RADON AND THE ENVIRONMENT

RADON AND THE ENVIRONMENT

Edited by

William J. Makofske **Michael R. Edelstein**

Institute for Environmental Studies
Ramapo College of New Jersey
Mahwah, New Jersey

Library of Congress Catalog Card Number: 87-35242
ISBN: 0-8155-1161-2
Printed in the United States

Published in the United States of America by
Noyes Publications
Mill Road, Park Ridge, New Jersey 07656

10 9 8 7 6 5 4 3 2 1

Library of Congress Cataloging-in-Publication Data

Radon and the environment / edited by William J. Makofske and Michael R. Edelstein.
p. cm.
"Proceedings of a conference that was sponsored by the Institute for Environmental Studies at Ramapo College of New Jersey on May 8, 9, and 10, 1986"--Pref.
Bibliography: p.
Includes index.
ISBN 0-8155-1161-2
1. Radon--Environmental aspects--Congresses. 2. Radon--Environmental aspects--United States--Congresses. I. Makofske, William J. II. Edelstein, Michael R. III. Ramapo College. Institute for Environmental Studies.
TD885.5.R33R298 1988 87-35242
363.1'79--dc19 CIP

Preface

This volume is the proceedings of a conference that was sponsored by the Institute of Environmental Studies at Ramapo College of New Jersey on May 8, 9, and 10, 1986. As the reader will discover, we have tried to go beyond the conference material to provide a thorough overview of the radon issue. We did not have to go far, however, because the conference was fortunate in attracting many of the key people in the unfolding radon issue.

The conference, and this volume, reflect a number of objectives regarding how the radon issue should be seen. First, the analysis is interdisciplinary, reflecting our long term collaboration in the Environmental Studies programs at Ramapo College, as a psychologist and physicist, respectively. In our view, radon is clearly not a problem to be bounded by narrow disciplinary perspectives. Second, we have tried to combine the objectives of communication among professionals and communication to the public. The conference speakers addressed their remarks to this combined audience, and this dual perspective is carried forth here as well.

The conference emerged from our realization that radon is a substantial threat about which the public is poorly informed. Media attention has been sporadic and often inaccurate. Until recently, there has been a paucity of government materials. At the same time that radon was ill-understood as a public policy and health issue, we also realized that there have been important new developments in the understanding of the topic. The conference "Radon and the Environment" was designed to bring together many of the key contributors to our current knowledge of radon in a forum which would clarify the policy issues. This volume takes up the same cause, we hope in a manner that is of equal use to the radon professional, the government official and the concerned citizen.

We have edited the contributed papers, solicited several additional papers and written focused section introductions which add some of our own conclusions while integrating the substance of each conference session. Discussion sessions from the conference have also been edited to highlight key points. Several appendices are added to provide useful information to the reader. The reader less familiar with radon will hopefully find the introductory chapter and the section introductions to be a pathway into the specific papers. A glossary of terms is appended which should help the reader deal with any jargon problems. We have converted most discussions of radon measurements into the more commonly used "picocurie per liter" or pCi/l measure for consistency.

Many people deserve credit for helping in our endeavor. New Jersey's Department of Environmental Protection, through the collaboration of Robert Tucker, Gerald Nicholls and Jeane Herb and the New Jersey Department of Health, through Judith Klotz, contributed strong support to our effort. Conference presenters disengaged themselves from complex schedules to fully participate. In countless hours of telephone calls prior to the session, the community of radon experts helped to educate us about the best way to approach our task. Particularly of assistance were Harvey Sachs, Richard Guimond, Kay Jones and Jeff Morales. We also enjoyed strong support from our colleagues and institution from top to bottom. In particular, we acknowledge the assistance of Patricia McConnell, Coordinator for Environmental Education at the Institute for Environmental Studies, for her help in coordinating and planning the conference. Deana Sanderson, a Ramapo student, was instrumental as conference manager during the session. Sherrill Cox, the Institute's secretary, spent many hours on the conference, as did Rebecca Manus, one of our students.

We also want to thank the following people who served as chairpersons for the conference sessions on which the various Sections of the book are based: Howard Horowitz, Ramapo College of New Jersey, and Patricia McConnell (Section 2–The Geographic Distribution of Radon); Robert K. Tucker (Section 3–Transmission and Mitigation of Radon); Gerald Nicholls (Section 4–Testing and Measurement); Theodore Sall, Ramapo College of New Jersey, and Robert A. Michaels, Consultant in Environmental Toxicology and Risk Assessment (Section 5–Radon and Health); Caron Chess, New Jersey Department of Environmental Protection (Section 6–Perception of Risk and Psychosocial Impacts of Radon Exposure); Jim McQueeny and Jeffrey Morales, Office of Senator Frank Lautenberg (Section 7–Socioeconomic Impacts of the Radon Issue); and Patricia McConnell (Section 8–The Role of Government in Responding to Radon Gas Exposure).

The Institute was fortunate to attract good outside support as well. Grants from the Fund for New Jersey, the Ramapo College Foundation and the New Jersey Department of Environmental Protection made the conference possible. A grant from the Separately Budgeted Research Fund at the college supported our editorial work on this volume. Many outside organizations, beginning with Orange Environment, Inc., and growing to the lengthy list which follows, gave support of various kinds to the conference: Orange Environment, Inc.; Sigma Xi; Bureau of Radiation Protection and the Office of Science and Research, New Jersey Department of Environmental Protection; New Jersey Department of Health; Biology Club of Ramapo College; Respiratory Health Association; Passaic River Coalition; Pennsylvanians Against Radon; Citizens Clearinghouse for Toxic and Hazardous Wastes; League of Women Voters,

Orange County (New York) South; League of Women Voters, New Jersey; New Jersey Audubon Society; New Jersey Sierra Club; New Jersey Environmental Lobby; Association of New Jersey Environmental Commissions; New York State Association of Conservation Commissions; New Jersey Association of Realtors.

The arduous task of checking the entire final manuscript for sentence structure, grammar and clarity was performed by Mary Makofske. We are grateful for her many changes and suggestions to improve the readability of the manuscript.

The actual transcription and typing of this manuscript was done by Liana Hoodes. We are greatly indebted to her for many hours of thoughtful labor.

The Institute is also indebted to the School of Theoretical and Applied Science, and to its Director, Edward Saiff, for providing a supportive structure and home for the Institute during its first year of existence.

We hope that the reader will learn as much from this proceedings as we did in putting it together and that this volume will contribute to the goal of limiting harmful levels of radon exposure to the general population.

M.R.E. and W.J.M.

Institute for Environmental Studies
Ramapo College of New Jersey
Mahwah, New Jersey October 1986

NOTICE

To the best of our knowledge the information in this publication is accurate; however, the Publisher does not assume any responsibility for the accuracy or completeness of, or consequences arising from, such information. This book does not purport to contain detailed user instructions, and by its range and scope could not possibly do so. Final determination of the suitability of any information, procedure, or product for use contemplated by any user, and the manner of that use, is the sole responsibility of the user.

Mention of trade names or commercial products does not constitute endorsement or recommendation for use by the Publisher.

The book is intended for informational purposes only. The reader is warned that caution must always be exercised when dealing with radon and other materials, or procedures, which might be hazardous.

Contents and Subject Index

SECTION 1
INTRODUCTION

SECTION 2
THE GEOGRAPHIC DISTRIBUTION OF RADON

SECTION 4
TESTING AND MEASUREMENT

SECTION 6
PERCEPTION OF RISK AND PSYCHOSOCIAL IMPACTS OF RADON EXPOSURE

THE PENNSYLVANIA RESPONSE

Section 1

Introduction

OVERVIEW OF THE RADON ISSUE

William J. Makofske and Michael R. Edelstein*

INTRODUCTION

To someone just learning about radon, the amount of information and its complexity may seem overwhelming. There are numerous unfamiliar scientific and technical terms. The radon issue spans many disciplines and has far-reaching implications for society. There are also many uncertainties stemming from a variety of sources. These include the often misleading and inconsistent media coverage of the topic, the newness of the issue, the lack of detailed scientific information and the way people preceive and respond to risk. The following section attempts to provide the radon novice with a brief overview of the issue and at the same time to systematically integrate the conference conclusions in order to provide a useful summary document.

RADON SOURCES AND DISTRIBUTION

Radon is a colorless, odorless, radioactive, chemically- unreactive gas which is formed directly from the decay of radium. Both radium and radon are part of the naturally-occurring decay chain as uranium radioactively decays to lead (See Figure 2.7). While radium has a half-life (the time for half of a given number of atoms to decay) of about 1600 years, radon decays with a half-life of just 3.8 days, and polonium-218, lead-214, bismuth-214, and polonium-214, the immediate "daughters" or "progeny" of radon, have half-lives on the order of thirty minutes or less. These elements all exist, to a greater or lesser extent, as components of rock and soil throughout the earth. As a result, radon may be found in varying concentrations in ambient air everywhere. Because of their particular geological history, certain types of soil and rock have enhanced concentrations of uranium and, therefore, radium and radon. Granites/gneisses, limestones (particularly dolomitic limestone), black shales and phosphate rocks are particularly linked to the potential for high radon concentrations. The wide distribution of radon emitting

*William J. Makofske is Professor of Physics and Director of the Institute for Environmental Studies. Michael R. Edelstein is Associate Professor of Psychology. Both served as conference co-organizers.

rocks and soils contradicts the current public belief that high radon levels are isolated in such areas as the the Reading Prong formation in the Eastern U.S. In fact, geologic formations with enhanced radon potential are quite common and may be found in every state, and, indeed, throughout the world (See for example Otton and Schutz & Powell, Section 2).

The main mechanisms by which radon moves through soil and rock are diffusion and convective flow. Radon may also move with ground water as a dissolved gas. In order to escape the rock and soil, the radon must migrate relatively quickly before it decays and its progeny combines chemically with surrounding elements. In areas where soil has higher porosity and permeability or is in proximity to fractured rock and fault lines, greater quantities of radon may reach the earth's surface. Likewise, radon formed in close proximity to ground water may dissolve in the water and travel with it.

RADON IDENTIFIED AS A HAZARD

The recognition of the hazard associated with radon exposure has been long in coming. Until recently, most concern over radon has been focused on occupational exposures to miners. For centuries, it has been known that hard-rock miners suffer excess deaths from a disease which we now identify as lung cancer. As scientific knowledge increased, radon was identified as the cause of these excess lung cancers. A second clue to the potential radon threat in the United States might have come from abroad. Several countries, most notably Sweden, had confronted the radon issue a decade before it emerged as a major national concern in the United States. A third development in the history of radon exposure involves the so-called "man-made" exposures. The extensive use of uranium after World War II, and, in particular, the inadvertent use of uranium mill tailings in the construction of foundations in homes, caused higher radon levels in dwellings in a number of western communities. These were discovered in the 1960's. Studies on these exposures, and on homes built on phosphate-mined land in Florida and, more recently, on industrial radium-enhanced soils in New Jersey, suggested the need to assess the threat from "normal" or naturally-occurring radon levels in homes. Finally, interest in this country over radon has intensified in the wake of a greater general attention to indoor air pollutants, an outgrowth of the effort in the 1970's to make homes more energy efficient and to understand the behavior of dwellings. Although high levels of radon were reported in American homes by several researchers in the late 1970's and early 1980's, it was the Watras incident[1], and the subsequent discovery of extremely high levels of radon in homes in the Reading Prong area of Pennsylvania, that brought the nation's attention to the problem of naturally-occurring radon gas.

HOW RADON ENTERS HOMES

Radon can enter homes from a number of sources:

(a) BUILDING MATERIALS -- While it represents one possible source of radon, studies have shown that the contribution from building materials under most circumstances is slight. Therefore, we will spend relatively little time on this source.

(b) GROUND WATER -- The second entry route for radon occurs when certain radium-bearing rock types are in proximity to ground water and significant amounts of radon are dissolved in the surrounding water. When this water is drawn from a well into a home, the dissolved radon outgasses and becomes airborne. In some areas of the country, particularly New England, high levels of radon in water have been found to contribute significantly to radon levels in air (see Hall, Section 2). Where water is stored for some time before reaching the user or where it originates from a surface source, radon levels are usually relatively small.

(c) SOIL GAS -- In contrast to the above two points of radon origin, soil gas has emerged as the most common source for radon entering American homes (See Sachs, Section 2). Because of combustion devices, venting and wind, houses often operate at a lower relative pressure compared to the surrounding soil. Soil gas is drawn in via cracks and sumps in the basement by this pressure differential. If the surrounding geology supplies radon and the soil type allows its easy movement, significant amounts of radon may be drawn in with the soil gas; the house acts as a collector of radon from the surrounding ground. This "convective mass flow" has been found to be the major mechanism for radon entry into homes (See Sextro, Section 3). The soil gas route, therefore, receives the greatest amount of attention in this volume.

VARIABILITY OF RADON LEVELS

Radon levels in homes undergo rather large fluctuations depending on many variables, some of which are not yet completely understood (See Oswald, Section 4). In light of current knowledge, several of these variables can be listed. First, the ability of a house to draw in radon from the surrounding soil depends on variable soil conditions which change with the seasons. Second, it is clear that changes in ventilation or combustion use will change the pressure differential and the pressure-driven convective flow of soil gas into the house. Third, wind and major barometric changes have been associated with large changes in radon levels as well. Some measurements have differed by a factor of 10 or more over a few day period. Fourth, even yearly average measurements have been found to vary by a factor of two, perhaps attributable to differences in soil moisture content. Fifth, radon in well water will contribute to large hourly fluctuations in air levels according to the pattern of water use throughout the day. The variability of radon in air has led to the EPA setting certain protocols or standard conditions of measurement to try to provide reasonable reproducibility of measurement results. The health risk, of course, is associated with the long-term average radon concentration, and this is the desired measurement to make in homes.

MEASUREMENT UNITS

The most commonly encountered measurement of radon gas is picocuries per liter (or pCi/l), The picocurie is a measure of activity or rate of decay such that 0.037 decays per second of radon take place. The Law of Radioactive Decay states that the rate of decay is proportional to the number of atoms. Accordingly, pCi/l represents the concentration of radon gas in air (the number of atoms of radon in the air per volume of air).

Ranges of radon in air vary considerably from a few tenths of a pCi/l or less in outdoor air to almost three thousand pCi/l inside the highest homes found so far. The average level in indoor air in the U.S. has been estimated to be about 1 pCi/l. The current EPA guideline for environmental exposure in homes is 4 pCi/l. The implication of this guideline is that residential radon levels above 4 pCi/l should be reduced to this level. The current standard for mines is about 60 pCi/l but is undergoing review and may be reduced to a lower value. Recent testing indicates that there are many homes in the U.S., possibly in the hundreds of thousands, which exceed the current standards for mines, and many millions which exceed the suggested guideline of 4 pCi/l.

Another relevant measurement has historically been used in discussing radon exposures in mines. Because the health effects in mines are due to the radon daughters and not the gas itself, it has been the practice in miner studies to measure "Working Levels" (WL), a unit of radon daughter concentration, instead of radon gas concentration (pCi/l). The working level is a measure of the total energy deposited by all the subsequent decays of the daughter nuclei.[2] Because the daughters are chemically active and may deposit (or "plate out") on surrounding materials, the radon gas and the radon daughters do not achieve a total equilibrium (defined by an equilibrium factor of 1). Based on actual measurements in homes, an equilibrium factor of at most 0.5 has been identified. The relation between radon gas in pCi/l and radon progeny or daughters in WL at an equilibrium factor of 0.5 is 1 WL = 200 pCi/l. With this assumption, the guideline of 4 pCi/l is equivalent to 0.02 WL. For the convenience of the reader, most references to WL in this volume are accompanied by the pCi/l equivalent.

The units of pCi/l and WL represent a rate of exposure, respectively, to either radon or its progeny. The length of exposure must also be considered if one wishes to estimate the potential health effects due to a given level of exposure. One accumulates a certain radioactive dose to lung tissue at a certain rate. Because the health effects depend on the cumulative dose received, the rate of exposure must be integrated over the time of exposure. For miners, the cumulative exposure is defined by the working level month (WLM). In simplist terms, this is the integrated product of the variable WL amount by the time of exposure in months. For occupational exposures, the month is taken to be 170 hours. For residential environmental situations, the equivalent type of cumulative measure for radon gas exposure is usually specified by the radon gas concentration in pCi/l over a defined time, usually a lifetime. By making assumptions about home occupancy and equilibrium ratio, these different units may be directly related.

TESTING

There are a variety of testing instruments and methods that measure either radon gas or radon progeny (See George, Section 4). These measurements may be made over differing time intervals; some are essentially instantaneous while others can take as long as a year or more. Different methods are suitable for different objectives. Some of the basic testing objectives are screening for high radon values in houses, characterizing the average radon level in a house, confirming the results from initial testing, diagnosing radon entry points as a first step in mitigation, assessing the success of mitigation, and certifying the house as "radon safe" for real estate transactions. In order to assure meaningful test results, the choice of a testing device needs to balance the objective with both the inherent limits found in each different testing method (discussed in Section 4) and with the methodological challenge from the large variability of radon levels in homes.

The comparability of tests has important implications for the ability to systematically apply the 4 pCi/l guideline and aggregate test results from many homes for the purposes of looking at health outcomes. For this reason, the EPA has developed sets of protocols or recommended standard conditions for making various kinds of measurements (See Ronca-Battista, Section 4). In order to assure reasonable reproducibility and accuracy, suggested conditons include testing in the wintertime under closed house conditions, assuring that closed house conditions have been maintained for at least 12 hours prior to the measurement, and determining appropriate locations in the home for placing detectors.

The two devices which are most used by homeowners and government agencies are the charcoal canister and the alpha track detector. Both measure average radon gas concentrations in pCi/l. Typically, the charcoal canister integrates over a period of a few days to a week, while the alpha track detector integrates over a period of a few months to a year. While the charcoal canister may be suitable for screening for high values of radon, its short integrating time coupled with the large variability of radon levels makes its one-time use problematical as a method for house characterization. This limitation can be overcome if four charcoal canister measurements, one at each season of the year, are taken and then averaged. The other and perhaps simpler method is to measure the average level directly with an alpha track detector over the desired time interval.

Radon levels in water are measured with more specialized instruments and generally require laboratory analysis of a water sample. The range of water values found varies from a few hundred pCi/l in surface water supplies to over a million pCi/l in several private wells in New England. There is no current recommendation or guidance on allowable values of radon in water. However, values on the order of 10,000 to 20,000 pCi/l in water are considered to be problematic. A rough rule is that each 10,000 pCi/l in water contributes about 1 pCi/l to the radon level in air. Of course, this depends on such factors as the amount of water used and local venting at points of water use.

HEALTH EFFECTS

The health effects due to radon exposure have been well studied in mining populations and provide firm evidence of the link between radon exposure and lung cancer. It appears that the radioactive daughter products, formed in the breakdown of radon gas, become deposited in the upper respiratory system. In some cases, these progeny are attached to dust particles and in other cases they are unattached. In either case, as further decay occurs, an alpha dose to the upper lung tissue causes lesions that are associated with an increased risk of lung cancer. At a lifetime exposure of 4 pCi/l, this risk has been estimated to lie between 0.9 and 5 per 100. In other words, at this rate of exposure, between 1 and 5 people out of 100 will develop lung cancer due to radon. EPA estimates that 5000 to perhaps 30,000 lung cancer deaths per year in the U.S. may be attributable to environmental radon exposures. The enormity of this risk can be seen when compared to risks that are associated with most other environmental exposures. For example, risk levels for toxic chemicals are typically on the order of 1-in-a-million or 1-in-one-hundred thousand.

Table 1.1 presents data from the conference which compares the risk of lung cancer from radon exposure alone (at a lifetime exposure to 4 pCi/l or 0.02 WL) to the risks of lung cancer to the general population, to non-smokers and to smokers.[3] Note that recent EPA figures project the radon risk to be as high as 5 in 100. It can be assumed that radon is a major component of the background lung cancer risk for the general population. Furthermore, in the case of radon exposure and smoking, the risks can be seen as at least additive (see Howe and L'Abbe, Section 5).

POPULATION	ESTIMATED LIFETIME RISK (per 100 people)
pop. exposed to 0.02 WL (4 pCi/l) for a lifetime	1.6 to 3.3 (excess risk from radon alone)
general population	3.9 to 5.9
non-smoking population	1.0 to 2.5
smoking population	9.9 to 25.0

TABLE 1.1: APPROXIMATE LIFETIME RISK OF LUNG CANCER, US POP.

The risk levels associated with radon are indeed alarming. How sure are we of this risk? Doubters frequently cite the lack of "residential" data. Miners, they argue, are not the equivalent of families. However, this argument is not convincing. Even though the studies show a range of risk values, the data is comparatively consistent and strong when compared to studies of other exposures. The data is from human populations, rather than generalized from non-humans. Statistical controls have successfully accounted for major variables. Furthermore, in many cases, the occupational mining data require no extrapolation to environmental exposures. There are many homes which are higher than the occupational standard of approximately 60 pCi/l. Furthermore data presented at the conference for the first time, indicates excess cancers down to a value of 16 WLM (see Howe and L'Abbe, Section 5). This is a level accumulated by people living in an average home! In other words, the linear extrapolation usually made for low-level radiation exposure, where assumptions are made about the incidence of cancers at low levels of exposure based upon data from high exposure, does not have to be invoked. The mining data is now showing sensitivity to excess cancers at typical residential environmental exposures. Table 1.2 shows a comparison of cumulative exposure in WLM in occupational, residential and school settings at different average WL concentrations.[4] Based upon what we know about radon levels in buildings, there is a significant population living at levels greater than allowed for miners. No wonder a major concern of government officials has been a possible panic reaction to the radon situation.

AVERAGE CONC. IN WL	EXPOSURE DURATION		
	1 year	4 years	10 years
Occupational			
5	60	240	600
1 (former standard)	12	48	120
0.33 (current standard)	4	12	40
0.1	1	5	12
0.02	0.2	1	2
Residential			
0.5	20	80	200
0.1	4	16	40
0.02	1	3	8
School			
0.1	1	3	10
0.02	0.2	0.5	2

TABLE 1.2: CUMULATIVE DOSE IN WLM FOR RADON PROGENY
(Note: 1 WL = 200 pCi/l)

There are some hopeful aspects to the current findings as reported in Section 5. For example, while the linear model would predict continually increasing risk from increasing cumulative exposure, it would appear that there is a point of saturation for radon where greater exposure does not increase risk. This is of great significance to people living in so-called "hot houses". Furthermore, children do not seem to be at higher risk compared to other segments of the population. Also, the risk appears not to be totally cumulative. If exposure is decreased or eliminated, it appears that additional future risk is not only eliminated, but some of the previously accumulated risk is reduced as well (depending upon the age of the person).

WHAT IS AN ACCEPTABLE RISK?

Given the known health risks of radon, an important question for the homeowner is what level of exposure is acceptable. As always, the issue is "acceptable from whose perspective?" From a governmental perspective, it can be argued that a guideline of 4 pCi/l is too low. Some question the tradeoff between excess cancers and the costs of bringing radon levels down to 4 pCi/l. Other countries have opted to tolerate greater levels of radon. An influential expert body, The National Council for Radiation Protection, recommends a guideline of 8 pCi/l. On the other hand, from the homeowners' perspective, it can be argued that the guideline is too high. For example, the American Society of Heating, Refrigeration and Air-Conditioning Engineers recommends a guideline of 2 pCi/l. Quantitatively, the risk at the guidance level of 4 pCi/l is 1000-10,000 times greater than risks associated with most other environmental exposures. If we will only tolerate one excess cancer per million in setting standards for PCB's, volatile organics and trihalomethanes, it can be asked why we should allow so high a risk for radon? Of course, there seems to be a practical lower limit for radon levels, the ambient concentration in outside air which is typically a few tenths of a pCi/l. It would appear that the relatively high risk associated with radon exposure even at these low levels will have to be accepted.

The government has been careful to note that they do not view 4 pCi/l as a safe level of exposure. As Christie Eheman, a scientist with the Centers for Disease Control, points out in this volume, the choice of 4 pCi/l was based on economics rather than health risk. It was thought that 4 pCi/l is the lowest exposure to which houses could economically be brought. Harvey Sachs, a radon scientist, criticized this decision as having been based on inadequate data about what radon levels could be attained in houses. Given the extent of the risk, the eventual ability to reduce homes to below 4 pCi/l may open up further debate aimed at lowering homeowner exposure. In the meantime, it is left to the homeowner to decide what radon level justifies expenditures for mitigation.

HOMEOWNER AND GOVERNMENTAL RESPONSIBILITY

One may well ask, "If radon is such a threat, why isn't government providing funds to test and remedy the problem?" While the federal government, under Superfund, has played a major role in attempting to address radon exposures that were human caused (such as in Essex County, New Jersey), government was quick to limit their responsibility for "cleaning up" naturally-occurring radon. Testing programs for characterizing geographical areas are often done at no cost to the homeowner, but these are usually random surveys which reach relatively small numbers of homeowners. Neither the federal government nor the state agencies involved in radiation issues are taking responsibility for large-scale radon testing programs. This is in part due to the fact that naturally-occurring radon cannot readily be blamed on some sector of society or easily be regulated by law. It also occurs in a domain which has traditionally escaped federal regulation, the private home. Radon has also been recognized as a widely occurring phenomenon. Hence, government has sought to limit its responsibility to conducting research projects from which findings can be generalized by the private sector. The major costs associated with the testing and mitigation of homes will generally be left to the individual homeowner. For a discussion of the government's definition of its responsibility, see Section 8 on The Role of Government.

MITIGATION OF HOMES

The target of the current Federal effort is to encourage the reduction where necessary of residential radon levels to below the EPA guidance value of 4 pCi/l. Recommendations by the EPA provide time ranges for action depending on the value found in the home. So, for example, homes having between 4 and 20 pCi/l have a recommended action time of a few years, those between 20 and 200 pCi/l have an action time of several months, and those having radon levels greater than 200 pCi/l require an action time of several weeks. The basic assumptions are clear. Higher radon values carry a greater risk. To avoid a more rapid radiation dose accumulation, occupants of homes having substantial radon levels should act more quickly to lower exposures than those with lower exposures.

The good news about mitigation is that it is possible, often not terribly expensive, and can sometimes actually be done by the homeowner. The bad news is that some houses are likely to be considerably more difficult than others, and the cost will correlate with the difficulty. Also, it is going to be difficult to get trained mitigators until the industry develops and matures.

There are several basic approaches to mitigation depending on factors such as the source of the radon, the radon level, and house construction. If the radon source is soil gas entering through the basement, the options are (1) seal the basement, (2) increase ventilation, (3) divert the radon by a collection and exhaust vent system and (4) prevent it from entering by depressurizing around the foundation. While these approaches may seem simple, the difficulty is in actually installing the appropriate working system at reasonable cost for a particular house (see Section 3 on transmission and mitigation of radon). The importance of accurate house characterization and diagnostic testing to find radon sources and entry points cannot be overemphasized. If the source is from radon outgassing from water, the GAC (granular activated charcoal) filter system will eliminate the radon at modest cost (see Lowry, Section 3). Unlike the more tricky task of soil gas mitigation, this strategy will work for any house. For some houses, both soil gas and water can be significant contributors to the radon in air.

Unfortunately, as was mentioned, houses tend to be somewhat individualistic in their behavior and construction, and hard and fast rules about the simplest and most economic way to mitigate any particular house are not yet available. A great deal of work has been done in the past several years by researchers, and this information is being distributed through an EPA-funded mitigation training workshop that is being given at a number of locations in the northeast (see Reese, Section 8). Also, an informative brochure on do-it-yourself mitigation has recently been published by the EPA. Efforts are continuing to spur the growth of a mitigation industry that is professionally qualified to handle the large number of homes potentially needing mitigation. However, based on the number of people testing and finding higher levels of radon, there is likely to be a "mitigatory gap" due to the insufficient number of skilled and trained mitigators, at least in the short term.

SOCIOECONOMIC IMPACT

The discovery of unacceptable levels of radon poses certain social and economic costs for individuals and for society as a whole. Since government has basically limited its responsibility for the naturally-occurring radon, certain segments and sectors of the society will need to address the financial issues. The obvious first cost is that the homeowner will have to pay for testing and mitigation. Just screening for hot houses with detectors at a low to modest per home cost of between $12-50 per test invokes a considerable cumulative expenditure, considering that there are several million homes in high risk areas such as northern New Jersey. While there are no firm estimates of exactly how many houses will be over 4 pCi/l, screening samples to date have indicated that the total number in the U.S. could be in the millions. While mitigation costs will vary drastically from house to house, many experts are quoting about $2000 as a typical amount. Total mitigation costs could easily be several billion dollars.

Other socioeconomic impacts will involve sectors such as the housing and construction industry, banks and mortgaging interests, the real estate industry, and the legal industry. There are certain critical times when homeowners and home buyers are directly involved with these sectors: when a new home is built and purchased and when an older home is bought and sold. While the full legal ramifications for such property transfers are as yet unclear, it appears that builders, sellers, and real estate brokers may be held liable for selling high radon houses, in some cases even when they were unaware of the risk (see Sykes and Toomey, Section 7). At the same time, there is a recognition that it is impossible to adequately characterize radon levels in houses during the short time frame of a typical housing transaction. Furthermore, such situations are vulnerable to the possibility of fraud. Until better short run testing methods are perfected, one way to address these problems is through use of an escrow account to provide for a longer test period under the buyer's control (see Sachs, Section 7).

There are also significant economic, not to speak of social, costs in NOT dealing with the radon problem. Clearly of major concern are the health costs of an estimated 5,000-30,000 additional cases of lung cancer per year. With a vigorous remediation effort and with the development of radon standards in new housing, these costs can be lessened over time. Other economic costs will include: building code modification, increased costs for new construction, lawsuits over liability in the transfer of houses, and regulation and research by government agencies.

One other major socio-economic impact is the creation of a potentially multi-billion dollar radon industry. Already there are hundreds of companies who do testing and a considerably smaller number in the mitigation area. How are they to be regulated to prevent unqualified contractors from entering the field? How will they be trained to provide the skilled testors, diagnosticians and mitigators needed to remediate the large number of homes above 4 pCi/l? A paper presented by Schutz and Greenberg of the American Association of Radon Scientists and Technologists (see Section 7) describes how some of these problems are being handled by a professional organization. Reese (see Section 8) describes the response of a governmental body, the New York State Energy Office.

PSYCHOLOGICAL, PERCEIVED RISK AND SOCIAL ISSUES

In addition to the socioeconomic issues outlined earlier, there is another class of problems and issues associated with radon. These include the psychological effects on families who find high levels of radon in their homes, the stigmatizing of neighborhoods and regions which become associated with radon problems and the stress of dealing with the accompanying health and economic uncertainties. The health risk invites "an inversion of home" whereby the place we most associate with security and identity is converted to a place of uncontrollable danger. The impact on families living under these conditions cannot be easily appreciated by others. The article by Kay Jones and Kathy Varady (see Section 6) outlines some of these impacts.

Another major issue is one of perceived risk. A unique situation is created by the very nature of the radon issue. The homeowner is left with the responsibility to act, so that the perceptions of the homeowner about the radon threat become central to the ability of society to address the issue. Radon has provided a major laboratory within the field of public health in which to address complex questions surrounding risk communication. The specific challenge is how to communicate about the radon risk so as to engender enough concern to encourage effective action while not inducing so much fear as to create panic. Based on several studies presented in the section on Perceived Risk and Psychosocial Impact (Section 5), it appears that people are not as frightened of radon as the risk from exposure dictates they should be. This is due to various factors: factual errors in the media and governmental presentations of the issue; inherent uncertainties and complexities of the radon issue; limitations in the public's ability to readily appreciate so complex an issue; the characteristics of radon itself as a colorless, odorless and natural hazard; and finally, the possible denial that so threatening a substance might be affecting one's home and family. As a result of this underappreciation of radon, a significant number of homeowners may be avoiding actions important for protecting their family's health.

THE ROLE OF GOVERNMENT

While the federal government has not accepted major responsibility for radon testing and mitigation on behalf of individual homeowners, they have responded by trying to pioneer work and develop information that can be used by the individual states. Thus, the EPA and the Centers for Disease Control have acted to set guidelines for exposures. The EPA and the Department of Energy also have funded several research programs on how to mitigate homes, and the EPA has provided information booklets to homeowners and mitigators on the issue. The EPA has developed a set of testing protocols to ensure accuracy and reproducibility of test results. They have also issued standards for calibration of equipment and quality control within the industry to ensure that there is uniformity to professional standards in the emerging radon industry. The federal role has also involved support for individual states as they begin to address the radon issue.

At the state level, in the absence of firm federal guidance, there is currently a noticably uneven response to radon, with the fastest action occurring where a crisis is identified (see Edelstein, Section 8). The level of action depends greatly on the perceived magnitude of this crisis, which translates into funding appropriated by the state legislature. Moreover, the states tend to be idiosyncratic in terms of their response to the radon issue, as several papers in the role of government section illustrate. Responsibilites are often divided differently throughout the array of state agencies. As noted, the level of concern seems to correlate with the level of radon found and the number of potential homes at risk. In many states, the data is still too incomplete to identify the location and number of homes at significant risk. Studies by the U.S.G.S. for the EPA are currently identifying those areas which are likely candidates for further sampling and characterization. The scope of state response is likely to broaden as additional problem areas are identified and as the generally widespread nature of the hazard is further confirmed.

THE ROLE OF EDUCATION

The nature of the radon problem together with the governmental response would seem to indicate that it will be the responsibility of the public to act on the issue. However, given the complexity of the issue and the persistent myths which have evolved as the media persists in disseminating outdated information, it is clear that the public does not yet have the knowledge to make informed decisions. The government and educators will clearly have to work together to communicate and educate the public on the issue. The first step in this process is a decision about what needs to be communicated. In general, media has so far been unable to convey the facts coherently while government has been reluctant to provide a clear message. The result has been complacency rather than panic. It would clearly be in everyone's best interest if a direct and forthright educational program is developed so that the response is appropriate to the public health risk involved and that action in testing and mitigation is based on scientific understanding rather than misunderstanding. It is hoped that this proceedings will provide an important step in this direction.

FOOTNOTES

[1]When Stanley Watras, an employee of the Limerick Nuclear Power Plant, set of radiation alarms on his way into work in December, 1984, exploration for the source of his contamination led to the discovery of extremely high levels in his home and neighborhood.

[2]To make it easier for the reader to keep track of pCi/l as a consistent measure of radon levels, most references to WL in this volume of material are matched with the corresponding pCi/l measurement in parenthesis.

[3]Presented in a slide by C. Eheman, Radon and Health Session of the Conference, May 9, 1986.

[4]Presented in a slide by J. Klotz, Radon and Health Session of the Conference, May 9, 1986.

Section 2

The Geographic Distribution of Radon

INTRODUCTION AND SUMMARY OF SECTION 2: THE GEOGRAPHIC DISTRIBUTION OF RADON

Patricia McConnell*

The public's perception of radon from a geological and geographic perspective has often been reduced to simple cause-effect relationships. One example of that is the belief that radon is found only in granite-gneiss formations such as the "Reading Prong" connecting Pennsylvania to Connecticut. But in fact, emerging evidence suggests a much more complex and widespread distribution.

Although relatively little testing has been done until recently on the actual radon levels in houses, there exists considerable work on the distribution of uranium, the ultimate parent of the radon gas. This section deals with both the geologic formation of radon through its precursors, uranium and more immediately, radium, and the actual geographic distribution in houses based on the latest testing results from New Jersey and Pennsylvania.

Harvey Sachs (Radon Consultant), in "Where Is All the Radon From," presents an overview of how our present understanding of radon evolved. Early recognition of the radon hazard in certain types of mining was followed by studies of natural background doses from radon. Energy conservation work in the 1970's provided an impetus for studying building materials as sources of radon and indoor air pollution in general. By the early 1980's, several studies in Maine, Pennsylvania, and New York showed that much of the radon in homes came from outside the building, either from ground water or from entering soil gas associated with specific types of geological formations. The distribution of radon in houses was found to follow a lognormal distribution law which implies that most houses will have relatively low levels but a significant fraction (perhaps 5-10%) will have unacceptably high values. As the result of this work, it was determined that the major mechanism for producing higher levels of radon in houses in most instances is convective mass flow caused by pressure differentials between the house and surrounding soil. Although a significant body of knowledge currently exists on radon, much more work still needs to be done.

Donald Schutz (President of Teledyne Isotopes, Westwood, NJ), in "The Influence of Geologic Environment on

---

*Environmental Staff Coordinator for the Institute for Environmental Studies at Ramapo College and Co-Chair of the session, The Geographic Distribution of Radon.

the Distribution of Uranium/Radon in New Jersey," enumerates the important reasons for studying geology, particularly uranium distribution as it relates to radon. On a broad scale, such information can identify areas of high and low uranium, providing greater efficiency in designing radon survey programs. Longer term studies, which would include combining geophysical, ground water, and soil data, would probably allow enhanced prediction for identifying specific areas likely to evidence high radon emanation.

James Otton (Geologist with the United States Geological Survey), in "Potential for Indoor Radon Hazards: A first Geologic Estimate," describes four decades of research and interest in radon by the USGS and summarizes current information regarding the geology and geo-chemistry of radon. Recently, the USGS has expanded its studies to include radium and uranium as they relate solely to radon in the environment, and it is currently assisting the Environmental Protection Agency by researching areas of the country with the potential for high radon values. Otton suggests that the two critical elements which determine the level of radon in a home are geologic factors and house factors. While the controls of radon in rock may seem simple, they are actually complex. Included in these geologic factors are radium content, porosity, and emanation power of the surrounding soil and bedrock. Another geologic factor is the permeability of the soil which depends on grain size, the variation in that grain size, and moisture content. So far, the rock types in the United States responsible for high radon levels include: a) rock with high uranium content (granites/gneisses, black shales, limestones/dolomites, phosphate rock), b) fractured rock, c) combinations of the above and d) highly porous and/or highly permeable rock and soil. Attempts to correlate geological research with levels of radon in houses are confounded by the type of house construction.

Schutz provides examples which illustrate that the geologic provinces of New Jersey are not simple formations but actually complex assemblages formed in different geologic environments. The Pre-Cambrian Reading Prong, "a complex assemblage of igneus rock," is in some cases highly radioactive and in other cases actually barren of radiochemical elements. These rocks were metamorphosed by heat and pressure from sedimentary and volcanic rock to the present formations. The Paleozoic Sedimentary rocks of the Valley and Ridge Province include the Jacksonburg limestone which has produced the high radon values found in the Clinton, New Jersey area. The Mesozoic Sedimentary rocks of the Piedmont Province include the Lockatong (a black shale) and the Stockton, basically a sandstone which concentrated its uranium from ground water. Finally, the possibility of glacial material as a source of radon is discussed.

Francis Hall (Professor of Geology, University of New Hampshire), in "Geologic Controls on Radon Gas in Ground Water," discusses the chemical and geo-chemical features that lead to radon gas in water, particularly focusing on the Concord and Two-Mica granites of New England. Water contamination by radon results from enriched parent material (i.e., uranium or radium bearing rock) in direct contact with active ground water circulation. Furthermore, it is also believed that wells play an important role in the occurrence of radon in water by increasing the circulation flow, which in turn increases oxidation. Slow ground water velocity and the short half-life of Radon-222 ensures that radon usually does not travel more than 100 feet in distance around the well. A summary of some of the values for radon found in ground water in commonly found rocks in New England is given. Using 20,000-70,000 pCi/l to indicate a "high" value and greater than 70,000 pCi/l to indicate a "very high" value, many wells in New England have been found in the "high" range, and some have been found to be "very high".

Bernard Cohen (Professor of Physics, University of Pittsburgh) and Nickifor Gromicko (Program Manager of the University of Pittsburgh's Radon Project), in "University of Pittsburgh Measurements in New Jersey and Eastern Pennsylvania," report the results of tests mostly done at the request of homeowners for a charge of $12. A number of no-charge measurements were also randomly selected in certain areas to assure a distribution of tests. All measurements were done using 3-inch charcoal canisters sent as kits through the mail.

The data for New Jersey showed that Hunterdon and Warren counties had the highest mean levels, with Somerset and Sussex counties somewhat lower, and Morris county ranking 5th of these five. Bergen, Passaic and Essex counties appear to have below average mean radon levels. It is interesting to note that while all these counties are traversed by the Reading Prong, the mean radon levels seem to decrease with increasing distance from the Pennsylvania border. This does not mean that there will not be houses with high values in these areas, but only that there will be fewer of them.

The data for eastern Pennsylvania counties show mean radon values which are considerably higher than the average New Jersey values. It is interesting to note that many of the counties in Pennsylvania with high mean values do not lie on the Reading Prong. Evidently, Pennsylvania appears to have a much greater radon problem than New Jersey, and the Reading Prong is not the main geological area at high risk.

In summary, some of the major conclusions from this session were:

1) The numbers and types of basic geological rock contributing to radon release are not fully known. At present, there are four known basic types: granites/gneisses, limestones (especially dolomitic limestones), Triassic sandstones, and black shales. In addition, phosphate rock, found mostly in the state of

Florida, is also a major formation capable of producing high indoor radon levels.

2) The Reading Prong, composed of granites and gneisses, often does yield high radon levels, although not always uniformly. However, there are other formations, such as the dolomitic limestones found in Clinton, New Jersey, that have produced high levels in homes and surrounding soil.

3) There are other factors besides rock type that contribute to radon in a house. These include the porosity and permeability of the soil and rock, the presence of faults and/or fractures, and the soil type overlying the bedrock. For example, clay is an excellent barrier material, as it is non-porous, making it difficult for radon to penetrate into a home basement.

4) All of the above factors may be mitigated or enhanced by the house construction type. A poorly constructed basement in a relatively low radon environment may actually have a higher reading than a properly constructed one sitting directly over a uranium source.

5) In some cases, ground water may be a significant contributor to the radon gas level in homes, and this will depend on the specific geological conditions that surround the well.

6) The counties of eastern Pennsylvania have on the average considerably higher mean radon values than New Jersey counties. There are counties off the Reading Prong in Pennsylvania with high values while there are counties traversed by the Reading Prong in New Jersey with below average values.

WHERE IS ALL THE RADON FROM?

Harvey M. Sachs*

INTRODUCTION

Background

The past year has seen greatly increased awareness of the public health problems associated with environmental exposure to excess radon in houses. From the public health perspective, radon is now recognized as the most significant indoor pollutant.

It is ironic that this conference occurs only two weeks after the terrible reactor accident in Chernobyl, in the Ukraine, because the maximum number of early fatalities from that accident is much less than estimates of annual US deaths due to environmental radon: 1,000-20,000 per year[1]. By contrast, the Three Mile Island accident released relatively little radioactivity before being brought under control. Still, news analysis by Drachler suggests that the Three Mile Island accident in 1979 so overwhelmed the Pennsylvania Department of Environmental Resources that its officials ignored or suppressed the information that came to them concerning the threats from radon in eastern Pennsylvania[2]. They continued to ignore the problem until publicity about the Watras house forced a response. If they had instead taken timely action and raised the level of concern in federal and other state agencies, the recent "Reading Prong" panic might have been avoided.

My goal is to illustrate how the technical community has achieved its present level of understanding that geological and building science knowledge are the keys to solving the radon problem in houses. However, this paper is not a formal literature review with a comprehensive bibliography. Excellent reviews of various aspects can be found in Nero[1], BEIR 1980[3], NCRP 77[4], and The Natural Radiation Environment III[5].

THE CONCERN DEVELOPS

Antecedents

---

*Harvey M. Sachs is a radon consultant and former faculty member at Princeton University's Center for Energy and Environmental Studies.

It has been known for a long time that some groups of underground miners die of lung cancer at much higher rates than expected. In fact, this knowledge predates the discovery of radioactivity. Even after the discovery of the role of radon progeny, radon was considered to be exclusively an occupational problem.

Following World War II, a small group of scientists accelerated the investigation of the "natural radiation environment" (the study of the amount of radiation we are all exposed to) and how it varies geographically. Radon progeny were recognized as components of the population radiation budget, and some estimates were made. As an example, BEIR estimated that approximately 1/3 of annual exposure was the lung dose largely attributable to radon decay products. These estimates were based on very limited data, and show little knowledge of the wide variability of radon concentrations in buildings.

Most investigations during the 1970's concentrated on the role of building materials as sources of radon in buildings (e.g., papers in [5]). Concurrently, early work on the building science aspects was begun in conjunction with the discovery of high radon concentrations in houses in uranium mining areas (Canada) or built upon tailings from mining operations (Grand Junction, Colorado). In particular, Arthur G. Scott undertook a noteworthy series of investigations for the Atomic Energy Control Board of Canada. He began to evaluate many of the mitigation methods used today, and determined that the high radon values in some houses were due to "natural" radon sources in the ground -- an early indication of the role of geology.

Energy Consciousness and Increased Activity

The oil price "shocks" of 1973 and 1979 focused concern on methods for reducing energy consumption in houses and other buildings. Work at Princeton University, Lawrence Berkeley Laboratory, and other institutions demonstrated that the uncontrolled infiltration of outside air that needed to be heated accounted for about a third of the heating budget of houses. From this came the development of methods to "tighten" buildings in order to limit and control their ventilation energy losses.

As shown in Figure 2.1, reducing the ventilation rate will proportionally increase the concentration of any preexisting pollutants. It is critical to stress that the modest ventilation changes achieved in existing houses do not create new pollution problems, but they can "amplify" existing sources. However, there was a new concern about the contamination of the indoor environment. Work began to inventory indoor air pollutants, discover their sources and develop appropriate standards and control strategies.

By about 1980, the scientific community had the essential elements required for rapid progress:

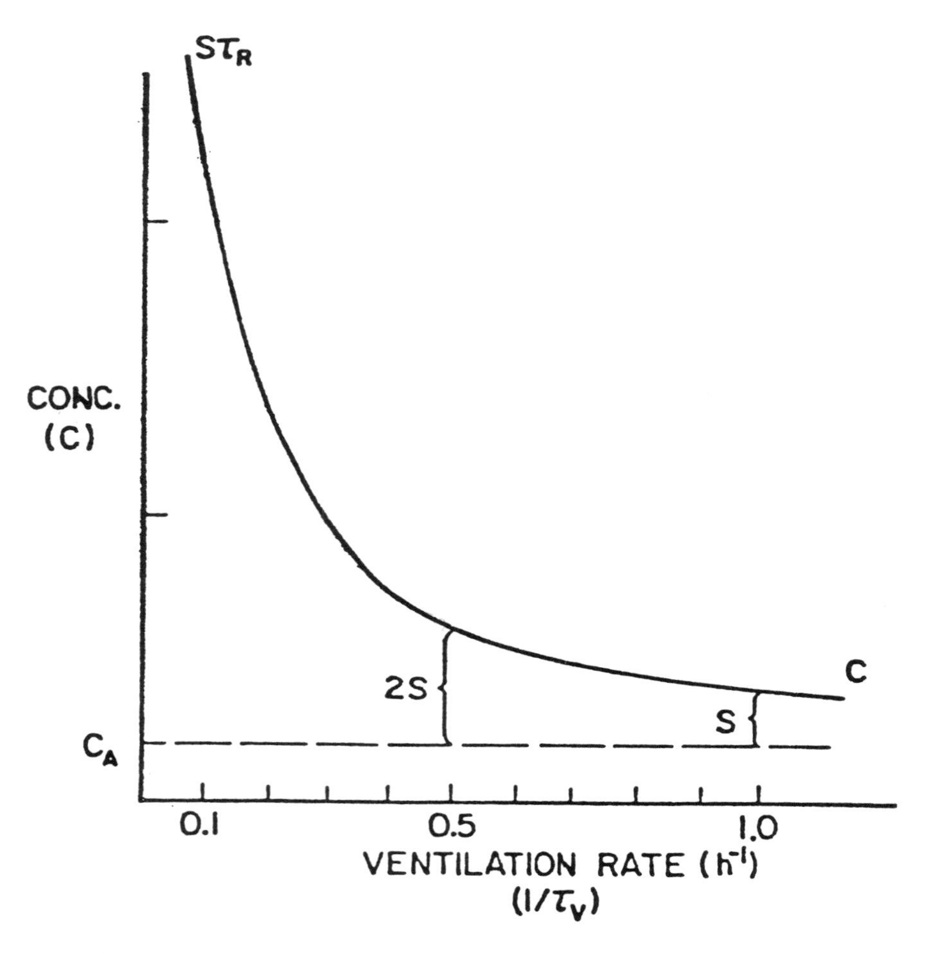

Figure 2.1 - Radon Concentration and Ventilation Rate for an Arbitrary Source Strength S. T_V is the building ventilation time constant, and T_R is the decay constant for Rn-222. C_A is the outdoor concentration. Y-axis has a linear scale with arbitrary divisions. From (7).

* Awareness of the health effects of radon progeny.

* Knowledge that much of the radon in houses came from outside the building.

* A political environment in which determining the extent of the indoor air pollution threat was important.

THE ROLE OF GEOLOGY

From 1980-1982, Hess and his co-workers in Maine showed that the amount of radon in well water supplies of houses was strongly associated with the degree of metamorphism of the rocks in which the wells were drilled: more granitic rocks were associated with more radon[6]. They also showed that the amount of radon in the air of the houses was correlated with the amount of radon in the ground water.

About that same time, the Pennsylvania Power and Light Co. (PP&L) carried out a study of indoor radon concentrations in 36 houses in their service region. At the suggestion of A. George of the U.S. Department of Energy, PP&L asked the Center for Energy and Environmental Studies at Princeton University for interpretation and analysis of the data. The PP&L data were the first to show clearly the association of specific outcrop bands of sedimentary rock with anomalous frequencies of high radon concentrations[7]. Subsequent work with another relatively small data set pointed toward the Beekmantown and other specific formations and suggested that older houses in the region had more ventilation and less radon on average[8].

Under New York State Energy Resources and Development Administration (NYSERDA) and Niagara Mohawk sponsorship, W.S. Fleming and Associates carried out a study of indoor air pollutants in about 60 houses in New York. Interpretation of these data by William Lilley and by Sachs again showed the association of high radon values with particular sedimentary rock sequences, in this case specific black shales. Similar investigations are underway by the Bonneville Power Administration and by other groups.

The distribution of radon in houses follows a lognormal law. This means that most houses will have very low concentrations, but that some (perhaps 5-10%) will have unacceptably high values. We have inferred that high-valued houses will tend to be found in restricted regions corresponding to specific geological conditions. Where the combination of bedrock and soil development yields high permeability and high radon concentrations in the soil, too much radon can move into the houses.

On the one hand, this is a message of genuine hope: the number of areas where high concentrations are frequent is likely to be small, and it should be possible to identify these areas. On the other hand, we do not yet have enough knowledge to be able to predict accurately the "bad" areas. Large scale, carefully constructed surveys are required. Such a study is being undertaken in New Jersey. Unfortunately, federal agencies did not begin such work four years ago, when the need became obvious. Sweden, Canada, and Great Britain have all carried out much more extensive programs.

BUILDING SCIENCE AND MITIGATION

Since radon comes from the ground in almost all cases, radon control requires isolating the house from the soil. Houses (or mobile homes) that are elevated, with a ventilated crawl space between the house and the soil, are very unlikely to have radon problems; tall buildings with high ratios of floor area to soil contact area are also unlikely to have problems.

However, the realities of construction methods and of air pressure distribution in houses make realization of this isolation difficult in most housing, as Art Scott, our group, and others have shown. Conventional construction practices leave many openings between the building volume and the exterior. These range from designed-in features such as french drains, sumps, and plumbing pass-throughs to incidental features such as the porosity of concrete block foundation walls. In addition, as houses age, their structures move and crack, introducing additional passages.

Nonetheless, attempts have been made to seal all openings to exclude radon. No successes have been documented for houses with very high concentrations of radon, but this approach recognizes that the primary transport mechanism is convective mass flow, rather than diffusion through the materials.

The air pressure distribution in houses causes the convective flows. In general, houses are frequently at lower air pressure than the soil beneath them, so pressure gradients drive soil gas into the buildings. Several groups of workers have attempted to exclude radon by building elaborate passive ventilation schemes comprising extensive perforated manifolds and tall stacks. These are expensive, and their performance varies with wind and temperature gradients.

In our experience, the simplest solution is to install an active (fan-forced) system to lower the pressure in the subslab gravel bed, where one is present. This "traps" the radon-rich soil gas before it gets into the house and vents it harmlessly outside. These systems are inexpensive and often very effective. A simple flag at the outlet suffices to show that the system is functioning, and estimated operating costs are modest, comparable to the cost of operating a basement dehumidifier. I believe that this method was first used by A.G. Scott; it is now offered commercially by at least one Swedish company. In addition to its simplicity and low cost, this method has two additional virtues: the metaphor of a sump pump for soil gas is like the sump pump used to control basement water, so it is easily accepted. Just as important, the tools and materials needed are familiar to tradesmen, so the work can be done efficiently.

Additional research needs to be done, however, to build good rules for the number and size of ventilation components required for specific construction methods, climate types, and soil conditions. In addition, when subslab ventilation is not successful, other measures may be required. Costs will rise, both for diagnostics and for implementation.

DISCUSSION

The stage is now set for rapid gains in our understanding of the radon problem. The basic paradigms are in place for distribution and mitigation:

* We know how to design and carry out the studies that will pinpoint "problem" areas.

* We know how to carry out the experimental studies that will lead to "routine" diagnostics and mitigation.

* We know how to carry out the experimental programs that will lead to changing building codes in problem areas. We are ready to test the hypothesis that minor construction changes can assure low-radon housing.

Unfortunately, the social will lags. Political pressure will be required to release agencies from the pressures of their internal agendas and to find budget resources for the required research. Because the benefits of this research cannot be "captured" for private gain, this must be a government responsibility, and it will require more effort than has been allocated to date.

It is also important to learn from mistakes that were made as research findings in energy conservation were transferred to government agencies responsible for action and to the private sector, to assure that the right training is provided and the right evaluation methods are built in from the start.

If this conference helps to build an agenda for action, it will have been a great success. If not, we will have more panic episodes, with repetition of the mistakes of the past.

FOOTNOTES

[1]Nero, A.V., Indoor Radiation Exposures from Rn-222 and its Daughters: A View of the Issue. Health Physics. V. 45, 1983, pp. 273-288.

[2]Drachler, Stephen, "Warnings on Radon Ignored," Allentown Call, September 22, 1985, p. 1.

[3]BEIR, The Effects on Populations of Exposure to Low Levels of Ionizing Radiation, National Academy of Sciences, Washington, D.C., 1980.

[4]NCRP, Exposures from the Uranium Series with Emphasis on Radon and its Daughters, National Council on Radiation Protection and Measurements, Report 77, Bethesda, MD, 1984.

[5]Gesell, T.F., and W.M. Lowder, Editors, Natural Radiation Environment III. V.1 and 2. Technical Information Center, U.S. Department of Energy, 1980.

[6]Hess, C.T., C.V. Weiffenbach, and S.A. Norton, Variations of Airborne and Waterborne Rn-222 in Houses in Maine, Environment International, V.8, 1982, pp.59-66.

[7]Sachs, H.M., T.L. Hernandez, and J.W. Ring, Regional Geology and Radon Variability in Buildings, Environment International, V.8, 1982, pp.97-103.

[8]Gross, S. and H.M. Sachs, Regional (Location) and Building Factors as Determinants of Indoor Radon Concentrations in Eastern Pennsylvania, Princeton University Center for Energy and Environmental Studies Report PU/CEES 146, 1982.

POTENTIAL FOR INDOOR RADON HAZARDS: A FIRST GEOLOGIC ESTIMATE

James K. Otton*

INTRODUCTION

The level of radon in a dwelling is dependent upon two principal factors: 1) the availability of radon in the ground; and 2) the structural characteristics of the dwelling. Of these, only the availability factor is a function of the natural environment and will be considered here. Although in specific buildings, high radon levels may be caused by the use of uranium-rich and radium-rich natural building materials or by the degassing of radon-rich waters in showers and other devices in the home, the principal natural source of indoor radon is soil gas that moves into the dwelling through the foundation by diffusion or by flow caused by air pressure differences. Subjacent bedrock is usually a major contributor of radon also, especially where soils are thin or where the bedrock is uranium- or radium-enriched and permeable. The availability of radon in soil gas is dependent on the radium concentration in the soil and bedrock, the emanating power of the soil (that fraction of the radium in mineral grains that produces radon in soil pores), the soil porosity and permeability, and the moisture content of the soils.

The radium content of soils and bedrock is a result of a series of complex water-rock interactions in the uranium geochemical cycle that involves leaching of uranium and its daughter products from their sources, transport of these elements in surface and ground waters, and their fixation in a host. Radon is the product of a decay chain of geochemically distinctive radioelements. Most of the radioelements in that chain are cations soluble in and largely transported by water, but radon is an inert gas that can be transported by either water or air.

Because a radon problem starts with its availability in the ground, an understanding of the geology and physical properties of soils ought to lead to predictions of where radon hazards may occur, but the interplay between radium concentrations in soils and subjacent bedrock and the porosity and permeability of the soils, although critical, has been little studied. Case studies do suggest, however,

*U.S. Geological Survey, Denver, CO.

that high uranium and radium concentrations in soils and subjacent bedrock can lead to severe radon problems (greater than 1 WL or 200 pCi/l) in indoor air no matter what the soil characteristics. High concentrations of uranium in bedrock are the causes of the highly publicized radon hazards in residences in the Boyertown area of the Reading Prong in Pennsylvania[1] and in residences built over certain granites and most alum shales in Sweden.[2] However, highly porous and permeable soils can lead to elevated radon concentrations (0.1 to 1 WL or 20 to 200 pCi/l) even if the uranium and radium concentrations in a soil are only average, simply because large volumes of soil gas may move into buildings through permeable soils. For example, houses constructed on glacial outwash gravels and eskers have been documented as having high levels of indoor radon in some areas of the U.S. and in many areas of Sweden. Conversely, heavy clay soils and sediments, especially where water saturated, can nearly block all soil gas migration.

GEOLOGIC SITUATIONS THAT MAY PRODUCE ELEVATED INDOOR RADON LEVELS

The existing information on uranium and radium geology and geochemistry, including the National Uranium Resource Evaluation aeroradiometric data and radon soil gas and indoor radon studies strongly suggest that terrane overlying uranium occurrences has the highest probability of producing severe radon levels in indoor air (greater than 1 WL or 200 pCi/l).[3] These data also show that certain rock types are known to contain uranium concentrations that are above the average for most rocks. These rock types are likely to contain elevated radon concentrations in soils and weathered bedrock across their outcrop area, but are not likely to produce severe levels. In some cases, unusually high permeability of the underlying rocks increases the radon hazard potential. Based on present information, the geologic terranes listed below are believed to have an increased likelihood for producing significantly elevated radon concentrations in homes.

1. Granitic and Metamorphic Terranes

Example A -- The Reading Prong of Pennsylvania, New York, New Jersey, and Connecticut. Locally anomalously uraniferous Precambrian metamorphic rocks and local uranium occurrences of this physiographic province have produced high average levels and locally severe radon levels in some homes in Pennsylvania. The national index home for radon (13 WL or about 2,600 pCi/l) is located in this province near Boyertown, Pennsylvania. Geologically similar terranes exist elsewhere in the eastern U.S., but these other areas are largely unevaluated.[4]

Example B -- Glaciated Paleozoic granitic and high-grade metamorphic terranes in Maine and New Hampshire.

Radon problems in ground waters associated with anomalously uraniferous granites and high grade metamorphic terranes were initially identified in Maine[5] and New Hampshire.[6] Subsequently elevated indoor air radon concentrations have been identified in Maine.[7] Similar areas occur throughout the eastern U.S., but many of them are not glaciated and instead are deeply weathered. The effects of deep weathering on radium geochemistry and radon movement in soils needs to be evaluated.

2. Phosphate Lands

Reclaimed phosphate mining areas in Florida were recognized in early studies as radon problem areas. Subsequent work has shown that unmined phosphate lands in these same areas have similar if not higher radon levels in homes.[8] Phosphate lands occur elsewhere in the U.S., but have been less thoroughly evaluated for radon problems.

3. Pleistocene and Holocene Glacial Outwash Gravels

These sediments are unusually permeable and form soils that also are unusually permeable. Elevated levels of radon are common in the east Spokane, Washington and northern Idaho areas that are underlain by such gravels, especially where they are derived from granitic rocks. In those areas, radon values in houses may approach severe levels.[9]

4. Glacial Till and Moraine, Loess, and Lacustrine Clay

Although these sediments are generally poorly sorted and low in permeability, they are locally highly permeable or are locally very fine-grained and have high radon emanating power. Thus, even though they are not always high in uranium or radium concentrations, soils over these sediments may have high radon availability. Radon levels could be elevated where these sediments are derived from uraniferous rocks such as granites or marine shales. A few indoor and soil-gas radon studies over these sediments in the north-central states suggest locally high values for radon.[10] A significant radon problem may also exist here.

5. Carbonate Terranes, Especially Karst Cave Areas

Caverns and caves in the National Park Service system have been evaluated for radon hazards and a significant number of them have hazardous levels of radon.[11] Some caves contain several working levels of radon. Although limestones and dolomites generally do not contain elevated levels of uranium or radium, they commonly contain phosphatic material, lenses of carbonaceous mudstone, or volcanic ash, and uranium contents are higher where these other materials are present. In addition, the clays and iron oxyhydroxides that form the principal weathering products of limestones and dolomites are very effective adsorbants of uranium and radium. The high permeability of weathered limestones and the fine-grain size of these weathering products also makes them very efficient radon emitters. The few studies of indoor air and soil gas that have been conducted in carbonate terranes suggest a significant percentage of elevated values and possibly locally severe values.[12] Carbonate rocks occur over wide areas of the central and eastern U.S.

6. Marine Black Shales, especially Devonian Shales of the Central U.S.

These rocks have long been known for their elevated uranium concentrations. One study of soil gas and indoor air in part of New York State suggests that elevated radon concentrations occur over these rocks.[13] Similar although more uraniferous rocks in Sweden have been identified as a major radon hazard.[14] Again, marine black shales are widely distributed and underlie parts of major population centers, especially in the central U.S. Other types of marine shales, for example, the Pierre shale of the western U.S., contain slightly elevated uranium concentrations and may create some radon hazards.

7. Arkosic and Vocaniclastic Alluvial Conglomerates and Sandstones

Such rocks exist in several areas of the country in terranes that range from Precambrian to Recent age. Areas with known elevated uranium concentrations include parts of basins adjacent to the Front Range of Colorado, parts of the Mesozoic basins of the eastern U.S., the areas underlain by the Ogallala and White River formations of the High Plains of Texas, Nebraska, and elsewhere, and many intermontane basins in the western U.S., notably in Wyoming and Montana. Many of these units host uranium occurrences and deposits. Locally, there may be severe radon levels in homes near uranium occurrences and generally there may be elevated radon values. A few studies of uranium in waters, and of radon in waters, soil gas, and indoor air in some of these areas suggest that concentrations of radon significantly above the national norm may be expected.

This list must be regarded as only a first compilation. It must be emphasized that not every building over one of these rock types has elevated radon levels. There are many sources of variability that affect radon levels in bedrock, soil gases, and inside structures. For example, granites vary in their uranium content. There is considerable variability attributable to housing factors. Much work is required on each of these types of terranes as well as on other terranes to establish whether radon hazards exist and, if they do, to determine their extent and their causes.

other rock types may also produce radon hazards and each of these potential sources must be evaluated.

FOOTNOTES

[1]Smith, Robert, Personal communication, 1986.

[2]Akerblom, G., P. Anderson, and B. Clavensjo, Soil Gas Radon - A Source For Indoor Radon Daughters. Radiation Protection Dosimetry, Vol. 7, No. 1-4, 1984, pp,. 49-54.

[3]Duval, Written Communication, 1986.

[4]Grauch, R.I., and Katrin Zarinski, Generalized Descriptions of Uranium-Bearing Veins, Pegmatites, and Disseminations in Non-Sedimentary Rocks, Eastern United States. U.S. Geological Survey Open-File Report, 76-582, 1976.

[5]Brutsaert, W.F., S.A. Norton, C.T. Hess, and J.S. Williams, Geologic and Hydrologic Factors Controlling Radon-222 in Ground Water in Maine. Ground Water, Vol. 19, 1981, pp.407-417.

[6]Hall, Francis, University of New Hampshire, Personal Communication, 1986.

[7]Lanctot, M., Maine Geological Survey, Personal Communication, 1986.

[8]Florida Department of Health and Rehabilitative Services, Study of Radon Daughter Concentrations in Structures in Polk and Hillsborough Counties. Florida Department of Health and Rehabilitative Services, 1978.

[9]Lawrence Berkeley Laboratory, Personal Communication, 1986.

[10]Fargo Forum, Fargo, North Dakota, October 20, 1985.

[11]Yarborough, K.A., Radon- and Thoron-Produced Radiation in National Park Service Caves, The Natural Radiation Environment III, U.S. Department of Energy CONF 780422, 1980, pp. 1371-1395.

[12]Sachs, H.M., T.L. Hernandez, and J.W. Ring, Regional Geology and Radon Variability in Buildings, Environment International, Vol. 8, pp. 97-103.

[13]Lilley, W.D., A Geological Assessment of Houses Monitored for Radon in Central New York State, Geol. Soc. America. Abs. with Programs, Vol. 17, No. 1, 1985, p.32.

[14]Akerblom, G., et al., Ibid.

OTHER REFERENCES

Phair, G. and D. Gottfried, The Colorado Front Range, Colorado, U.S.A. as a Uranium and Thorium Province, in Adams, J.A.S., and W.M. Lowder, Eds., The Natural Radiation Environment, Chicago: University of Chicago Press, 1964, pp. 7-38.

Sinnaeve, J., G. Clemente, and M. O'Riordan, The Emergence of Natural Radiation, Radiation Protection Dosimetry, Vol. 7, No. 1-4, 1984 pp. 15-17.

THE INFLUENCE OF GEOLOGIC ENVIRONMENT ON THE DISTRIBUTION OF URANIUM/RADON IN NEW JERSEY

D.F. Schutz and J.A. Powell*

The discovery of a potential hazard from radon in private residences and commercial and public buildings in New Jersey and adjacent areas has led to extensive publicity by newspapers, radio and television media. While the media provide a needed public information service and make a conscientious effort to be accurate, their requirement for presentation of simplified concepts described in nontechnical language has led to oversimplification of inherently complex scientific problems.

One of the most basic oversimplifications has been the concept that the rocks of the "Reading Prong" are the only significant source of radon in New Jersey. When other sources of radon <u>are</u> identified, they are treated as extraordinary or as exceptions to the "established" premise that the "Reading Prong" should be the area of major if not sole concern. This practice continues in spite of early evidence that other sources of radon are present and warnings by state officials that broader areas of northern New Jersey outside the "Reading Prong" may be expected to show high radon values.

The purpose of this paper is to examine the geology of New Jersey and attempt to shed light on some of the complexities of the problem of relating radon in houses to the uranium content of underlying geological formations. In sympathy with the news media, it must be confessed that the concepts and terminology will, of necessity, be oversimplified. This paper will, however, attempt, without becoming too technical, to provide a sharper view of the distribution of geologic environments favorable for uranium, and therefore radon, than has generally been possible in media accounts.

---

*This paper was presented by Donald F. Schutz, president of Teledyne Isotopes, Westwood, NJ and of the American Association of Radon Scientists and Technologists, Inc. Jonathan A. Powell is a geologist working on radon issues for Teledyne Isotopes.

Why Study Geology?

The first question we should ask is: Why is the study of geology relevant to radon? The main reason is that radon comes from uranium in earth materials such as rock, soil, and ground water. If, therefore, we map the distribution of these materials, we should be able to predict where the radon will appear. We find, however, that there are situations in which houses adjacent to each other have very different radon contents even though they are certainly in the same general geologic setting. The problem is that our techniques of geologic mapping are not precise enough to predict the source of high radon on a house-to-house basis.

In spite of evidence that knowledge of the general geologic setting may not lead to precise predictions of radon concentrations in a particular house, the study of geology is potentially fruitful on two counts: 1) it may make it possible to delineate areas that have unusually high or low uranium; and 2) once an area of high uranium is established, the mode of occurrence may be described in order to improve our predictive capabilities and to suggest strategies for remedial action.

Definition of favorable and unfavorable areas for uranium occurrence can then be used to improve efficiency in survey programs. For example, using existing geologic data, New Jersey was able to reduce the number of radon samples required in surveying the coastal plain area of the state.[1] While the validity of that decision must be verified, it nonetheless allowed for a greater concentration of resources in areas believed to have a higher probability for uranium occurrence.

If geologic data can be combined with existing geophysical, ground water, and soil data bases, then improved survey strategies may be developed at lower cost than broadly applied saturation sampling. This is important from an epidemiological standpoint because the public money simply will not be available to sample every house in northern New Jersey. Pinpointing favorable and unfavorable areas is certainly useful in designing large-scale surveys, but what good is such information to the mayor of a town or an individual house owner in an area with a generally high probability for uranium occurrence such as the Reading Prong?

To improve our ability to make predictions about smaller areas, subprovinces, or particular geologic formations or environments, we must determine the style of uranium occurrence characteristic of such areas. Is, for example, the uranium evenly distributed throughout a formation so that all of the houses situated on that formation could be expected to have high radon? Or is the uranium restricted to faults or other linear features so that only houses along those features would be likely to have high radon?

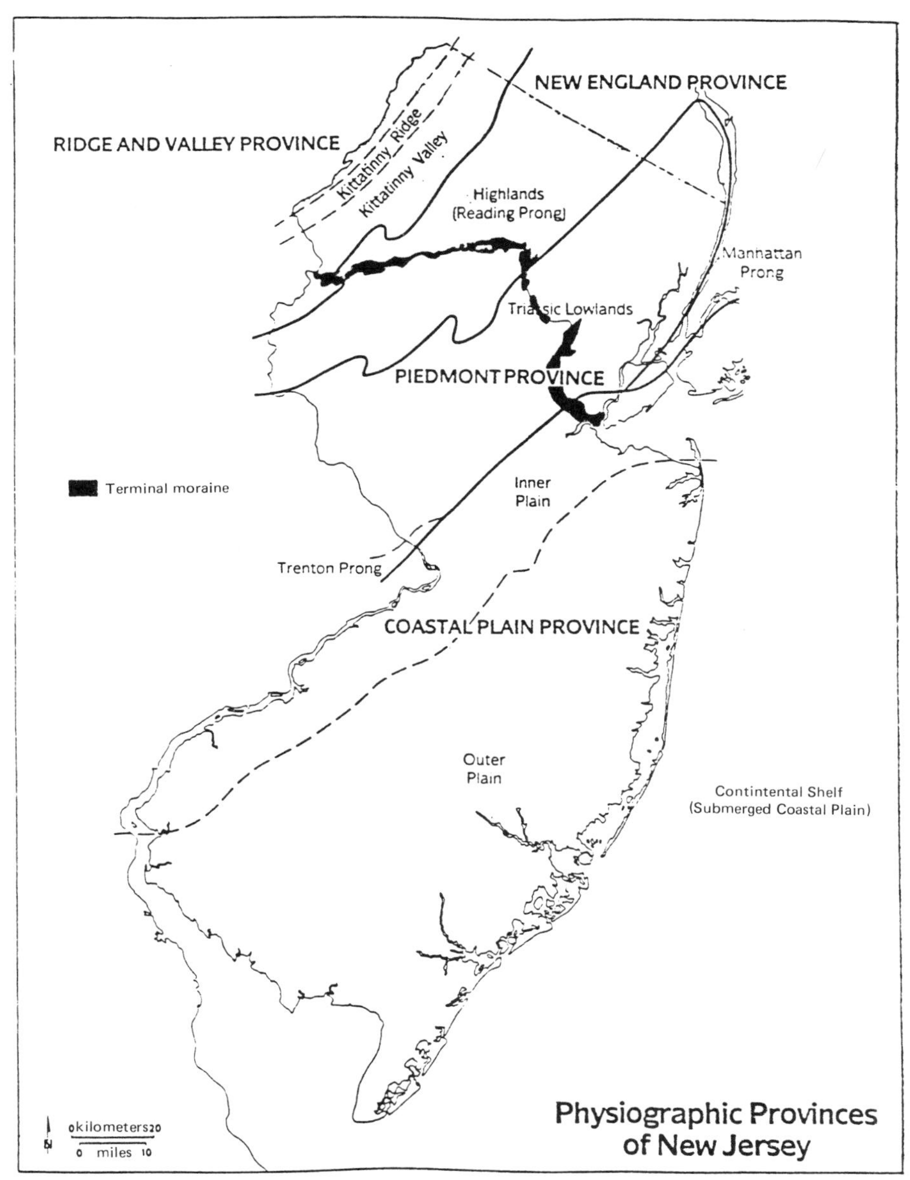

Figure 2.2 - Physiographic Provinces of New Jersey.

GEOLOGIC PROVINCES OF NEW JERSEY

In looking at the geology of New Jersey, the broadest and simplist subdivision of the State has been done according to broad terrain features or physiographic provinces as shown in Figure 2.2. Terrain features are, in a general way, related to rock type and structure. Thus, the igneous and metamorphic rocks of the Reading Prong make up the New Jersey Highlands, the folded Paleozoic sedimentary rocks make up the Valley and Ridge Province and the less resistant Triassic sandstone and shales make up the Piedmont Province or Triassic Lowlands. The coastal plain is underlain by gently inclined rocks that are not very resistant to erosion so the area is generally flat.

While the geologic provinces can be broadly defined, they are in fact complex assemblages of rocks formed over hundreds of millions of years in different geological environments. Because the uranium content can be a function of the chemical and physical conditions in the geologic environment at the time of origin, large differences in uranium content may occur. Not only are conditions at time of origin important, but subsequent conditions may lead to movement of uranium into new locations.

It is important to examine some of the complexities of different rock types which form units within the larger geological subdivisions and discuss the reasons for high uranium content of some units and its absence from others.

The Reading Prong: Igneous and Metamorphic Rocks of the New Jersey Highlands Province

The Reading Prong is now a widely recognized feature, the geographic location of which is shown in Figure 2.3. Contrary to the most common accounts, the Reading Prong is not a "formation"; it is not primarily composed of granite; and it is not uniformly high in uranium content. The Reading Prong is a complex assemblage of igneous rocks which were once molten and metamorphic rocks subjected to high temperature and pressure deep in the earth's crust. These rocks are of pre-Cambrian age (900 to 1,150 million years old). While some of them are highly radioactive, some are essentially barren.[2] The Reading Prong is itself only part of a larger geologic province which extends to the northeast and to the southwest.

The rocks of the Reading Prong were formed for the most part from previously deposited older sedimentary and volcanic rocks. Some of the sedimentary rocks were formations rich in iron and uranium of volcanic origin. The unusual concentration of uranium by sedimentary processes is believed to have been the primary cause of its high

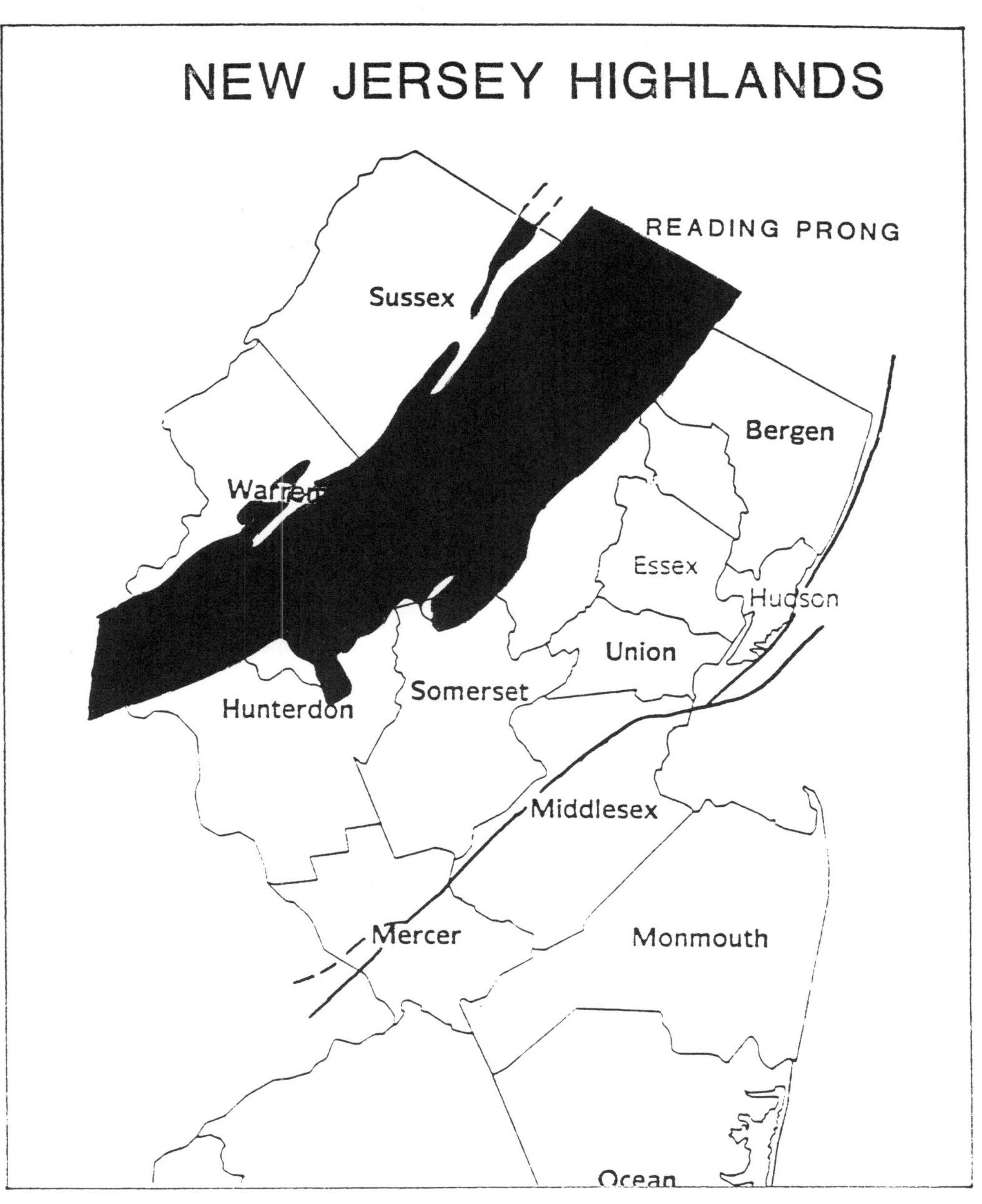

Figure 2.3 - Reading Prong.

concentration in the New Jersey Highlands and adjacent areas of Pennsylvania and New York. There is, therefore, a scientific basis for regarding part of the Reading Prong as the source of the radon problem, but it is incorrect to create the impression that the high concentrations of uranium are limited to the area of the Reading Prong itself.

The rocks of the Reading Prong are not uniformly radioactive. High uranium concentrations tend to be associated with magnetite deposits, which are metamorphic equivalents of iron-rich sedimentary formations,[3] and with pegmatites, which are rocks formed by precipitation from mineralizing solutions in the late phases of granite emplacement.[4] As a consequence of the mode of formation, the uranium bearing magnetite and pegmatite bodies tend to be long and narrow, and the resulting patterns of radon occurrence can be expected to be the same, a possible explanation for adjacent houses having very different radon concentrations.

Part of the rocks exposed in the Reading Prong were once molten or were subjected to such high temperatures and pressures that their uranium content was mobilized and transported to other environments leaving the source rock barren. It is also true that certain rock units never had high uranium and never provided environments that would cause the fixation of mobilized uranium that may have been present from other sources. Thus, two different mechanisms may be responsible for Reading Prong rocks which have little uranium and are unlikely to be significant sources of radon.

In addition to the inhomogeneity with the igneous and metamorphic rocks of the Reading Prong, the area includes inliers of sedimentary rocks of Paleozoic age (245-570 million years old) not related in origin to the rocks beneath them. The areas underlain by these rocks are shown in Figure 2.4. Some of these sedimentary rocks, especially recrystallized limestones, are potentially strong sources of radon, whereas others are not likely to be, because of the absence of uranium or lack of permeability to radon migration.

Paleozoic Sedimentary Rocks of the Valley and Ridge Province

Sedimentary rocks of the same age as the inliers in the Reading Prong occur throughout the Valley and Ridge province to the west and in small areas to the east of the Reading Prong. Such rocks may be barren, as is the case for most of the sandstones, or may produce high levels of radon as has been found for certain limestones. An example of an occurrence of high radon in a limestone is the Jacksonburg limestone which has produced high radon levels near Clinton, New Jersey. Similar rock types such as the Kittatinny limestone may be equally suspect as sources of radon, and areas of outcrop should be examined to determine whether the

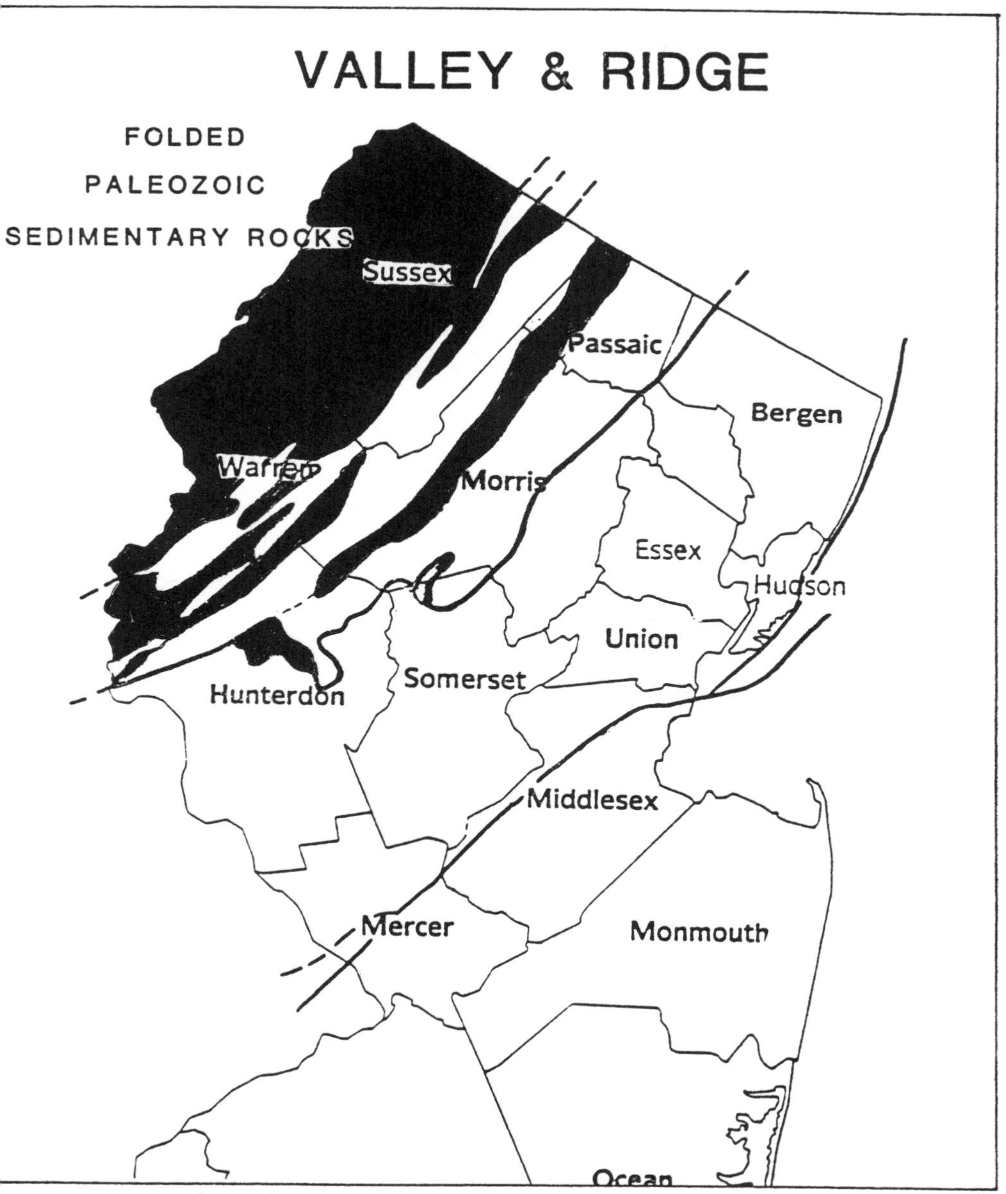

Figure 2.4 - Folded Paleozoic Sedimentary Rocks.

mechanisms producing high radon in homes near Clinton are peculiar to that area or whether they may be found to have been effective elsewhere.

Three mechanisms for producing high concentrations of radon appear to be operative near Clinton and should be looked for wherever similar limestone outcrops occur.[5] The first mechanism involves the presence of phosphatic fossil-shell fragments which have the power to concentrate uranium. The uranium may have been concentrated at the time of origin or later from percolating ground waters.

The second mode of concentration may be precipitation along faults and shear zones from fluids originating in the underlying uranium rich igneous and metamorphic rocks of the Reading Prong. One of the major shear zones in the Clinton area was found to contain over 400 ppm uranium oxide or over 100 times the average uranium abundance in all rocks near the earth's surface. A third mechanism for enhancing the concentration of uranium involves the accumulation of uranium in iron-rich residual soils formed by the action of rain water in dissolving the limestone surface. The role of soil as a concentrating mechanism could explain the unusually high percentage of houses in the affected Clinton neighborhood, a distribution pattern that would not be likely if higher concentrations along linear features such as faults and shear zones were the only operative mechanism. Recognition of the existence and importance of such soils might also permit the development of preventive remedial action by removing the soil from the surface before starting construction.

Mesozoic Sedimentary Rocks of the Piedmont Province

Rocks of Mesozoic age (66-245 million years old) are located in the Piedmont region to the east of the New Jersey Highlands. The two formations which have been found to have the highest uranium contents are the Lockatong formation and the Upper Stockton formation.[6] Two other formations, the Brunswick and the Hammer Creek conglomerate may have local concentrations of uranium in localities where sufficient organic matter existed to cause the chemical fixation of uranium.

The Lockatong formation is a black shale deposit formed in the bottom of a lake bed which extended over a large part of northeastern New Jersey during Triassic time about 230 million years ago.

The area covered by the Lockatong formation is shown in Figure 2.5. It is believed that the uranium in lake sediments was fixed on decomposed plant material near the sediment-water interface and resulted in thin laterally extensive low grade uranium deposits with concentrations ranging from 8,000 to 20,000 ppm uranium.[7] Thus, while the ore grade is lower than the ores of the Reading Prong pegmatite and magnetite associations, the occurrences are

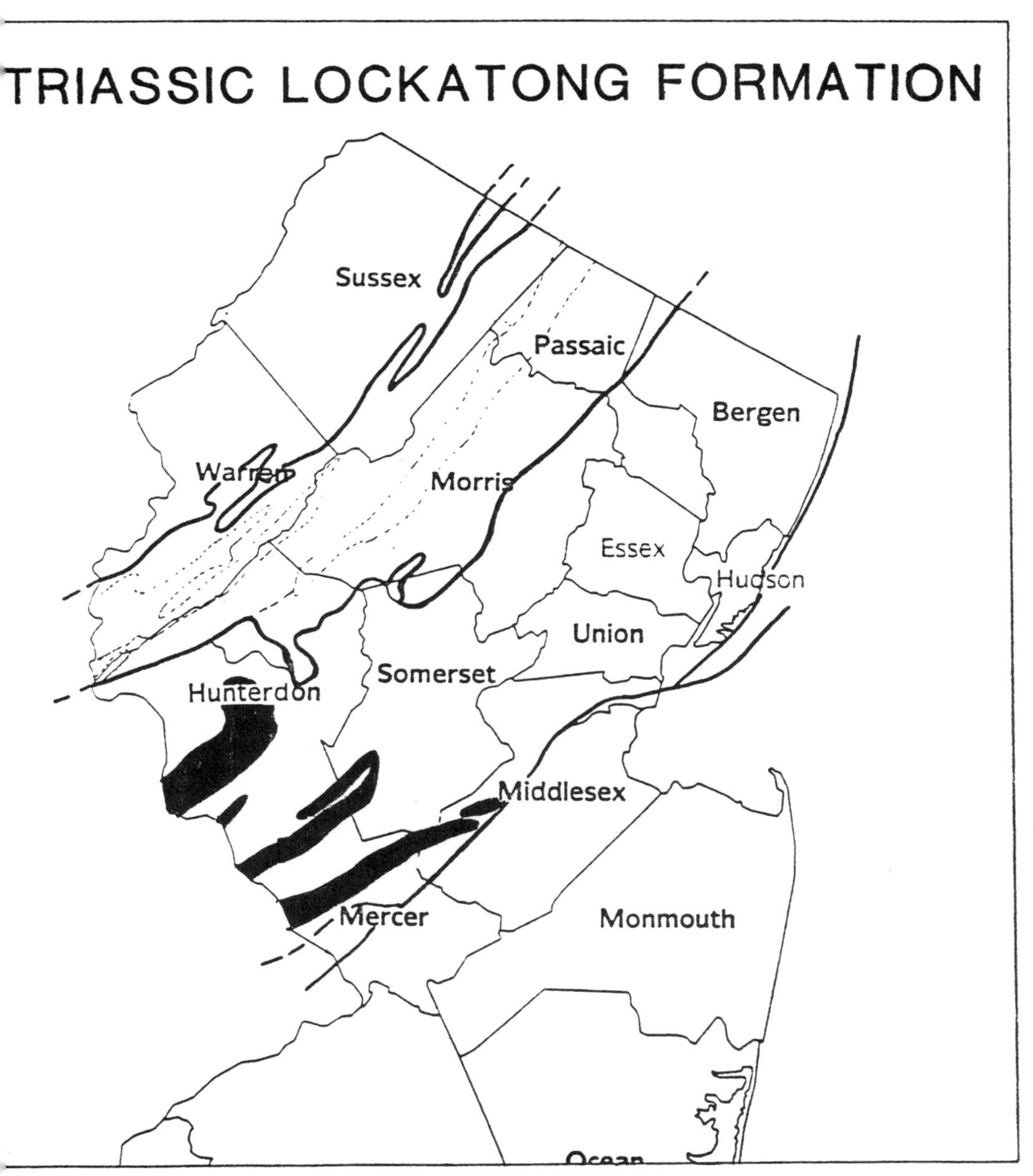

Figure 2.5 - Triassic Lockatong Formation.

more widespread and higher in grade than most black shales, including the well known Alum shale of Sweden.[8] Consequently, the uranium concentrated in the Lockatong lake environment has the potential for creating excess radon concentrations in a high percentage of houses over significant areas of the highly populated Triassic Lowland.

In addition to the occurrence of uranium in the offshore black mudstone of the Lockatong formation, uranium also occurs in the Upper Stockton formation which is a sandstone formed on the lakeshore. The distribution of Upper Stockton formation is shown in Figure 2.6. The uranium in the sandstone beds is believed to have been concentrated from ground water by the organic materials which originated in the Lockatong formation. Uranium occurrences contain 0.01 to 1.28 percent uranium which is sufficient to cause extremely high radon values in soil gas.

GLACIAL MATERIALS

As indicated in Figure 2.2, there is a band of glacial moraine crossing northern New Jersey which marks the maximum extent of the last glaciation about 20,000 years ago. North of the limit of glaciation much of the surface is covered by glacial deposits. Upland areas are covered with an unsorted mixture of sand, clay, and boulders. Low areas are filled with up to 350 feet of sand and gravel.

Glacial deposits may be of unknown provenance and may thus be derived in an unpredictable way from rocks high in uranium content. Outwash may be clean and well-washed sand low in uranium, but high permeability may allow radon to be transported over long distances and, thus, provide ample supplies of radon for accumulation in houses.

Not enough studies of radon in houses built on glacial material have been reported to make possible generalizations about the actual effects of glacial sub-soil.

GROUND WATER

Radon is highly mobile and readily enters ground water in saturated strata. While radon is not usually a problem in large aquifers serving public water supplies or in reservoirs open to the atmosphere, small local aquifers serving private wells can be significant sources of radon in houses even where the host rocks are not exposed at the surface. For this reason, radon may be found over broader areas than are indicated by the geologic outcrop maps shown in Figures 2.3-2.6.

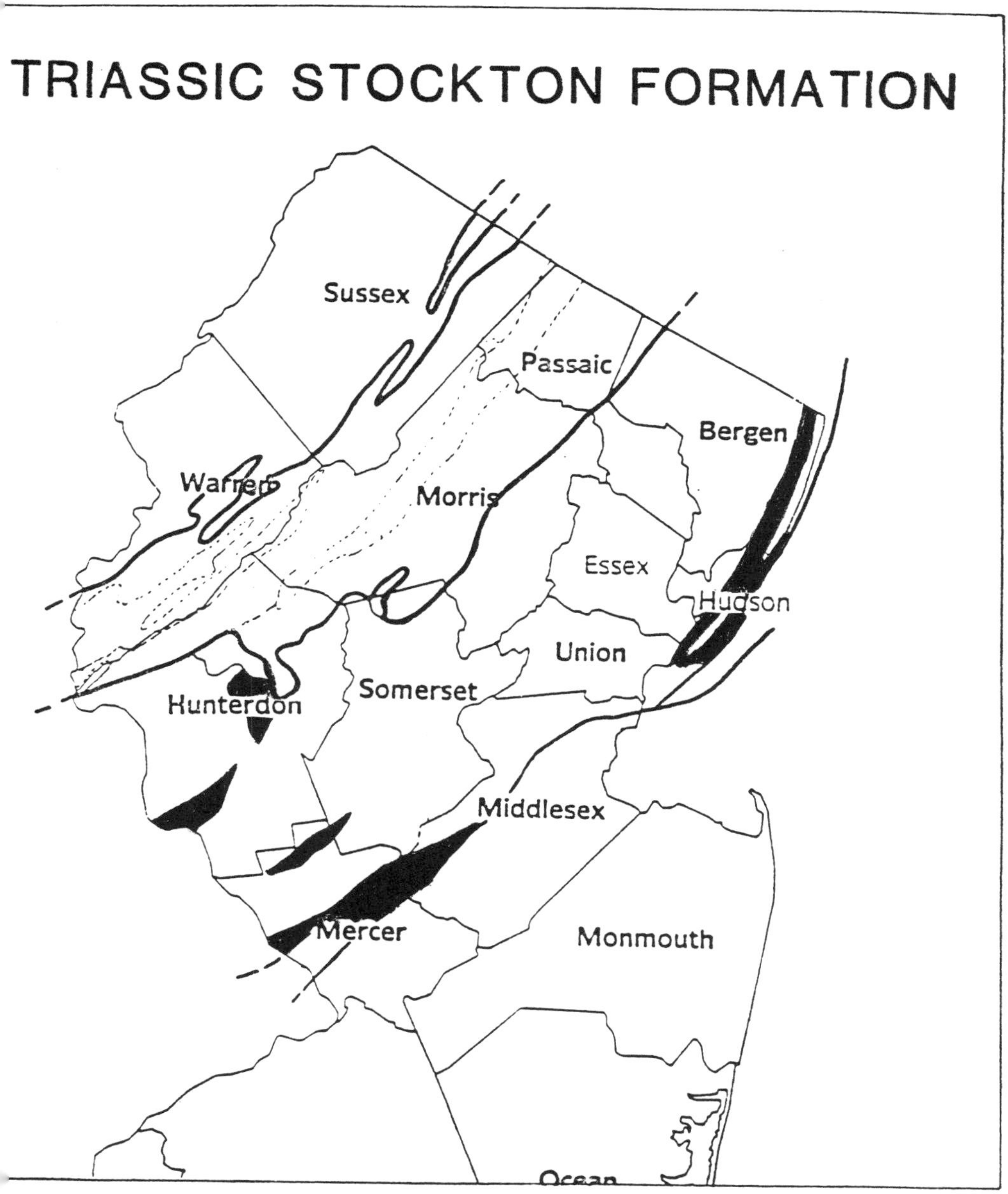

Figure 2.6 - Triassic Stockton Formation.

SUMMARY AND CONCLUSIONS

The geologic terrain we now call New Jersey has for over a billion years been the site of unusually high uranium concentrations. The highest such concentrations occur in the area of the Reading Prong, but the rocks underlying the Reading Prong have for hundreds of millions of years supplied uranium to a variety of geologic environments in adjacent areas.

Because of the variability of uranium distribution in the rocks of the Reading Prong, not all houses will necessarily have elevated radon concentrations and, in fact, higher population exposures may occur outside the Reading Prong in geologic environments where the uranium is more uniformly spread over large areas. Even though the peak concentrations in those areas may be lower than in the Reading Prong, a larger percentage of the houses might show elevated concentrations, and, coupled with higher population density, result in a larger total human exposure.

Areas of particular interest outside the Reading Prong include the Paleozoic limestones which provide chemical environments for concentrating uranium and making the radon decay products available to enter houses now situated above its outcrop areas.

In addition, throughout the highly populated Triassic Lowlands, there are areas underlain by the Lockatong formation in which uranium was concentrated in black shale deposits formed in a vast lake which existed for millions of years.

On the lake shore, sandstones were formed and provided another environment which later became host for migrating organic materials which caused the fixation of uranium now found in outcrops of the Upper Stockton formation.

In later times the penultimate stage of redistribution came when the glaciers covered the northern part of the area and created a pattern of occurrences not now subject to coherent description. The final stage of the mobilization of uranium decay products is still in process as ground waters make available dissolved radon to owners of private wells.

In all these environments the uranium in the rocks only provides the potential for radon to be present. Soil conditions, house construction, heating and ventilation systems and source of water are crucial in determining whether radon present in the underlying rocks will actually enter and accumulate in the enclosed spaces of our dwellings.

Because of the significant implications for health and for property values, careful geological analysis should precede any statements that could, through inaccuracy or oversimplification, place an unwarranted stigma on a given area or, conversely, give a false sense of security with regard to a particular locality.

FOOTNOTES

[1]Request For Proposal for Statewide Scientific Study of Radon for the Bureau of Radiation Protection X-391. Issued by the State of New Jersey Department of the Treasury, Division of Procurement and Central Services, February 13, 1986.

[2]Popper, G.H.P. and T.S. Martin, National Uranium Resource Evaluation Newark Quadrangle Pennsylvania and New Jersey. Bendix Field Engineering Corporation, PGJ/F-123(82), March 1982.

[3]Gundersen, Linda C., Geology and Geochemistry of the Precambrian Rocks of the Reading Prong, New York and New Jersey - Implications for the Genesis of Iron - Uranium - Rare Earth Deposits. USGS Research on Energy Resources, 1986, Program and Abstracts, Edited by L.M.H. Carter, V.E. McKelvey. Forum on Mineral and Energy Resources, U.S. Geological Survey Circular 974, 1986, p. 19.

[4]Popper, G.H.P., op.cit.

[5]Tanner, A.B., Personal Communication, 1986.

[6]Turner-Peterson, C.E., "Sedimentology and Uranium Mineralization in Triassic-Jurassic Newark Basin, Pennsylvania and New Jersey," in Turner-Peterson, C.E., Ed., Uranium in Sedimentary Rocks: Application of the Faciew Concept to Exploration: Short Course Notes. Rocky Mountain Section of Society of Economic Palentologists and Mineralogists, 1980, pp.149-175.

[7]Turner-Peterson, C.E., "Sedimentary Framework and Uranium Potential of the Newark Basin, Pennsylvania and New Jersey." Abs. Society of Economic Geologists, Fall Meeting, November, 1976, p. 742.

[8]Turner-Peterson, C.E., P.E. Olsen, and V.F. Nuccio, Modes of Uranium Occurrence in Black Mudstones in the Newark Basin, New Jersey and Pennsylvania. Proceedings of the Second U.S. Geological Survey Workshop on the Early Mesozoic Basins of the Eastern United States, U.S. Geological Survey Circular 946, IN PRESS, pp. 120-124.

GEOLOGIC CONTROLS ON RADON GAS IN GROUND WATER

Francis R. Hall*

INTRODUCTION

Radon is a short-lived, inert gas which is soluble in water. The concern with radon is that it is carcinogenic to humans by inhalation from air and by ingestion of water. It is essential to differentiate between two main sources of radon gas. One source which is discussed herein occurs mainly in private dwellings which use drilled wells in bedrock for water supply. Larger municipal systems either use surface water which is low in radon or ground water which has adequate time for radon to degas or decay before reaching the consumer. The other source, which is from a direct flux into buildings from surrounding soils and rocks, is discussed elsewhere in this volume. The two sources can operate together in private dwellings, but the direct flux can occur whether or not a bedrock well is involved.

Health standards have not been set for radon in ground water, but there seems to be a general consensus that a value on the order ot 10,000 - 20,000 picocuries per liter (pCi/l) is worrisome. Also, it seems generally accepted that values about 100,000 pCi/l are definitely a cause for concern. For present purposes, 20,000 pCi/l is arbitrarily taken to indicate high radon and values in excess of 70,000 pCi/l to indicate very high radon.

CHEMICAL FEATURES

Radon-222 (Rn-222) is of primary concern, and it is a product of the uranium-238 (U-238) decay series (Figure 2.7). Note that half lives are given which represent the time required for an original amount to decay to one half that amount. Therefore, the larger the half life the longer lived the isotope. There are two other much shorter-lived forms of radon from the thorium-232 and uranium-235 decay series which will not be considered further. The key features of the U-238 decay series are:

---

*Francis R. Hall is Professor of Geology at the University of New Hampshire, Durham, N.H. 03824.

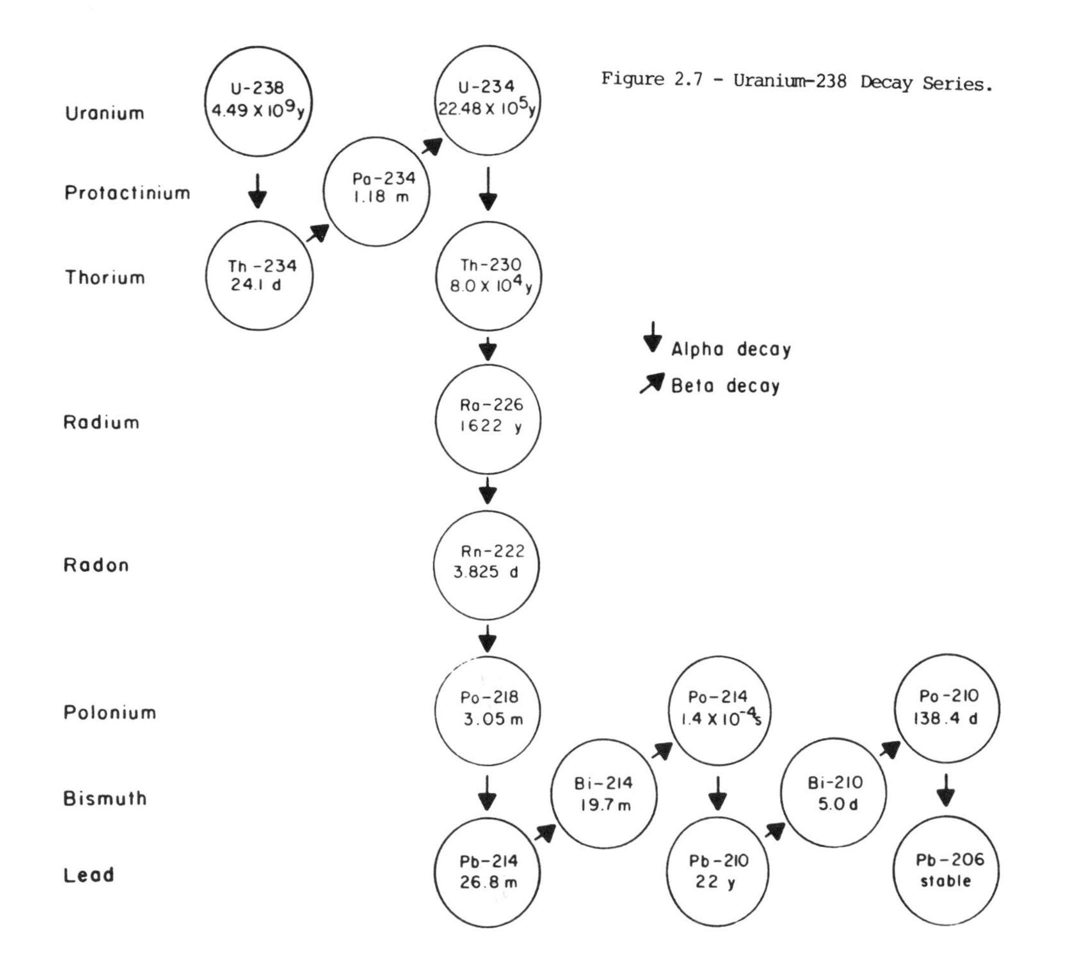

Figure 2.7 - Uranium-238 Decay Series.
Uranium
Protactinium
Thorium
Radium
Radon
Polonium
Bismuth
Lead
U-238 4.49 X 10^9y
U-234 2.48 X 10^5y
Pa-234 1.18 m
Th-234 24.1 d
Th-230 8.0 X 10^4y
Alpha decay
Beta decay
Ra-226 1622 y
Rn-222 3.825 d
Po-218 3.05 m
Po-214 1.4 X 10^{-4}s
Po-210 138.4 d
Bi-214 19.7 m
Bi-210 5.0 d
Pb-214 26.8 m
Pb-210 22 y
Pb-206 stable

1. The primary, very long-lived U-238 decays to U-234.

2. U-234 decays to thorium-230 (Th-230) which decays to radium-226 (Ra-226).

3. Ra-226 with a half life of 1,622 years decays to Rn-222 with a half life of 3.82 days. Thus, Ra-226 is the immediate parent of Rn-222, whereas U-238 is the ultimate parent.

4. Rn-222 decays through intermediate steps until the longer-lived lead-210 (Pb-210) is reached (Figure 2.7). The further decay to stable Pb-206 will not be discussed. The important feature is that the decay products, mainly alpha particles, down to Pb-210 pose hazards to human health.

Therefore, in order to have radon gas in ground water, it is necessary not only for the U-238 decay series to operate, but for there to be a process whereby Ra-226 is in contact or near contact with the water. Furthermore, because of the short half life of Rn-222, the radium must be fairly close by. That is, Rn-222 can not move very far in ground water in the 20 days or so required for nearly complete decay. In fact, most of the radon decays in five to six days. The process leading to radon gas seems fairly straightforward, but chemical and rock properties combine to create a rather complicated situation in the field.

GEOCHEMICAL FEATURES

Uranium is widespread in rocks of the earth's crust with an average abundance of 2.7 parts per million (ppm).[1] This abundance varies by rock type with averages of 4.8 ppm in granite, 0.6 in basalt, and 3.2 in shale. Carbonates and sandstones have values in the one to two ppm range.[2] It is also noteworthy that more than 99 percent of naturally occurring uranium is U-238. In order for the U-238 decay series to generate high radon values, it is necessary to have a concentration process such that the rock is enriched above average values and to have a secondary process to concentrate it even further under more localized conditions.

Because of its importance in New England, granite will be used as an example of how these processes work to cause high radon in ground water. Nevertheless, similar processes can concentrate uranium in a rock type such as shale. Also, uranium can be concentrated in one rock type and then migrate to another rock type. These latter possibilities are beyond the scope of this paper.

Uranium is an incompatible element in the sense that it does not readily enter into the crystal structure of many rock-forming minerals. This is due both to its large atomic radius (1.05 angstroms) and to its high valence state (+4 or +6 under natural conditions). The same is also true for thorium at 1.10 angstroms and +4, respectively. Therefore, during the emplacement of a granite after melting, the late stage melt or residual fluid will be enriched in uranium and thorium. The resulting granitic rocks and associated pegmatite dikes will also be enriched in uranium and thorium.

Good examples of the enrichment process are the Concord and other two-mica granites of the New Hampshire plutonic succession (shown as Hercynican Two-mica Granite on Figure 2.8) and Conway and related granites of the White Mountain plutonic succession (shown as Mesozoic White Mountain Plutonic-Volcanic Series on Figure 2.8).[3-5] It is of note that the two-mica granites have a high to very high content of radon in ground water whereas the Conway has somewhat lower values[6] although it is generally the more radioactive rock of the two granites.

The next required step is to mobilize and move members of the uranium decay series so as to increase local concentrations at shallower depths below present land surface. This can begin late in the formation of the granite, but it is enhanced as soon as circulating ground water reaches the granite and chemical weathering begins. This requires that the uranium be accessible to weathering and not be embedded in the crystal lattice of resistant accessory minerals. If the latter is the case, then the process is less effective. Uranium in the original rock is in the insoluble +4 valance state, so it is not mobile. The same is true for thorium which remains insoluble. If oxidizing conditions occur as with oxygenated ground water then the oxidized or +6 valance state U-238 becomes mobile, and it can be transported by ground water. Also, radium is mobile to a limited extent, so it can move away from its immobile parent Th-230. Under appropriate conditions, the uranium and radium will be redeposited mainly along fractures. Therefore, the migration of these constituents can be viewed as a stepwise process which is controlled primarily by local oxidation-reduction conditions and the ground water flow pattern.

RADON GAS IN GROUND WATER

The final aspect to be considered is how radon gas gets into ground water. This also turns out be fairly complex. For example, most high radon waters in granite are not in equilibrium or supported by parent isotopes in solution. There may be some dissolved uranium and even a little radium but little or no thorium, so these can not supply the observed radon. Therefore, the parent isotopes have to be contained within the crystal lattice of minerals or occur between mineral grains or as coatings along fractures.

The radioactive decay takes place by the alpha recoil mechanism where the new isotope moves in a direction opposite to the alpha particle. The distance of the movement is short, on the order of a few hundred angstroms.

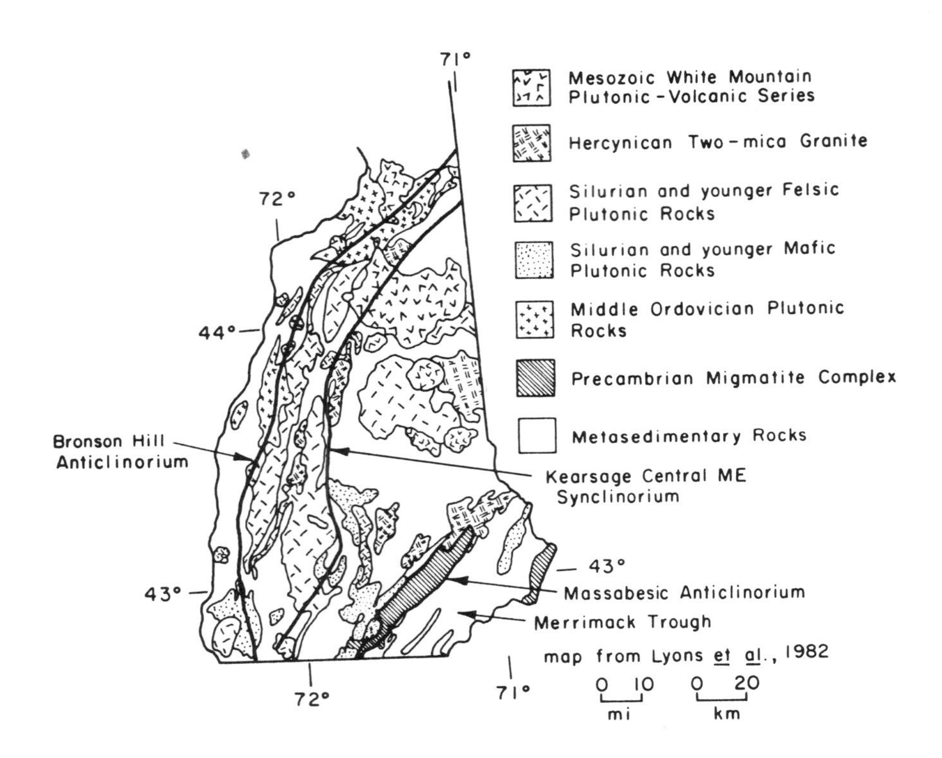

Figure 2.8 - Bedrock Geology Map of New Hampshire.

Therefore, if the parent, Ra-226, is in a mineral grain it must be very close to the outer boundary for the radon to leave the grain. Furthermore, if the radon encounters a dry void or another mineral grain then it will go into adjacent grains and be lost from our point of view. If the voids are filled with water, then the radon will be dissolved by the water. Obviously, this process is easier if the voids are coated with uranium and/or radium. Finally, the water-filled voids need to be part of a ground water flow system in order for the radon to move very far. Tanner has thoroughly reviewed these processes.[7,8]

Recent work by autoradiography shows that in the two-mica granites the radioactive constituents are found mainly along grain boundaries and in small fractures which implies close contact with ground water.[9] Similar work was not done for the Conway granite, but it is known to have much of its radioactivity tied up within accessory minerals, which implies less contact with ground water. Therefore, ground waters from two-mica granites can generally be expected to have a higher radon content even if the Conway is more radioactive in bulk composition.

ROLE OF WATER WELLS

There is increasing evidence that the water well itself plays a significant role in the occurrence of radon. This would appear to come about by some combination of the direct effect of pumping stress on the natural flow system and a more indirect effect from this stress by causing changes in geochemical conditions. For example, pumping moves more water more quickly from a greater volume of rock than would happen naturally. Thus, various fractures in a fracture pattern might have varying radon content; so what appears at the well is a function of pumping rate and length of time of pumping and whether pumping is intermittent or continuous. Another possibility is that a pumping well may induce inflow of lower radon water from nearby water bodies or overlying sediments. Also, by creating oxidizing conditions, the pumping could cause uranium and/or radium to move in closer to the well. Alternatively, reducing conditions could cause iron complexes which would be mobile and cause uranium to move. Then a change in conditions could cause a redeposition close to the well and so on. Some examples and possible explanations are given in Hall, et al.[10] and Wathen[11]. It should also be noted that because of the short half life of Rn-222 and generally slow ground water velocities, in most cases radon does not come from more than a hundred feet distance from the well.

RADON IN GROUND WATER IN NEW HAMPSHIRE

General ranges in pCi/l for various rock types in New Hampshire are: glacial deposits, less than 1,000; lower grade metamorphics, 700-5,500; higher grade metamorphics,

1,300 - 30,000; diorite, 700-8,600; quartz monzonite, 4,000 - 105,000; Conway granite, 300-70,000; and two-mica granite, 600-1,300,000. The lower values in the two granites if not the other igneous and metamorphic rocks probably represent either incorrectly identified rocks or cases where well water is diluted by nearby surface water or recharge from overlying sediments. Values in springs, headwater streams, and dug wells tend to be low even in two-mica granites whereas larger streams are quite low to below detection limits.

CONCLUSIONS

In order for ground water to have a high radon content where high is somewhat arbitrarily taken as 20,000 pCi/l, it is necessary to have the following processes:

1. The parent rock is enriched above the natural average uranium content.

2. A secondary process operates so as to cause a further generally more localized enrichment that is fairly close to land surface and that is in a zone of active ground water circulation.

3. The parent isotopes must be close to grain boundaries or occur along grain boundaries and fractures under conditions of water filled voids that are connected to the ground water flow system.

All of this should take place so as to cause enrichment of the parent isotopes within 100 feet or so of the well as radon decays too rapidly to move much further.

Examples are given for New Hampshire, where two-mica granites have very high radon concentrations due mainly to the causes listed in item 3 above. The Conway granite, which actually is more radioactive in bulk, has lower radon because the parent isotopes are found within resistant accessory minerals. Many wells have radon in excess of 20,000 pCi/l (depending on rock type) with values in excess of 1,000,000 pCi/l in rare cases.

FOOTNOTES

[1]Krauskopf, K.D., Introduction to Geochemistry, New York: McGraw-Hill, 1967.

[2]Hem, J.D., "Study and Interpretation of the Chemical Characteristics of Natural Water," 2nd ed, U.S. Geological Survey Water-Supply, Paper 1473, 1970.

[3]Armstrong, F.C. and E.L. Boudette, Two-Mica Granites. Part A: Their Occurrence and Petrography, Geological Survey Open-file Report 84-173, 1984.

[4]Armstrong, F.C. and E.L. Boudette, Two-Mica Granites. Part B: The Petrochemistry of the Major Nonmetallic and Metallic Oxides, U.S. Geological Survey Open-File Report 85-, 1985.

[5]Lyons, J.B., E.L. Boudette, and J.N. Aleinkioff, "The Avalonian and Gander Zones in Central Eastern New England," Geological Association of Canada, Special Paper 24, 1982, pp.43-65.

[6]Hall, F.R., P.M. Donahue, and A.L. Eldridge, "Radon Gas in Ground Water of New Hampshire," Proceedings of the Second Annual Eastern Ground Water Conference, National Water Well Association, 1985, pp. 86-100.

[7]Tanner, A.B., "Radon Migration in the Ground, A Review," The Natural Radiation Environment, J.A.S. Adams and W.M. Lowder, Ed., Chicago: The University of Chicago Press, 1964, pp. 161-190.

[8]Tanner, A.B., "Radon Migration in the Ground, A Supplemental Review," Natural Radiation Environment III, T.E. Gesell, and W.M. Lowder, Eds., Technical Information Center, U.S. Department of Energy, Vol. 1, 1980, pp. 5-56.

[9]Wathen, J.B., Factors Affecting Levels of Rn-222 in Wells Drilled into Two-Mica Granites of Maine and New Hampshire, Unpublished M.S. Thesis, University of New Hampshire.

[10]Hall, et al., 1985, op.cit.

[11]Wathen, 1986, op.cit.

UNIVERSITY OF PITTSBURGH MEASUREMENTS IN NEW JERSEY AND EASTERN PENNSYLVANIA

Bernard Cohen and Nickifor Gromicko*

The University of Pittsburgh Radon Project measures radon levels by the diffusion barrier charcoal adsorption method using one week exposures. The collector is a metal can, 3 inches in diameter and 1 inch high, containing 25 g of charcoal. Entrance is through a diffusion barrier which controls the rate at which radon can enter or leave the charcoal, and thereby gives the radon level averaged over several days. Exposures are started by removing a tape covering the entrance, and are terminated by replacing that tape. The collector is then put back into its box and sent back to the Laboratory using a postage paid mailing label which is part of the kit. Its radon content is measured by detection of three gamma rays characteristic of radon decay. Twenty measuring systems are used giving a Laboratory capacity of about 600 measurements per 24 hour day. Current operations average about 300 measurements per day.

Two programs are in progress, $12 measurements and random, no-charge measurements. The former measurements are done for all requestors at a charge of $12 and measured radon levels are correlated with information obtained from questionnaires. These measures are useful for finding areas that might have high average radon levels. The random, no-charge measurements are made on a computer-selected random sample of houses and are used to determine mean radon levels in selected counties and states and to make the data obtained from the $12 measurements more useful.

In general, the average radon levels obtained from $12 measurements are higher than those obtained from random measurements. There are several obvious reasons for this. For example, a person is more likely to purchase a measurement if high radon levels have been reported or suspected in his area or if his house has properties that are believed to be correlated with high radon levels.

The arithmetic (average) and geometric means of our measured radon levels for each county in New Jersey are listed in Table 2.1 as obtained from our $12 measurements during time periods roughly corresponding to the past winter and spring. Only counties where at least 10 measurements

---

*Bernard Cohen is Professor of Physics at the University of Pittsburgh, Pittsburgh, PA and Nickifor Gromicko is Project Manager for the University of Pittsburgh Radon Project.

Table 2.1 - Average and mean radon levels in pCi/liter for New Jersey Counties from $12 measurements during two time periods, March-June 1986 and November 1985 to April 1986. "Number" is number of measurements on which averages and means are based.

County	March-June 1986			Nov.85-Apr.86		
	number	average	mean	number	average	mean
Atlantic	2	3.5	3.44	0	-	-
Bergen	313	1.0	.77	388	1.4	.96
Burlington	7	0.9	.81	14	1.2	1.00
Camden	4	2.2	1.82	9	1.8	1.37
Cape May	1	0.7	.70	1	0.9	.90
Cumberland	1	0.8	0.80	-	-	-
Essex	128	1.4	.98	105	1.4	1.02
Gloucester	3	1.5	1.25	2	1.5	1.24
Hudson	12	1.2	.86	8	2.4	1.10
Hunterdon	803	13.0	3.72	596	13.7	3.66
Mercer	73	2.7	1.29	48	3.1	1.79
Middlesex	100	2.3	1.10	44	1.5	1.06
Monmouth	34	1.9	1.12	12	2.4	.98
Morris	2,081	3.6	1.66	2,625	3.4	1.71
Ocean	4	0.8	.69	5	0.8	.58
Passaic	337	1.7	1.13	485	2.5	1.28
Salem	1	1.2	1.20	0	-	-
Somerset	324	4.5	1.74	279	4.2	2.01
Sussex	964	3.1	1.81	883	3.7	2.07
Union	81	1.2	.90	64	4.2	1.36
Warren	203	5.3	2.71	270	8.2	3.41

have been made are worthy of consideration in drawing conclusions.

It seems evident that the highest mean levels are in Hunterdon and Warren counties, which border Pennsylvania and are traversed by the Reading Prong. Somerset and Sussex counties, which are also associated with the Prong, apparently rank third and fourth, and Morris County, which is well covered by the Prong, probably ranks fifth. All of these have (geometric) mean levels above the U.S. national mean which is believed to be about 1.0 pCi/l, but for purchased measurements would probably be about 1.5 pCi/l. However with the exception of Hunterdon and Warren counties, the excess seems to be rather negligible. Bergen, Passaic, and Essex counties, which are also traversed by the Reading Prong, seem to have below average radon levels. It seems that the radon problems associated with the Reading Prong decrease with increasing distance from the Pennsylvania border.

In many counties, the average is much greater than the mean (i.e., the arithmetic mean is much larger than the geometric mean). This indicates that there are some measurements very much higher than the mean. For example, a single measurement over 1,000 pCi/l in Hunterdon County increased the average by more than 1 pCi/l, but had a negligible influence on the mean.

Similar data for selected eastern Pennsylvania counties are shown in Table 2.2. It is immediately evident that mean radon levels are very much higher in Pennsylvania than in New Jersey. It is also evident that the problem is not limited to the Reading Prong counties, Berks, Lehigh, and Northampton; in fact, the other counties listed, Cumberland, Dauphin, Lancaster, Lebanon, and York counties, all seem to have at least comparable radon levels, and in all cases they are at least twice the levels in any New Jersey counties except Hunterdon and Warren, and substantially higher than the latter two. Clearly, radon is much less a problem in New Jersey than in Pennsylvania.

Some results from random, no-charge measurements are listed in Table 2.3. We see that the means for Monmouth, Morris, and Somerset counties (New Jersey) are about 30 percent less than in the comparable $12 measurements. The differences for the Pennsylvania counties vary rather widely, but seem to be somewhat larger. Table 2.3 also seems to indicate a much larger difference between spring and summer for the Pennsylvania data. In general, the statistical uncertainties in the random, no-charge data are still quite large, making it risky to draw strong conclusions from them.

Summarizing the New Jersey situation, radon is much less a problem than in Pennsylvania and the problem decreases as the distance from the Pennsylvania border increases. With the exception of Hunterdon and Warren counties, mean radon levels are near or below the U.S. national mean. This should not be interpreted as meaning

Table 2.2 - Average and mean radon levels (pCi/liter) for selected Pennsylvania Counties from $12 measurements. See caption for Table 2.1.

County	March-June 86			Nov.85-Apr.86			July-Oct. 85		
	number	average	mean	number	average	mean	number	average	mean
Berks	136	7.5	3.83	126	8.8	3.94	124	12.7	6.20
Cumberland	77	9.4	5.77	51	8.3	5.21	7	7.0	5.06
Dauphin	24	7.3	4.18	13	6.5	4.92	0	-	-
Lancaster	223	7.7	4.22	29	22.7	7.08	3	9.4	3.92
Lebanon	13	32.0	6.14	17	14.5	8.12	0	-	-
Lehigh	189	8.1	3.99	470	11.4	5.03	360	9.9	4.43
Northampton	125	10.7	3.45	300	27.8	6.56	335	14.8	3.73
York	35	18.7	4.98	15	7.5	4.57	5	14.6	7.58
All PA	1429	7.5	3.18	1716	12.1	4.05	910	-	4.18

Table 2.3 - Average and mean radon levels (pCi/liter) from random-no charge measurements in Pennsylvania and New Jersey Counties

County	March-May 1986			Jan.-Apr. 1986			June-Sept. 1985		
Pennsylvania	number	aver.	mean	number	aver.	mean	number	aver.	mean
Berks	3	3.8	2.1	-	-	-	16	1.3	1.0
Cumberland*	4	12.5	10.3*	28	7.7	4.9*	-	-	-
Lebanon	37	8.8	4.0	-	-	-	10	1.6	1.2
Lehigh	25	8.1	3.9	16	20.9	7.6	22	2.1	1.2
Northampton	42	8.5	3.2				49	2.0	1.2
All PA	268	6.5	2.9	121	6.8	2.7	257	1.7	1.1
New Jersey									
Atlantic	12	0.5	0.5	-	-	-	16	0.6	0.6
Monmouth	29	0.8	0.7	-	-	-	39	0.7	0.6
Morris	144	1.8	1.2	-	-	-	-	-	-
Somerset	56	1.8	1.3	3	2.5	1.8	23	0.9	0.7

*A random-no charge study in Cumberland County, PA during Winter 1984-85 involving 165 houses gave an average of 9.2 pCi/liter and a mean of 6.3 pCi/liter.

that there are not individual houses with severe radon problems. Some houses with over 100 pCi/l were found in Morris, Sussex, and Somerset counties as well as in Hunterdon and Warren counties. There is evidence from our work that this situation may be typical -- individual houses with very high radon levels have been found in many areas where there is no evidence of generally high mean levels.

HIGHLIGHTS FROM THE DISCUSSION: THE GEOGRAPHIC DISTRIBUTION OF RADON

COMMENTS OF GERRY NICHOLLS*, DISCUSSANT

G. Nicholls: I just have three comments on the presentations. My first comment is that the quality of background information which the panel has presented will serve us well in the next three days. It is unusual to go to a meeting where you don't learn something; I think this morning we are fortunate in that virtually everyone here, myself included, can take new ideas away from the discussion. And I use the term "new" advisedly, as an entry into my next comment. New in radon can mean either a few minutes ago when you heard a news broadcast or it can mean 1930 and 1935 when uranium miners were coming down with a disease called "Mountain Sickness" in Europe. For those of you who are hearing those stories for the first time, they are new, but one of the striking things about the indoor radon phenomena is that most of it is not new. The geology is fairly well known; it has been for quite some time. The fact that radon causes or at least increases one's risk of lung cancer is well established. The places where uranium exist are fairly well known and have been understood for quite a while. The thing that is new is that we have a sudden concern, basically prompted by the Watras event. You can trace most of the interest in radon in this country on the public level to the Watras event. There have been people who have been working on radon for quite some time, and they would have eventually uncovered the Watras-type houses because we know there are more than one of them around. It was an unfortunate event for the Watras family that became a very fortunate event for the public health of the citizens of the northeast, particularly New Jersey. When Stanley Watras walked through that monitor at Limerick, it brought to our attention something we did not know, which is that radon could build up to truly gigantic levels within a home. I think Watras still holds the record: something around 14 WL (2,800 pCi/l). We haven't seen anything that high in New Jersey yet. I sincerely hope that he keeps the record, because his home is remediated. The third item that I thought was interesting was Dr. Sachs' comment on the Chernobyl disaster. As he points out, the number of people who die attributed to radon exposure in this country is estimated at EPA to be something of the order of 10-20,000

*Gerald Nicholls is the Acting Head of the Bureau of Radiation Protection of the New Jersey Department of Environmental Protection.

people, or 5-20,000 people, and that is more folks in one year than are anticipated to die in the worst case estimates that we have heard so far out of the Ukraine. We truly have something here of an immense public health significance, and the next few days should be very informative.

RADON MEASURES AND GUIDELINES

Question: From a layman's point of view, please explain the difference between picocuries per liter of air and Working Levels.

G. Nicholls: Let's take an equivalence between picocuries per liter and Working Level such that 200 picocuries per liter is assumed to be equivalent to 1 Working Level. Now the assumption inherent in that equivalence is that there is a 50 percent equilibrium between radon and its decay products. That is probably far more conservative for real houses than most people realize. Real houses that we have seen in New Jersey typically have an equilibrium ratio of something on the order of 20 percent to 40 percent. So if you want a quick and dirty way to get from picocuries per liter down to Working Levels, divide the picocuries per liter by 200 and you will probably wind up in agreement with most of the people who have to make that off-the-top-of-their-head calculation on a day-to-day basis.

Question: The New Jersey DEP seems to recommend 4 picocuries per liter as a probable maximum safe level for indoor radon. Is this a correct interpretation?

G. Nicholls: 4 picocuries per liter is approximately equivalent to 0.02 WL. 0.02 WL is not a standard, but it is a number that has been used in the past in at least 3 specific situations as guidance to when remediation should be undertaken. It has been used in the state of Florida with regard to phosphate mining. It has been used in Grand Junction, Colorado with regard to housing structures that were built on mill tailings. And it has been used right here in New Jersey, in Montclair, Glen Ridge, and West Orange, where there is a "technologically-enhanced" radium problem. That is a euphemism for "industry put it there, and now government has to take it away." In those homes, the remedial action level of 0.02 WL, as I understand it, was a compromise between what could reasonably be attained by mitigation and what the health effects were.

H. Sachs: In early 1981, the U.S. EPA was ready to release guidelines which would have had the sanction of the US government based on epidemiological evidence. It was the best they could do at the time. We had an election and the new government officials decided that this was a business that government should not be in, in contrast with I think England, certainly Australia, Canada, Sweden, and much of the rest of the civilized world. We, therefore, do not have a federal standard which can be used to help the public, the industries, and for that matter state and local governments. We have not allowed our wise people to come up with a number that would bear the sanction of the federal government as being the best guide we could achieve now because it was on the agenda of this particular administration to stay out of that business. Presumably the private sector could do it.

VARIABILITY OF RADON LEVELS IN HOUSES

Question: If a house has a safe level of radon, will this level change in the future?

J. Otton: This is basically a housing question; there are other people in the audience who might best answer the question. It depends in part on what kinds of measurements you have made. If you have just one 5-minute grab sample, the likelihood is that the measurement does not represent the average radon level in your home. The length of the measurement is important. You have to measure a minimum length of time to take care of all the variations in the home. If you have people going in and out of the house during the course of a day, the radon levels might be low, simply because you are getting a lot of fresh air into the house. But at night, when everything is closed up tight and people have gone to sleep, the radon can seep in without a lot of air exchange and the radon level is going to go up. I am not the expert on housing situations, but I have seen recommendations of a minimum of a one week measurement in a situation where the radon level is likely to be high. I think the best recommendation I have seen is a three-month measurement in the indoor heating season when your house is normally closed up. This will give you a pretty good idea of what the average radon level is during that time of year when it is likely to be the highest.

D. Schutz: There is one interesting set of data from Naomi Harley that bears on the fundamental variability of indoor radon beyond these fluctuations from wind velocity, opening doors, and so forth. They had studied a house in Teaneck for 4 years with hourly measurements and from one year to the next there was a factor of 2 difference in the yearly average radon concentration. There was a fundamental change that was probably related to the overall moisture of the soil, but the actual source was not determined. You can expect fluctuations, apparently up to at least a factor of 2 in the fundamental input of radon in addition to all these other factors that change the shorter term variability.

H. Sachs: There is of course the additional question, "If it is low today, and I buy the house, what is the chance that it's going to be high in 5 years?" Maybe Richard Sextro has some data; I don't know of much. But if the foundation of this house changes, suddenly introducing new cracks, or the way the house is used changes, there are certainly many ways in which the radon level in the house could change. The general assumption has been that for a mature house, one that has finished the adolescent growth pains of settling and cracking, it is not likely to change much.

J. Otton: In looking at geologic correlations between indoor radon and geology, the longer term the measurement, the better the correlation that we see between those two factors. There is so much variability that short-term measurements aren't very meaningful for determining the average radon concentration.

ROCK SOURCES OF RADON

Question: Where there is rock with uranium, is it more hazardous if it is disturbed or if it is left alone?

D. Schutz: Any process of disturbing it which would tend to make the metal grains more available to give up their radon would certainly enhance the possibility of radon emanation. I would have to go with leaving things alone and not disturbing it. Building a foundation for a house is certainly something that can enhance the radon availability.

Question: Was the 1.28 percent uranium oxide sample in the Lockatong formation one specimen or a large representative sample result?

D. Schutz: I believe that came from the Stockton formation, and as far as I know it was one single specimen. It was probably the highest ever observed, but it does show the wide range that exists. I wouldn't expect that you would find large areas with that kind of composition.

H. Sachs: I would like to add an additional remark. Historically in central New Jersey, the Stockton was one of the most readily available quarry rocks that was used. In the era before concrete block availability, many of the foundations that we have looked at, including my own house, turn out to be made of weathered Stockton. There is not enough volume to be a very large source--I hope. It isn't in my house. But it is a contributor, and there is this irony that many of these houses are built, not on a geology of Stockton, but out of building materials of Stockton.

D. Schutz: Is that the source of the brownstones in New York?

H. Sachs: No, I don't think there is any Stockton there.

RADON AND GROUND WATER

Question: Why have radon concentrations in the ground water of the Triassic basin been relatively low, approximately 6,000 pCi/l, when such high uranium concentrations can be found in the Stockton and Lockatong formations?

D. Schutz: I don't know that the water sampling has been correlated with the location where you might find the uranium. However, 6,000 pCi/l is not that low. It is not dangerously high, but most ground water has been on the order of 1,000 pCi/l.

G. Nicholls: We have some data on wells in New Jersey now. For people who have asked us to do confirmatory testing, our highest number is about 39,000 pCi/l. We have heard reports of higher values, in the hundreds of thousands, but we do not know which aquifers or which formations those were associated with. That is one thing that we will be working on with our geologists.

D. Schutz: Most of the water supplies in the Triassic lowlands are surface water supplies. The presence of private wells is not as important as it is in the western part of the state where you can find higher values in the water. Certainly in the reservoir systems you don't need to be particularly concerned about radon.

RADON AND FAULT LINES

H. Horowitz: Dr. Otton, have you found any correlation between radon levels and fault lines in your work?

J. Otton: Yes, this is an issue I would have liked to address more. By fault, I simply mean any breakage of the rock where you get zones of opening so that air and water can move through. Faults and fractures are clearly zones where uranium concentrates through geologic time simply because ground water that contains uranium can flow through it trapping uranium into the fault zones. Also, these zones are of higher porosity and permeability. Penn State did a study in the Reading Prong where they looked at the influence of fractures on radon concentrations in soil gas, and they found that the fracture zones do enhance the radon concentrations that they see in soil gas. Another study of the black shales in the Cleveland area shows that the highest radon levels in soil gas were adjacent to fracture zones where they cut through. So fracture zones are very important. Even assuming that you don't have uranium mineralization concentrated in a fracture zone, the fracture zone may enhance the radon concentration in the soil gas by a factor of 3 or more, and make it a much more serious situation.

Section 3

Transmission and Mitigation of Radon

INTRODUCTION AND SUMMARY OF SECTION 3: TRANSMISSION AND MITIGATION OF RADON

Robert K. Tucker*

This session provided an overview of transmission mechanisms for the entry of radon gas into homes, source diagnosis and mitigation procedures. In the prior section, the occurrence of radon was described as being more widely distributed than has been commonly supposed. Furthermore, we now know that measured concentrations of radon can vary widely from house to house in the same area and that concentrations vary seasonally in a single house. Analysis of factors affecting the radon source strength, the movement of radon gas in soil, and its transmission into structures is a key component in understanding these variations. The papers in this section describe the behavior of homes in the transmission of radon, an approach to diagnosing radon sources and mitigation methods useful for reducing radon entry from both soil and water.

In "Radon in Dwellings," Richard Sextro (Co-Director of the Indoor Environment Program at Lawrence Berkeley Laboratory) describes five sources of indoor radon. These include (1) the surrounding soil, (2) building materials, (3) water, (4) natural gas and (5) infiltrating outdoor air. Of these, the predominant mechanism and contributor to houses with radon concentrations of 20 pCi/l or more appears to be the convective flow of soil gas carrying radon. The driving force for this convective flow is the pressure differential between the house and the soil caused by temperature differences, wind loading, combustion and ventilation. Radon entry points consist of cracks and penetrations through the building shell. Several parameters involving the building shell and soil determine the radon concentration. Principal among these are the air permeability of soils and the building structure. It is, however, the variations in radon source strength from house to house that appears to account for most of the variability in observed indoor radon levels.

Harvey M. Sachs (Radon Consultant) was a discussant in this session, and subsequently submitted a paper, "Rapid Diagnosis for Radon Sources in Homes," based on work that he and others had done at Princeton University. The method described is based on determining the radon mass balance of the house and comparing it with measured flux emanation

*Director, Office of Science and Research, New Jersey Department of Environmental Protection.

rates from suspected radon sources. After ventilating thoroughly, the house is closed up and radon ingrowth rates are measured over several hours. Grab samples are analyzed with a Lucas scintillation cell counting system. Simultaneous measurements of house source strength and ventilation rate, together with flux measurements from suspected sources using a variation of the Jonassen method, allows radon source strengths to be identified. Although the approach requires highly trained personnel, it requires only one visit to the site.

Arthur Scott (Mitigation Scientist with American Atcom currently directing EPA mitigation programs in Pennsylvania) details four basic mitigation methods in his paper,"Radon Mitigation Methods in Existing Housing." The four methods are to (1) increase the ventilation rate, (2) seal the openings, (3) reverse the pressure differential between soil and building and (4) reduce radon concentrations in soil next to the building. While the principles are straightforward, the implementation is often difficult and sometimes costly. Depending on the type of house construction, sealing is often too difficult and too expensive as a sole remedy. Ventilation can often be effective at radon levels less than 20 pCi/l. In the Reading Prong area, higher level houses (100 pCi/l or greater) have often required the use of active soil ventilation measures. In most cases, these have cost less than a few percent of the house price. In general, the decision about which method to apply depends greatly on the radon entryways, details of the house construction, and actual radon levels encountered. In complicated cases, several mitigation methods often need to be combined in order to reach a satisfactory solution.

While the previous selections concentrate on radon emanating from soil, in some cases water from individual wells can be a significant contributor to indoor radon levels. Since radon is quite soluble in water (more than 40 times more soluble than oxygen in water), ground water movement provides another pathway for radon transport. If radon contaminated water is pumped into homes from wells, its subsequent outgassing from the water can provide radon gas exposure to the inhabitants. Such outgassing occurs during normal water usage in such activities as showers and baths, washing dishes and washing clothes. It should be pointed out that surface water and reservoir supplies generally have low values of radon, typically less than 1000 pCi/l. Although there is no clear cutoff, values of radon in water greater than 10,000 or 20,000 pCi/l should be considered problematical.

Jerry Lowry (Department of Civil Engineering, University of Maine at Orono) and Sylvia Lowry (Scientist and Principal of Lowry Engineering, Inc., of Thorndike, Maine) discuss various methods of water treatment in their paper, "Techniques and Economics of Radon Removal from Water Supplies." They document that significant elevated levels of radon gas can occur in residences from water use. They contrast two methods for removing radon in water supplies: (1) the use of Granular Activated Carbon (GAC) Absorption/Decay and (2) aeration. GAC is extremely effective and economical for the small flow rates found in households, costing approximately $750 to $850 and removing 99.9% or more of the radon. Aeration methods are better suited and more economical for the higher flow rates found in community and municipal supplies.

In summary, understanding transmission gives us a better handle on ways to mitigate exposure to radon. The ability to reduce such exposure offers hope that we can prevent some of the considerable risk our health effects experts tell us is associated with such exposure. The ability to mitigate exposure changes risk from an involuntary one to a risk that people have control over. As discussed in the section on risk perception, people react quite differently to risks imposed on them than to those they choose. More effective mitigation methods will provide options for people to deal with the risk from radon decay.

RADON IN DWELLINGS

Richard Sextro*

INTRODUCTION

Radon[1] is a chemically inert, radioactive gas produced by the radioactive decay of radium. Radium, part of the ^{238}U decay series, is a ubiquitous component of all crustal material, and as a consequence, is found in soils and in building materials derived from rock or soil components. The radon gas formed by this process can migrate through these materials or soils into the interior of homes. Radon is present in virtually all dwellings, with observed concentrations ranging from less than 1 pCi/l to, in very rare cases, greater than 1,000 pCi/l.

As with other indoor air pollutants, the actual distribution of radon concentrations in homes throughout the United States is not known, since no survey of randomly selected homes across the U.S. has been conducted to date. However, a recent study[2] systematically reviewed the results of various investigations of indoor radon concentrations in single-family homes scattered around the country and compiled an aggregate set of data that provides an estimate of the frequency distribution of radon concentrations in U.S. homes. The resulting distribution is shown in Figure 3.1; the data appear to be lognormally distributed, with a geometric mean of 0.9 pCi/l, a geometric standard deviation of 2.8, with an average (arithmetic mean) of 1.6 pCi/l. The lung cancer risk associated with a lifetime exposure to this average radon level is about 0.3%, which is one to two orders of magnitude larger than that associated with many other environmental risks.[3]

Although there are currently no standards regulating indoor (non-occupational) radon exposures (and there are no proposals and no real regulatory framework to do so), several organizations and agencies have proposed guideline concentration values, including the American Society of Heating, Refrigerating, and Air-Conditioning Engineers (ASHRAE), the National Council on Radiation Protection and Measurements (NCRP), and the Environmental Protection Agency (EPA). These guidelines can then be compared with the apparent radon concentration distribution in homes noted

*Indoor Environment Program, Lawrence Berkeley Laboratory, University of California, Berkeley, CA 94720

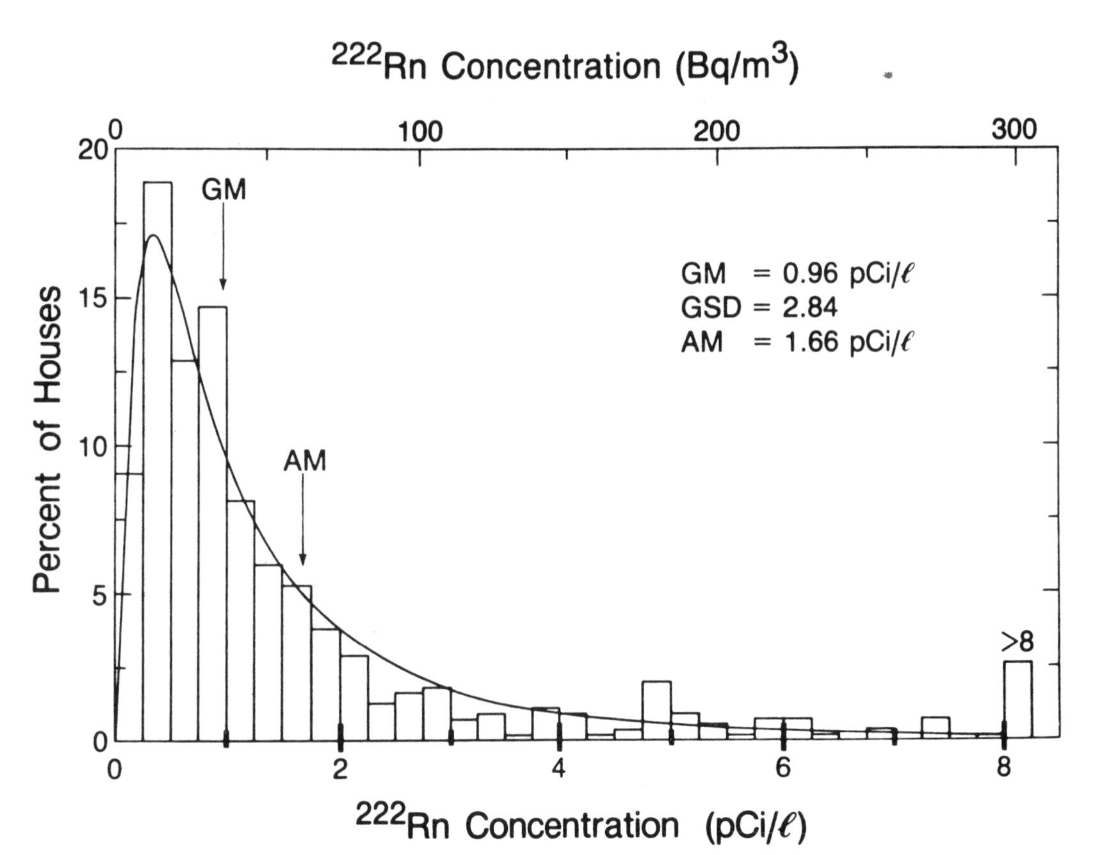

Figure 3.1 - Probability Distribution of Rn-222 Concentrations in 552 U.S. Houses. The smooth curve is a lognormal function with the parameters shown in the figure.

earlier, and an estimate made of the number of single-family homes that equal or exceed the guideline value. The ASHRAE guideline is 0.01 working level (WL), which is approximately 2 pCi/l of radon[4]. Almost 20 percent of U.S. homes (some 12 million single-family units) are estimated to exceed this value. The NCRP guideline is about 0.04 WL, which translates into a radon concentration of approximately 8 pCi/l. At this value, about 1 to 2 percent of the homes (or approximately 1 million units) can be expected to exceed this guideline. The EPA guideline, which as of this writing is not yet finalized, is expected to be in the vicinity of 4 pCi/l; approximately 7 percent of U.S. housing (approximately 4 million units) would equal or exceed this guideline.

In addition to the range of indoor radon concentrations, regional differences have been observed in the distribution of measured indoor radon levels. The geometric means of the individual data sets comprising the aggregation discussed earlier[7] are themselves lognormally distributed with a geometric standard deviation of 2.0. This distribution appears to arise from the geographic variability of the radon source strengths. This regional or geographic variation is perhaps best illustrated by the distributional data shown in Table 3.1, where comparisons are made for three radon concentration distributions. As can be seen, the recent data from that area of eastern Pennsylvania known as the Reading Prong has an average of 15.3 pCi/l, compared with the average of the aggregate data discussed earlier of 1.5 pCi/l and the average of 103 measurements in the Houston, Texas area of 0.59 pCi/l. Even more striking is a comparison of the expected number of homes with concentrations exceeding various indoor levels. A representative comparison is tabulated in Table 3.1. Using 4 pCi/L as an illustration, based on the aggregate data, 7 percent of U.S. homes would be expected to equal or exceed this level, while only 0.07 percent of the homes in the Houston area would. In contrast, more than half (54 percent) of the homes monitored in the Reading Prong (PA) exceed 4 pCi/l. In fact, based on these distributions almost 7 percent of the Reading Prong homes are expected to have concentrations greater than 50 pCi/l, while only 0.01 percent (about 6,000 homes) of U.S. homes would exceed this concentration, with a negligible number in the Houston area.

SOURCES OF INDOOR RADON

There are basically five sources of indoor radon, as illustrated schematically in Figure 3.2. These include the soil surrounding the building shell, building materials, the use of water or natural gas indoors, and infiltrating outdoor air. The average source strengths for each of the sources of radon are summarized in Table 3.2. These source strengths, when multiplied by the infiltration rate, will give the contribution of that source to the indoor radon

Regional Variation in the Distribution of
Observed Indoor Radon Concentrations

Indoor Radon Concentration (pCi/L)	Percent of homes within selected geographical area that exceed given radon concentration		
	U.S.[a]	Reading Prong (PA)[b]	Houston TX[c]
1	46	85	13
4	7	54	0.07
8	2	37	<<0.01
10	1	32	—
20	0.4	18	—
50	<0.01	7	—
Lognormal Distribution Parameters:			
Number of homes	552	~5000	103
GM (pCi/L)	0.9	4.8	0.47
GSD	2.8	4.58	1.94
AM (pCi/L)	1.5	15.3	0.59

[a] Data compiled by Nero *et al.* (1984)

[b] Data reported by George (1986)

[c] Statistical data from Prichard *et al.* (1982)

Table 3.1 - Regional Variation in the Distribution of Observed Indoor Radon Concentrations.

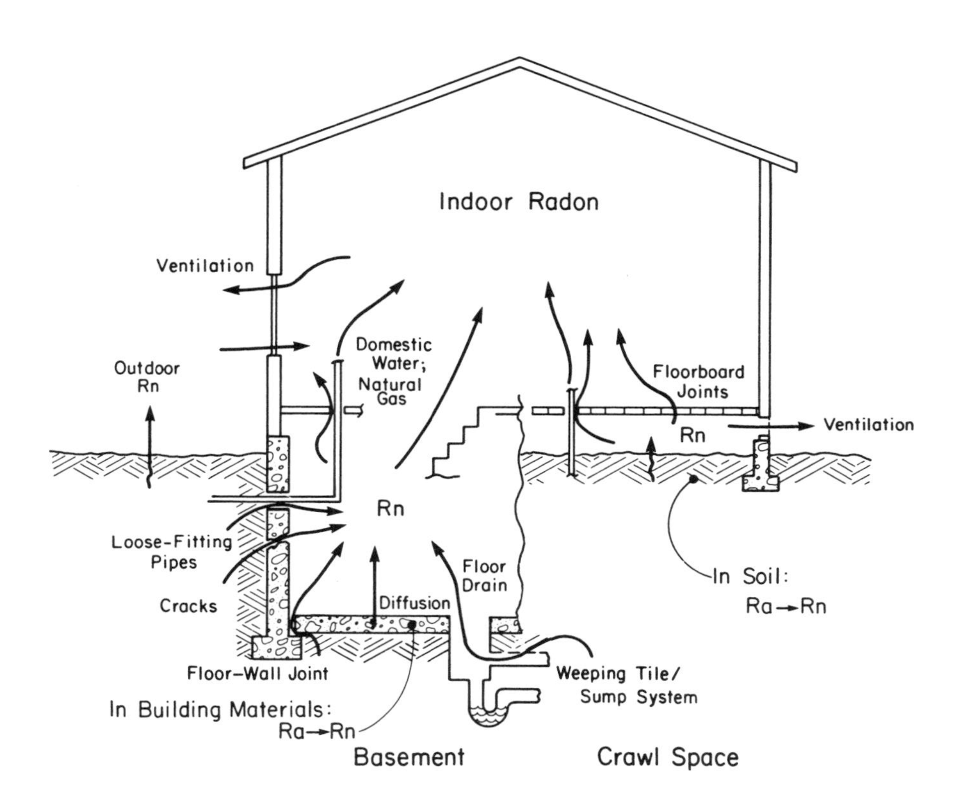

Figure 3.2 - Schematic Illustration of Radon Entry Pathways for Typical Residential Building Structures.

concentration. Each of these sources of radon is discussed in greater detail below.

Radon from the surrounding soil can enter a house through cracks and penetrations in the building shell. This may be due either to the settling and aging of the structure or to building design and construction practices used, such as the joint between the basement floor and wall. Radon entry occurs through two principal mechanisms: diffusion and convective flow. Convective flow will be discussed in the subsequent section. Radon diffuses from the soil through cracks and other openings or through concrete and other building materials. This is typically a slow process which, due to the radioactive decay of radon, generally makes only a limited contribution to indoor radon concentrations. The estimates shown in Table 3.2 of the radon source strength of soil are based on diffusion from uncovered soils, as if the basement floor of a house were open soil. With the addition of a concrete floor, the diffusion rates are retarded by approximately a factor of 20 or more.

Earth-based building materials, such as gypsum board, concrete, and brick, all contain trace quantities of radium, and thus become sources of radon emanation into the building interior. In almost all cases, however, the source strengths are usually small, as illustrated by the average radon source strength estimated in Table 3.2 for concrete. In a few cases, notably in Sweden, where for a time alum shale deposits containing significant concentrations of radium were used in concretes, building materials have been identified as important sources of indoor radon.

Radon produced by radium in subsurface rock formations can enter ground water and natural gas contained in fluid reservoirs in the rock or passing through the formation. Subsequent use of these fluids indoors will then release any remaining radon that was contained in the fluid into the indoor air. Usually natural gas use occurs some distance from the original well, and the transit time to the end user is large enough so that most of the radon will have decayed. The equivalent radon entry rate is, therefore, quite small.

In some areas of the country, high concentrations of radon dissolved in ground water have been reported[8]. However, the resulting radon concentration in air due to radon release from the water is approximately 1 in 10,000. That is, the radon concentration in water must be 10,000 pCi/l to contribute, on average, 1 pCi/l to airborne concentrations.[9] For public water supplies, derived either from surface water or ground water sources, average radon concentrations in water are 30 pCi/l and 300 pCi/l, respectively. Almost half the U.S. population uses water from surface water sources, and another approximately one-third of the population uses water from public ground water sources. Eighteen percent of the U.S. population obtains potable water from small public (less than 1,000 persons) ground water supplies or from private wells. In these

Typical Radon Source Contributions
for a Single-story Residence

Source	Average Source Strength ($pCi\ L^{-1}\ h^{-1}$)	Reference
Outdoor air	0.3	Gesell 1983.
Potable water		Nazaroff *et al.*, 1985a
surface water	0.004	
public groundwater	0.04	
private wells	0.7	
Natural gas	0.01	Johnson *et al.*, 1973
Concrete floor[b]	0.06	a.
Soil		
diffusion through floor	0.04	Nero and Nazaroff 1984.
uncovered soil	0.9	Wilkening *et al.*, 1972.
Compare with:		
Average entry rate	1.8	b.
Entry rate for house with average indoor concentration of 20 pCi/L	22	

[a] Assumes half the flux from a 100 m^2, 20 cm-thick concrete floor enters the house (Ingersoll 1983).

[b] Arithmetic mean indoor radon concentration (Nero *et al.*, 1984) divided by an average ventilation rate of 0.9 h^{-1}.

Table 3.2 - Typical Radon Source Contributions for a Single-Story Residence.

cases, there is usually no aeration of the water, and the total transit time between the water source and end-use is small, providing little opportunity for the radon to be released from the water or to decay. The data on radon concentrations in private well water is quite limited, although there is wide variation in the observed concentrations. In some cases, the concentrations exceed 10,000 pCi/l and the water will contribute measurably to the radon concentrations in indoor air. It appears that these situations are isolated and are highly dependent upon local geological conditions.

Finally, outdoor air contains small concentrations of radon gas that has diffused from the earth. These concentrations are typically in the range of 0.2 to 0.6 pCi/l, depending upon the nature of the soils and rocks in the vicinity.[10] This radon can be carried indoors by infiltrating outdoor air, but steady-state concentrations due to this source will not exceed the outdoor levels.

RADON ENTRY INTO DWELLINGS

From the information presented in Table 3.2, it is clear that, on average, outdoor air, potable water, natural gas, and building materials make only minor contributions to average radon entry rates based on the aggregate U.S. concentration distribution. In the case of houses with elevated radon concentrations, e.g., 20 pCi/l or more, the average contribution from these sources is insignificant.

On the other hand, convective flow of soil gas appears to be a significant if not predominant mechanism for radon entry into buildings. Some of the routes of soil gas entry into the structure are illustrated in Figure 3.2. The driving force for this convective flow is the pressure differential that develops across the building shell due to the difference between the indoor and outdoor temperatures and due to the wind loading on the building superstructure. Additional depressurization of the building shell may be caused by the operation of an exhaust ventilation system or by the use of an appliance such as a clothes dryer that exhausts air to the outside. Similarly, a combustion appliance, such as a gas- or oil-fired furnace will draw air from the house as the hot combustion gases rise and escape through a vent pipe.

All these effects can combine to produce pressure differences across the lower part of a building shell, for example near the basement floor, on the order of 10 Pa. This pressure difference is quite small, but persistent. The effect is usually largest in the winter, when the indoor-outdoor temperature differences are greatest. This is also consistent with the general observation that indoor radon concentrations are greatest in winter.

The radon entry points illustrated earlier provide a means of coupling the building shell with the soil. As a result, the negative pressure in the basement sets up a

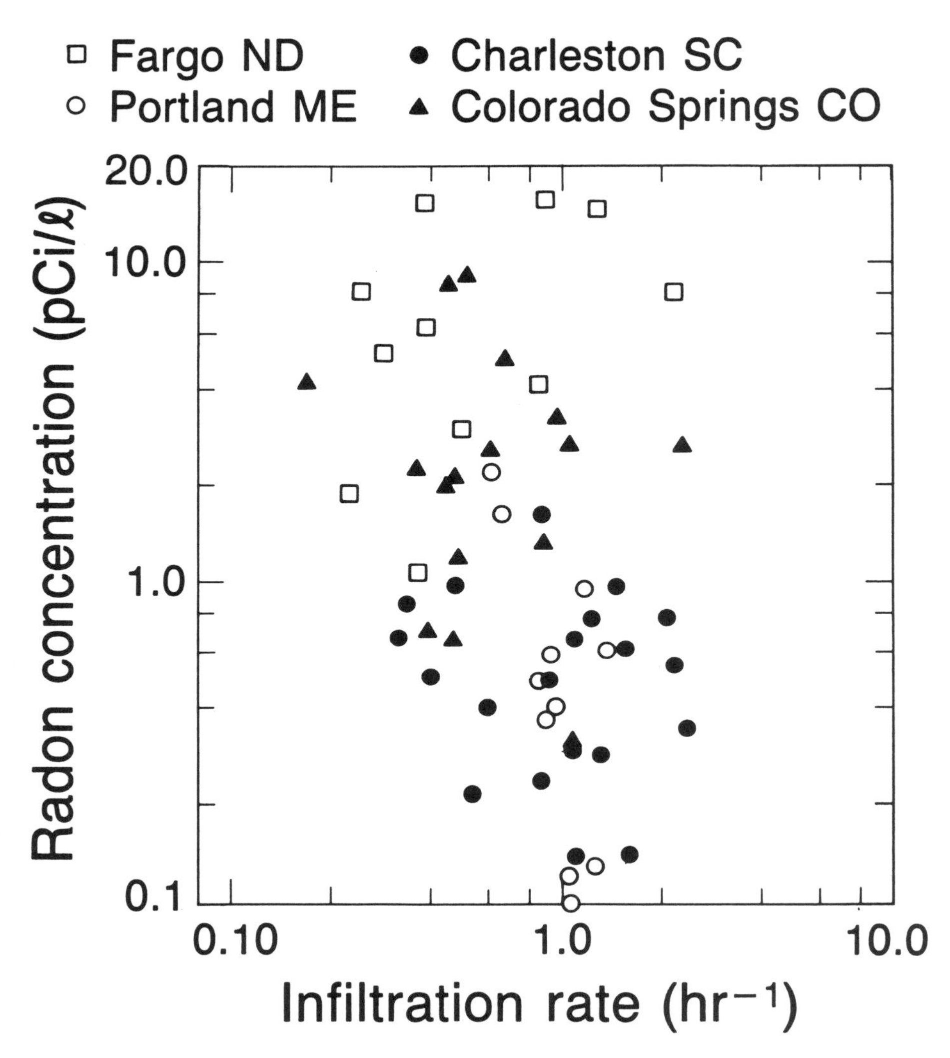

Figure 3.3 - Scatter Plot of Time-Averaged Radon Concentration and Air Infiltration Rates Measured in 58 Houses in Four U.S. Cities.

negative pressure field in the surrounding soil. The existence of this pressure field has been demonstrated experimentally, and the resulting migration of soil gas has been observed with the use of tracer gases injected into the soil.[11] These experiments provide direct physical evidence for pressure driven flow of soil gas through the soil and into buildings. Thus, a house does not simply act as a passive accumulator of radon diffusing from the soil but has an active role in actually pumping radon-bearing soil gas into the building interior.

The interaction between the building shell and the soil depends upon a number of parameters. These in turn appear to have an influence on the radon concentration observed indoors. Principal among these is the air permeability of soils and the "leakiness" of the building structure. In fact, it is the variations in the radon source strength from house to house that account, for the most part, for the variability in the observed indoor radon concentrations. Figure 3.3 is a scatter plot of time-averaged indoor radon concentrations and simultaneous measurements of air infiltration rate from houses in several cities.[12] As can be seen, there is no apparent correlation between indoor radon concentration and infiltration rate. If there were a relationship between these two parameters, the indoor concentrations would drop as infiltration rates increase.

GENERAL OBSERVATIONS AND CONCLUSIONS

While a general picture of radon entry into houses has emerged, many of the details are based on empirical observations in a few homes. Continuous, multi-parameter data have shown that the variations in environmental conditions, such as changes in outdoor temperature, wind speed and direction, or changes in building operation, all influence indoor radon concentrations in a complicated way. Additional detailed experiments will provide further insights into understanding radon entry. Such results also have implications for designing and deploying radon mitigation strategies.

As noted earlier, there are potentially one million or more homes that may exceed any of the suggested indoor radon concentration guidelines. Locating such homes is not an easy task. Even in areas where a significant number of homes have excessive indoor radon concentrations, most of the homes surveyed have approximately average indoor radon concentrations. Thus, there is a need to assemble a predictive approach to help identify areas with the potential for having homes with elevated indoor concentrations.

FOOTNOTES

[1]Most discussions of radon usually refer to the ^{222}Rn isotope, which has a half-life of 3.8 days. Another isotope, ^{220}Rn, may also be present in indoor atmospheres, although its accumulation indoors is limited by its short, 55 second half-life. Nevertheless, the average dose to the lung from the ^{220}Rn decay products has been estimated to be about 25 percent of that from the ^{222}Rn progeny (UNSCEAR, Ionizing Radiation: Sources and Biological Effects, United Nations Scientific Committee on the Effects of Atomic Radiation, United Nations, New York, NY, 1982). While much of the discussion in this paper is applicable to either radon isotope, most of the details apply to ^{222}Rn (referred to as radon in this paper) and its progeny.

[2]Nero, A.V., M.B. Schwehr, W.W. Nazaroff, and K.L. Revzan, "Distribution of Airborne Radon-222 Concentrations in U.S. Homes." Lawrence Berkeley Laboratory, report LBL-18274, revised 1985, to be published in Science, 1984.

[3]NCRP, Exposures from the Uranium Series with Emphasis on Radon and its Daughters. National Council on Radiation Protection and Measurements, NCRP Report 77, Bethesda, MD, 1984.

[4]ASHRAE, ASHRAE Standard 62-81, "Ventilation for Acceptable Indoor Air Quality." American Society of Heating, Refrigerating, and Air-Conditioning Engineers, Atlanta, GA, 1981.

[5]Unlike the other two proposed radon concentration guidelines, the NCRP guideline is based on exposure and is 2 working level months per year (2 WLM/yr). Translating this into an equivalent radon concentration at 100 percent home occupancy gives 8 pCi/l. A better estimate would be based on an occupancy of 80 percent, which yields a concentration equivalent of 10 pCi/l.

[6]NCRP, 1984, op.cit.

[7]Nero, et al., 1984, op. cit.

[8]Hess, C.T., C.V. Weiffenback, and S.A. Norton, "Environmental Radon and Cancer Correlations in Maine." Health Physics 45, 1983, pp. 339-348.

[9]Nazaroff, W.W., S.M. Doyle, A.V. Nero, and R.G. Sextro, "Potable Water as a Source of Airborne Radon-222 in U.S. Dwellings: A Review and Assessment." Lawrence Berkeley Laboratory, Report LBL-18514, (to be published in Health Physics), 1985.

[10]Gesell, T.F., "Background Atmospheric ^{222}Rn Concentrations Outdoors and Indoors: A Review." Health Physics 45, 1983, pp. 289-302.

[11]Nazaroff, W.W., S.R. Lewis, S.M. Doyle, B.A. Moed, and A.V. Nero, "Experiments on Pollutant Transport from Soil into Residential Basements by Pressure-Driven Flow." Lawrence Berkeley Laboratory, Report LBL-18374, submitted to Environmental Science and Technology, 1985.

[12]Doyle, S.M., W.W. Nazaroff, and A.V. Nero, "Time-Averaged Radon Concentrations and Infiltration Rates Sampled in Four U.S. Cities." Health Physics 47, 1984, pp. 579-586.

RAPID DIAGNOSIS FOR RADON SOURCES IN HOMES

Harvey M. Sachs, Gautam Dutt, Thomas Hernandez and James Ring*

Where radon levels in homes demand remedial action, and low cost remedies are sought, an inexpensive method for diagnosing radon sources is important. Several alternative approaches to diagnosis have been utilized. These include "the iterative empirical approach," involving the implementation of recommendations by an experienced field engineer followed by retesting. The "local concentration approach,"[1] involves the placement of passive concentration monitors at all suspect entry points to look for high radon concentrations in that locality. The "charcoal flux canister approach"[2] uses activated charcoal canisters attached to surfaces for monitoring radon flux over several days; analysis is done in the laboratory. Limitations exist for all of these approaches. The iterative approach may involve much trial and error. The local concentration approach fails to measure fluxes. Both it and the superior charcoal flux canister approach require multiple visits to the subject house.

A fourth approach was utilized in research by the Center for Environmental Studies at Princeton University. It is based upon a method developed by Jonasson to measure the radon mass balance. It can be used for specific sources as well as the entire house and requires only one visit. It is somewhat less accurate, however, than the charcoal flux canister approach for measuring radon mass balance and requires the use of highly trained personnel.

Prior to arrival at the test house, the field engineer reviews radon values and additional house information drawn from an occupant questionnaire in order to formulate hypotheses. At the house, 10-30 minutes are spent discussing the work with the occupants and sketching floor plans, sump locations and other salient features. Grab air samples are then taken in each room, after which the house is completely ventilated, driving the radon concentration to a low level. The house is then closed up so that successive tests can be taken over a three-hour period.

---

*This paper is based upon a paper entitled "A Rapid Diagnostic Procedure for Radon Sources in Houses," Working Paper 73, Center for Environmental Studies, Princeton University, March, 1985.

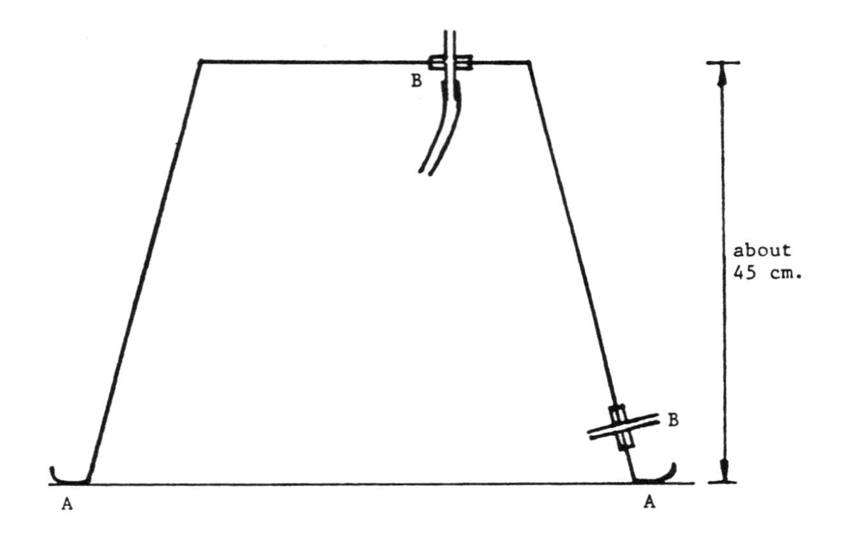

Figure 3.4

Jonassen flux bucket cross section showing:

A. flat-lip seal to surface being tested;

B. pass through sampling ports for taking grab samples.

Table 3.3

Building Materials Emanation Measurements, In Situ(1)

House N, April 8/9, 1982

Sample numbers	Material tested	Emanation rate, (this study)	Charcoal Canister flux measurements (A. George, Spring '82)
		(picocuries/ m^2/ hour, pCi/m^2/h)	
1 - 3	hexagonal floor tile	none detected	(3)
5 - 8	exposed brick arch (2)	8000	
9 - 12	wall, storage room (2)	9000	
13 - 15	square floor tile, storage room	none detected	(3)
17 - 19	concrete wall, storage room	none detected	
21 - 24	finish plaster	none detected	
25 - 28	white kitchen tile	none detected	
29 - 32	finish plaster	none detected	
33 - 35	concrete floor slab, mechanical room	none detected	(3)
37 - 39	concrete roof plank	none detected	<< 1
41 - 44b	hearth brick	none detected	5
45 - 47	closet wall, master bedroom	none detected	

Notes: (1) Long term average radon levels measured in five different rooms of the house, prior to our visit, ranged from 45 to 50 pCi/liter.
(2) Radon later shown to be originating in subsoil space and diffusing through wall.
(3) Two unspecified floor measurements were made; both were < 1 pCi/m^2/h.

Radon source strength is measured using a variation of the Jonassen "flux bucket," a modified plastic trash can into which replacement air can be pulled (see Figure 3.4). Successive grab samples are taken to monitor radon ingrowth rates from multiple potential sources, ranging from building materials to slab cracks and sub-floor sumps. The samples are analyzed using a Lucas scintillation cell counting system. A flux balance is computed by comparing measurements from within the home to contributions measured from potential sources. This is important because of the large temporal variability of radon concentrations and imputed source strengths. For example, Table 3.3 shows results from "House N" studied by Sachs, et al. While at the beginning of the initial site visit, radon concentrations were only 17 pCi/hr despite the house being closed up for 24 hours, the average radon concentration was 50 pCi/l (based upon the average of five long-term track-etch measurements).

To determine the local emanation rates from suspected radon sources, such as specific building materials, sumps, cracks or wall areas, a relatively small container is sealed to the feature of interest and successive samples taken through access ports.[4] Thus, Table 3.3 shows these measurements from some 10 materials. Because only two of these showed large emanation rates, the measurements indirectly led us to the principal radon source. This is further illustrated by Table 3.4, which shows results from "House W" and other sites in eastern U.S. locations where radon emanation was found from sump pumps in the cellar and other locations. When appropriate, field personnel carry common hand tools and supplies to test mitigation strategies. Ingrowth from "fixed" sources can then be tested.

Figure 3.5 also illustrates how radon ingrowth measurements can lead to the isolation of the primary radon source at a given test site. It was suspected that sub-slab heat distribution ducts were a potential pathway for radon to enter the house from soil underneath. Radon ingrowth measurements were conducted to compare concentrations with (a) the heat distribution ducts open in their normal operating position and the circulation fan operating and (b) with the ducts sealed from the living space and the circulation fan off. The house was thoroughly ventilated before each experiment. The concentration increased much faster in case (a), where the ducts were open, suggesting that this as a significant pathway for radon entry.

While the source strength of the house is being determined by the ingrowth method, a simultaneous measure of ventilation rate is made for a primary chamber of the house, such as a basement area, where air is internally well mixed and concentrations are highest for the building. Ventilation rate is measured by the tracer gas dilution method[5] employing sulfur hexafluoride as a tracer gas

Table 3.4

Results From Radon Diagnostics and Mitigation in a Number of Houses(1)

House ID	Study year	Rapid diagnostic measurement	Remedial action taken	Radon concentrations picocuries/liter (2) Pre retrofit Living area	Pre retrofit Cellar	Post retrofit Living area	Post retrofit Cellar	Refs.
W	1982	Source strengths (pCi/hr): Cinder block wall: none detected Sump in basement -- 120,000 and 810,000 (Feb 29 & Mar 4 data)	Sub-slab depressurization	6.1 (3.3) n=9	21.0 (5.0) n=9	0.6 (0.4) n=2	0.9 (0.5) n=6	(3)
N	1982	Source strengths (pCi/l-hr): Circulation fan off = 6.8 May 23 Circulation fan on = 9.2 May 24 Circulation fan on = 10.6 Jun 9 Circulation fan off = 4.5 Jun 10 After retrofit = 1.7	Sub-slab depressurization	>45 n=5	no cellar	<2	no cellar	(3)
F21	1983	Source strengths (pCi/hr): Floor over crack 75,000 Floor near sump 80,000 Ventilation (in cellar, ACH): 0.2, 0.2, and 0.4 (on 3 days)	Sealed and ventilated sump; sealed sill and floor-wall crack; painted concrete block wall.	10.2 (6.4) n=4	49.8; 14.8	0.9; 0.6	1.4	(4)

(Subslab depressurization was successful in summer. However, long-term monitoring by W.S. Fleming Associates established partial control failure in winter. Additional wall sealing was implemented by Fleming to maintain control.)

Notes: (1) Units and abbreviations: radon concentrations are given in picocuries per liter (pCi/l); radon source strengths are given in picocuries per hour (pCi/hr) or picocuries per liter per hour (pCi/l-hr); radon source fluxes from surfaces are sometimes expressed as picocuries per sq. meter per hour (pCi/m^2-hr); ventilation data were collected using tracer gas decay and are expressed in air changes per hour (ACH).

(2) Where three or more measurements were taken, numbers in parentheses refer to standard deviations and the number of measurements is shown as "n="; where only one or two measurements were made, these are shown individually.

(3) Sachs, Hernandez, and Nobel, 1984.

(4) ECRI, 1983; Fleming, 1985.

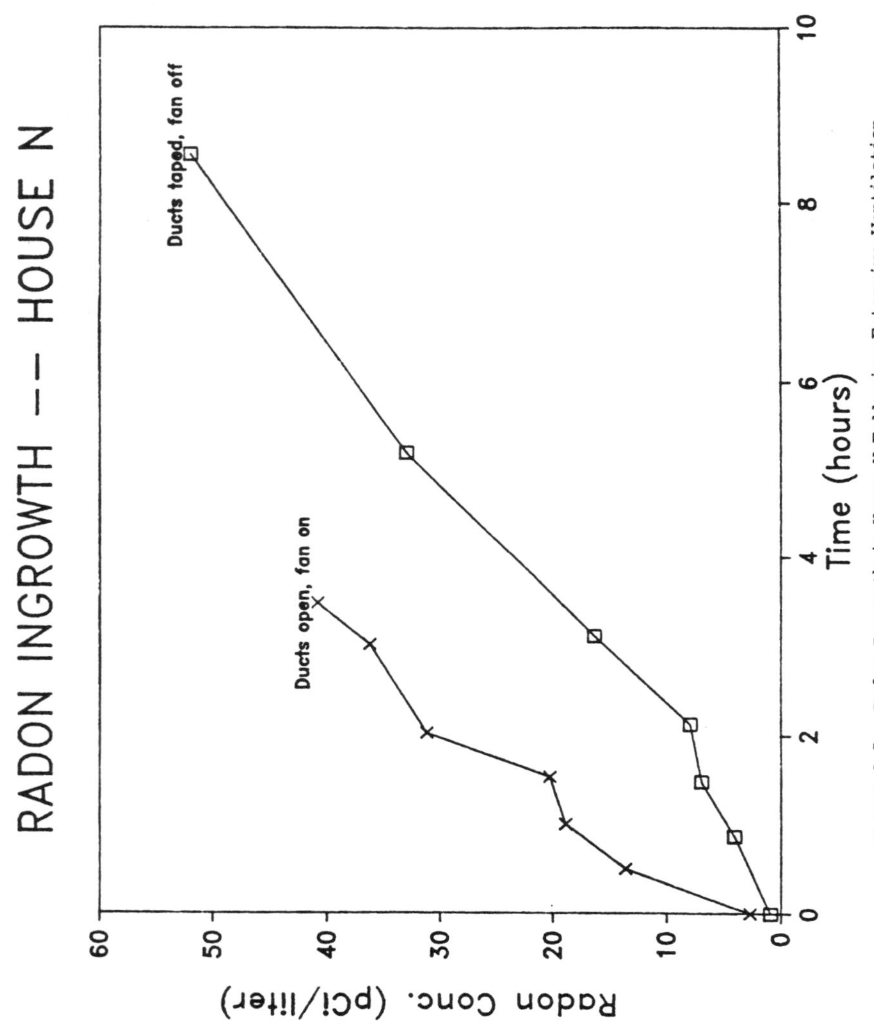

Figure 3.5 - Radon Ingrowth in House N Following Extensive Ventilation.

because it is inert, readily available, easily detectable and has very low background concentrations. This measure provides for an independent assessment of the source strength of radon in the building if prior average radon concentrations are known. This is because the equilibrium concentration of radon in the house is a function of the average source strength divided by the air changes per hour (ventilation rate). Ventilation rate also assists in prescribing the most cost-effective radon mitigation approach. If the building is unusually "tight," the air change can be multiplied several times without a substantial energy cost penalty. In this case, enhanced ventilation may be the most effective radon control method. Alternatively, if the ventilation rate is relatively high, augmenting ventilation further will make the house drafty, uncomfortable and expensive to heat.

The final product of this effort is an engineer's report typically including a contract specification for corrective measures to be implemented by a local building contractor. Follow-up monitoring is recommended subsequently to verify the efficacy of the remedy.

CONCLUSION

Certain building conditions point toward limitations in the method described here. As houses become more complex, diagnosis becomes more time consuming. This is found where relatively low radon concentrations (less than about 10 pCi/l) may not have "point sources" large enough for reliable ingrowth measurements by the technique described. The technique is also limited where basement walls made of weathered and wet fieldstone present extremely irregular surfaces which prevents the sealing of the Jonassen chamber or where radon entry is through porous concrete block walls which have a low area-specific emanation with possible "back-diffusion" of radon.

However, as Table 3.4 illustrates, the method described can serve generally as a cost-effective diagnostic method for large scale use. When soil gas entry occurs via discrete paths, this method can lead directly to the recommendation of appropriate remedial measures such as sub-slab depressurization. In all the cases in Table 3.4 but one, remedial action led to a reduction in radon level.

FOOTNOTES

[1]Wadach, J.B., W.A. Clarke, and I.A. Nitschke, "Testing of Inexpensive Radon Mitigation Techniques in New York State Homes." Paper presented at the 29th Annual Meeting of the Health Physics Society, New Orleans, LA, 1984.

[2]Countess, R.J., "^{222}Rn Flux Measurement with a Charcoal Canister." Health Physics, Vol. 31, Nov., 1976, p.455.

[3]Sachs, H.M., T.L. Hernandez, and J.J. Nobel, "Residential Radon Control by Subslab Ventilation," Preprint for the 77th Annual Meeting of the Air Pollution Control Association, San Francisco, CA., 1984.

[4]Jonassen, L., and J.P. McLaughlin, "Exhalation of Radon-222 from Building Material and Walls." In Gesell, T.F. and W.M. Lowder, Eds., Natural Radiation Environment III, Parts 1 and 2, 1980, pp. 1211-1224.

[5]Hunt, C.M., J.C. King, and H.R. Trechsel (Eds.), Building Air Change Rate and Infiltration Measurement, American Society for Testing and Materials, ASTM - STP 719, 1980.

ADDITIONAL REFERENCES

Dietz, R.N., and E.A. Cote, "Air Infiltration Measurements in a Home Using a Perfluorocarbon Tracer Technique." Environment International, Special Issue on Indoor Air Pollution, 8, No. 1, 1982, pp. 419-433.

ECRI, "NIEI/ECRI Project Report 414-027 to W. S. Fleming and Associates, Syracuse, NY.", ECRI, Plymouth Meeting, 1983.

Fleming. W.S. and Associates, "Indoor Air Quality, Infiltration, and Ventilation in Residential Buildings: Final detailed technical report." W.S. Fleming Associates, Syracuse, NY, 1985.

George, A.C., M. Duncan, and H. Franklin, "Measurements of Radon in Residential Buildings in Maryland and Pennsylvania, USA," presented at the International Seminar on Indoor Exposure to Natural Radiation and Related Risk Assessment. Capri, Italy, 1983.

Hernandez, T.L., and J.W. Ring, "Indoor Radon Source Fluxes: Experimental Tests of a Two-Chamber Model." Environment International 8, 1982, pp. 45-57.

Hess, C.T., C.V. Weiffenback, and S.A. Norton, "Variations of Airborne and Waterborne Rn-222 in Houses in Maine," Environmental International 8, 1982: pp. 59-66.

NCRP, Exposures from the Uranium Series with Emphasis on Radon and Its Daughters. National Council on Radiation Protection Report #77, Bethesda, MD, 1984.

Nero, A.V., "Indoor Radiation Exposure from Rn-222 and its Daughters: A View of the Issue," Health Physics 45, No. 2, 1983, pp. 277-288.

Sachs, H.M., T.L. Hernandez and J.W. Ring, "Regional Geology and Radon Variability in Buildings, Environment International. 8, 1982, pp. 97-103.

Sachs, H.M., D.Harrje, H.W. Prichard, K. Gadsby, and D.I. Jacobson, "Radon Concentrations and Ventilation Rates in Eastern Pennsylvania Houses." Symposium on Management of Atmosphere in Tightly Enclosed Buildings., J.E. Jansseen, Ed., ASHRAE, Atlanta, GA, In Press.

Sherman, M.H., D.T. Grimsrud, and R.C. Diamond, "Infiltration Pressurization Correlation: Surface Pressure and Terrain Effects." ASHRA Transactions 2, 1979, pp. 458-479.

RADON MITIGATION METHODS IN EXISTING HOUSING

Arthur G. Scott*

A few years ago radon mitigation was of interest only to a few specialists who were largely involved in uranium mine or processing plant waste clean-up operations. A number of mitigation methods were developed then, based on an intuitive understanding that the cause of elevated radon levels was the mass flow of soil gas containing radon into the building. The flow was driven by the pressure differentials between the soil and the building produced by wind and thermal forces and entered via openings in the foundation structure.

Given this, it is plain that there are only four basic methods to reduce radon concentration in a dwelling. They are:

1. Increase the ventilation rate to dilute the radon that enters the dwelling.

2. Seal all the openings through which the soil gas enters to reduce the flow rates.

3. Reduce or reverse the pressure differential between the soil and the building to reverse the flows (e.g., soil ventilation).

4. Reduce the radon concentration in the soil gas adjacent to the building.

The theoretical and experimental work which has been carried out since those early days has confirmed that soil gas is indeed the cause of significantly elevated radon levels in houses. We now know and understand more than we did even a year ago, and there is some hope that within a few years there will be good methods to predict what areas might cause elevated radon levels in conventional housing.

The difficulties of radon mitigation are not that the principles are poorly understood, but only in the difficulty in applying them in actual houses and at a reasonable cost.

---

*Arthur Scott works for American Atcon, Inc. and is presently researching mitigation methods in the Reading Prong under an EPA contract.

The principles are perhaps misleadingly simple and can compare to the instructions for crossing Niagara Falls on a tightrope:

1. Keep walking.
2. Don't fall off.

The difficulty is in the performance.

In this context it should be pointed out that theoretical studies, although vital to providing an understanding of the situation, cannot be expected to produce new mitigative methods. Once it is understood that soil gas contains radon and flows under pressure from the soil to the house, then the principles of effective mitigation methods are so obvious that additional thought is unlikely to produce new insights. If homeowners are to reduce radon levels, then they need implementations of these principles that are known to be effective in a variety of housing styles. As life is frequently stranger than thought, these implementations will have to be developed and their effectiveness demonstrated by installation in real houses. As not all methods will be effective, it is unfair to expect the homeowners to pay for all the experimentation out of their own pockets. The EPA program in the Reading Prong that I am directing is an attempt to develop and test lower cost methods of radon mitigation and is carried out on an experimental basis at no direct cost to the homeowners whose houses were selected.

Most of the houses in the Colebrookdale area of the Reading Prong are single story ranch style houses built on a basement with walls of hollow concrete block (often called cinderblock--but cinders have not been readily available for aggregate since the railways stopped using steam locomotives). The walls are hollow, the block voids are exposed to the house interior at the top of each wall, and all the voids are interconnected; it is very difficult to seal all these openings from the house interior. In addition, many walls are cracked as the result of settlement, which can be expected to continue, and this will crack any rigid sealant. The only sealant for block walls is, therefore, a flexible membrane applied to the exterior wall. This requires excavating around the house with a large amount of expensive hand digging, so the cost of this work is bound to be high. The floor/wall joint and any other openings in the floor still have to be sealed at additional cost. If we restrict the planned cost of mitigation to less than a few percent of the house value, then sealing is not a valid option. It would cost far too much.

If we reject sealing, then the mitigation methods remaining involve some form of ventilation, either house or soil. Ventilation implies gas movement, and that requires pressure differences to drive and control flows. These pressures can be produced either "actively" by a small fan or "passively" by a stack through the building roof which uses the same wind and thermal pressure differentials that produce the soil gas flows into the building. As the forces are comparable in size, the passive system must have a large low resistance collection system to effectively compete with the building for soil gas. In contrast, as the fan in the active system can produce almost any level of suction required, active systems can trade off collection system size against fan size. Installation of the collection system is the most expensive part, so active systems are the least costly. The cost of installing a passive stack will generally exceed the cost of a fan and its running cost for the life of the house.

The easiest mitigative measure is to increase ventilation. This can be done passively by leaving the windows open on all sides of the house to provide a cross draft. This is very cheap and effective but is unpopular in the cold months. To increase ventilation without large heat losses, the airflows must be limited; either a small fan, or a heat recovery ventilator (air-to-air heat exchanger) can be used. Radon concentrations will be reduced in the area in which fresh air is discharged by factors of 2 to 5.

As increasing the ventilation requires no knowledge of where the soil gas enters the house and installation causes a minimum of disruption, it is a reasonable cost-effective choice for mitigation in houses with lower radon levels (20 pCi/l).

However, in the higher level (100 pCi/l plus) Reading Prong houses, the reduction needed is higher than can routinely be produced by ventilation. We have, therefore, concentrated our efforts on preventing soil gas entry, using active soil ventilation methods. This seemed to be the only way to get the high reductions needed at moderate cost.

Time is money, so low cost mitigation means that the work must be done quickly. If homeowners are to be able to do the work, then special skills or materials cannot be required.

To reduce time and hence cost to the minimum, the decision was made not to install collection systems but to adapt existing buildings systems and structures. This limits applicability of some methods but is the lowest cost approach.

Some houses have a "weeping tile drain"--a perforated pipe laid in crushed stone around the footing. Its purpose is to drain water away from the basement. On sloping sites the water is discharged at a low point, on flat sites the water is often led into an internal sump so that it can be pumped away from the house. The major house-to-soil connections are the junction of the solid concrete footings and the first course of the hollow concrete blocks of the basement wall, and the joint between the floor slab and the interior face of the basement wall. As the weeping tile runs in just this area, there is a good chance that simply connecting a small exhaust fan to the tile will draw air from the house through these connections and so prevent soil gas entry. Our experience is that this works very well in those houses where the tile goes completely around the house. On sloping sites where the tile is accessible by a small excavation, the cost can be as low as a few hundred dollars. Most of the money is in the digging cost. Unfortunately, only about 30% of the houses have weeping tiles or this would be a general solution.

The radon source in the Reading Prong may be the heavily weathered and fractured granite bedrock. This lies only a few inches below the basement floor slab, so a subslab ventilation system can be expected to intercept much of the soil gas that moves out of the rock.

We have used minimal systems with either one or two central pipes through the floor and more elaborate systems with one or two pipes placed through the floor close to each wall. Results have been mixed, for some concrete floor slabs are poured directly on the rock, and as a result it is very difficult for the fan to draw gas from more than the immediate area. In other houses, there is a good layer of gravel or crushed stone beneath the slab and the fan suction extends beneath the entire slab even with only one or two suction points. This reverses the airflow through all openings in the slab and prevents the entry of soil gas with radon.

If the fan suction is high enough, air will be drawn down through the soil adjacent to the house walls, reducing the radon concentration in the soil gas there. When this leaks into the house through openings in the walls, the amount of radon brought in will be lower than before. Subslab ventilation systems can, therefore, reduce the radon entry rate through walls, but the residual flows may still be high enough to cause concern.

If this is the case, then the voids in block walls can be turned into a collection plenum by closing the opening at the top of the walls either by mortaring them closed or by caulking the sill plate to the top of the block wall. This closed volume can then be ventilated with a small fan, or pipes can be run from each wall to a single fan. Again, if the pressure in the wall is less than that in the house, then soil gas cannot enter the house.

As closing the tops of concrete block walls is slow work, wall ventilation is definitely the highest cost option. Initial work suggested that the walls were the major route of soil gas entry in many of these houses, and wall ventilation alone gave encouraging results. However, these results were not typical, and in general it now looks as if wall ventilation will usually have to be combined with a floor treatment. In that case, the best approach to soil gas exclusion in houses without weeping tile is to install a subfloor system first, and only if that does not produce a sufficient reduction in radon levels proceed to wall ventilation.

In conclusion, it is possible to produce very large reductions in radon concentration by soil ventilation at costs of less than a few percent of the house price. However, it is a sobering thought that there are houses where the costs of soil gas exclusion could be a large fraction of the house value. In those cases, the only economical solution will be improved ventilation coupled with pressurization of the building basement.

TECHNIQUES AND ECONOMICS OF RADON REMOVAL FROM WATER SUPPLIES

Jerry D. Lowry and Sylvia Lowry*

INTRODUCTION

There are three major sources of airborne radon in households and other buildings. These are:

1. Soil Gas
2. Ground Water Supply
3. Construction Materials

The actual contribution by each of these sources in a given household is a combination of many factors. While most researchers have identified soil gas as the major source of radon in homes, in general, it is not uncommon for the water supply to be a major contributor to the airborne radon concentration. In extreme cases, such as the one shown in Figure 3.6, the water supply can be essentially the only source of extremely elevated and significant levels of airborne radon inside a dwelling. The data in Figure 3.6

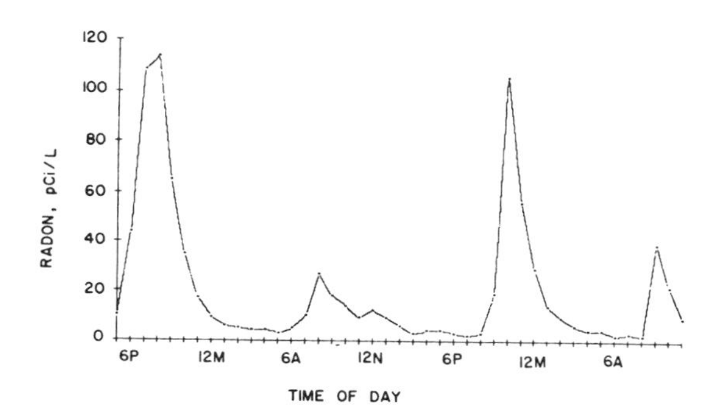

Figure 3.6 -Airborne Radon Levels in Kitchen Area

*Jerry D. Lowry is a faculty member in Civil Engineering at the University of Maine, Orono, Maine 04469. Sylvia Lowry is a principal of Lowry Engineering, Inc., RFD 2 Box 2400, Thorndike, Maine 04986.

show the fluctuations in airborne radon measured in the kitchen resulting from water use in a home located in Maine.

In homes that have elevated air and water radon levels, water treatment can be an effective mitigation technique. In several instances, the authors have found water treatment to be the only mitigation required to reduce airborne radon to acceptable levels. In other cases, where the soil gas contribution is significant, water treatment only provides a part of the necessary reduction. A comprehensive analysis of both air and water radon must be made to determine if water treatment represents a potentially effective mitigation technique. Unfortunately, both air and water radon levels can fluctuate widely and multiple sampling is required, but rarely practiced. Grab samples for water radon can vary by an order of magnitude from one day to the next, making repeated sampling desirable. Another complicating factor in determining whether water is likely to be a major contributor, even when it is relatively high, is that the actual transfer of radon from water to air is a function of water use habits in a given household. This is further complicated by the many variables that determine the household ventilation rate and in turn the ultimate airborne radon concentration that results. Although some obvious indication of the potential benefit of water treatment can be made by comparing water and air radon levels in a home, the only way to accurately document the benefit is to monitor the reduction in air radon before and after water treatment is implemented.

One of the few cases where before-and-after air radon and water treatment data have been collected is a location in Leeds, Maine. After water treatment, the kitchen air radon levels, presented in Figure 3.6, were reduced to below 1.5 pCi/l. Levels as high as 2,000 pCi/l airborne radon in the bathroom were reduced to the same level. While the authors believe that this may be a rare and fortunate case, in terms of mitigation ease, it does demonstrate the importance of waterborne radon and the effectiveness of water treatment to reduce the airborne concentrations.

REMOVAL TECHNIQUES

There are two cost-effective water treatment methods that can be used to remove radon from water supplies: 1)Granular Activated Carbon (GAC) Adsorption/Decay and 2)Aeration. The use of decay/storage is also possible; however, it does not represent a viable economic alternative when compared to GAC and aeration processes.

GAC Treatment

The senior author first identified the potential of GAC adsorption/decay for radon removal from household water supplies seven years ago. Since those initial experiments, further research has enabled us to determine the relative effectiveness of specific GAC products, model the design/operation of any GAC system to accurately predict its effectiveness, and to study the buildup of radioactivity within the GAC bed. Today, it is possible to properly select and size a GAC system to yield any level of removal desired for any level of influent radon encountered. Recently, we have designed a system that reduces radon from over one million pCi/l to less than 500 pCi/l, for a 99.9+ percent removal efficiency.

The key to the effectiveness of GAC treatment is the adsorption/decay steady state that occurs for radon and its short-lived daughters. This has been detailed elsewhere[1]. The result of the steady state operation is contrasted to normal GAC operation for a non-decaying adsorbate in Figure 3.7. The net result of this unique operation, brought about by the short half lives of the adsorbate and the extremely small mass accumulation it represents, is that a GAC bed will last for decades with respect to radon and its short-lived progeny. This has a significant influence on the economics of GAC vs. aeration at small (household) flows, making the operation and maintenance costs negligible.

An example of the effectiveness of GAC treatment is shown in Figure 3.8. This well had an average radon level of greater than 700,000 pCi/l, and the radon was reduced to less than 1,000 pCi/l. The detrimental effect of routine backwash is illustrated by the first 180 days operation. Operation without backwash is illustrated by the remaining period of operation.

GAC adsorption/decay can be modeled by a first order differential equation as follows:

$$dC/dt = K_{ss}C$$

where C is the bulk solution radon concentration and K_{ss} is the steady state rate constant for radon adsorption/decay. Nine GAC products have been tested to date and the range of K_{ss} ranges from 1.25 to over 3.0 per hour. Therefore, the GAC selected is very important in the design process. The equation above can be put into its more useful form by integrating, and:

$$C_t = C_o \exp(-K_{ss}t)$$

where C is the radon concentration at any empty bed detention time (EBDT). Thus, for a given GAC, any level of removal can be provided by selecting the appropriate GAC volume, in terms of EBDT.

The kinetics and degree of radon and daughter levels that build up during the attainment of steady state (approximately 10 days) have been detailed elsewhere[2] and are beyond the scope of this paper. However, from a public health standpoint, this small build up is not significant in the vast majority of treatment cases.

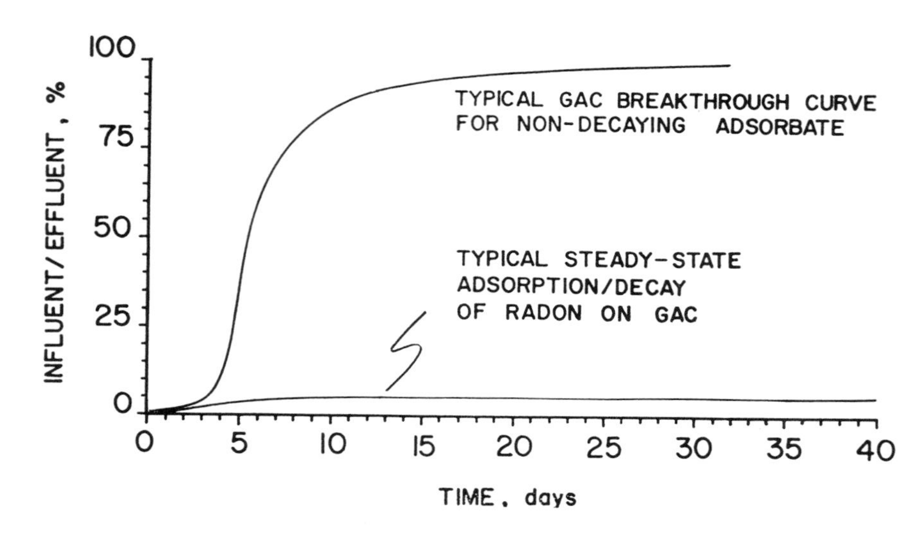

Figure 3.7 – Comparison of Adsorption/Decay Steady State with Typical Breakthrough for Non-Decaying Adsorbate

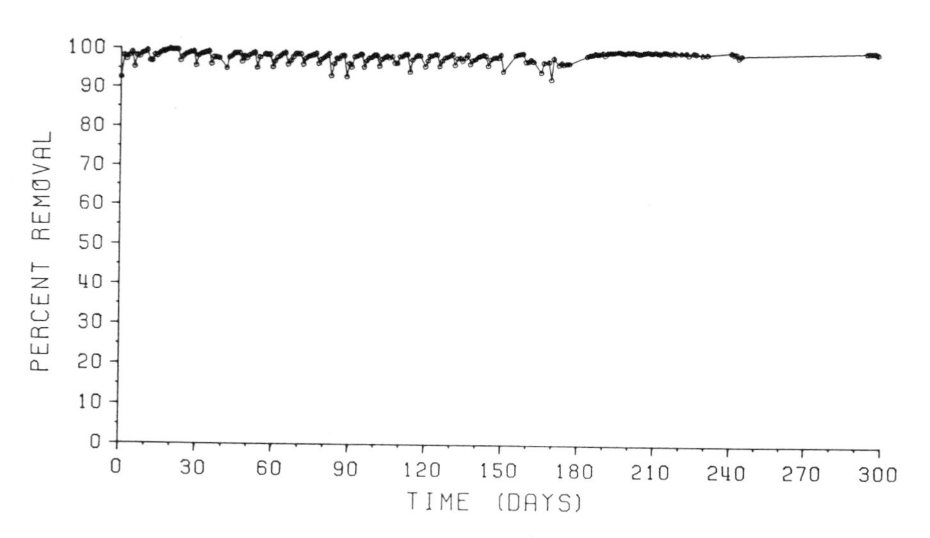

Figure 3.8 –Steady State Radon Removal for an Extremely High Radon Water Supply Containing 750,000 pCi/L

Aeration Methods

There are three major methods of aeration: 1) diffused bubble, 2) spray, and 3) counter current packed tower. All have been used to remove radon from water supplies over a range of flow covering household to industrial applications. The first two are applicable to smaller flows at sites, such as households, that are not practical for the installation of a packed tower. Large scale diffused bubble aeration has been used at a location in England for over twenty years[3]; however, it is generally recognized that packed tower aeration for large flows is a more cost-effective alternative.

For household applications, two commercially available aeration units have been developed. The State of Maine Department of Human Services, Division of Health Engineering, constructed a novel spray aeration device that has been marketed. It has been used to remove up to 93 percent of the radon from an extremely high radon well. A second alternative has been developed by the authors and is based on multi-staged diffused bubble technology. This system has recently been tested at a site containing 250,000 pCi/l of radon and routinely removes radon to background levels of 50 pCi/l. It is capable of radon removal to background for any household water supply.

A disadvantage of aeration for household flows is that aeration must be done under atmospheric conditions and the treated water must be re-pressurized. Re-pressurization represents a significant portion of the capital, operation and maintenance costs for these alternatives.

ECONOMICS

The use of GAC vs. aeration at a given site will be determined primarily by the capital and operating/ maintenance (O&M) costs. Other factors such as the need for pretreatment, discharge of radon to the immediate atmosphere, and the buildup of radon and daughters in the GAC bed will also be possible considerations in the selection process.

The economics of radon removal begins with the selection of the most cost effective treatment alternative. A reasonable perspective of relative cost-effectiveness between the alternatives can best be attained by dividing all applications into three categories, based upon the magnitude of flow treated (as measured in gallons per day or gpd):

1. Household Supplies
   (50 to 500 gpd)

2. Multi-Unit and Small Community Supplies
   (500 to 20,000 gpd)

3. Municipal Supplies (greater than 20,000 gpd)

For household supplies, GAC treatment is the most economical alternative. It costs approximately $750 to $850, installed, and has negligible O&M costs. It is mechanically simple, with no operating parts, and treatment occurs under pressure. In contrast, installation of spray aeration and diffused bubble aeration cost approximately $2600 and $2000, respectively. In addition, they have associated O&M costs and are mechanically more complex. While they avoid the slight build up of radon and daughters associated with GAC treatment, it is very doubtful if they will become popular as a result of their high relative cost.

It appears that aeration will become more cost-effective than GAC as the design flow increases to between 10,000 to 20,000 gpd. A current EPA/Cincinatti Demonstration Grant in New Hampshire is investigating the economics of aeration vs. GAC at two public water supplies to provide data to determine the crossover point more clearly.

At flows greater than 20,000 gpd, packed tower aeration is the most cost-effective alternative. The authors have designed an aeration and GAC system for a flow of 350 gpm (504,000 gpd) and found that GAC treatment was approximately ten times more costly than packed tower aeration.

FOOTNOTES

[1]Lowry, J.D. and J.E. Brandow, "Removal of Radon From Water Supplies,"Journal of ASCE, Environmental Engineering Division, Vol. 111, No. 4, August 1985.

[2]Ibid.

[3]Hoather, R.C. and R.F. Rackham, "Some Observations on Radon in Waters and Its Removal By Aeration," Proceedings of the Institution of Civil Engineers, Great George Street, London, S.W. 1., Dec. 7, 1962, pp. 13-22.

HIGHLIGHTS FROM THE DISCUSSION: TRANSMISSION AND MITIGATION OF RADON

RELATIONSHIP OF RESEARCH AND MITIGATION

H. Sachs: To me, the most fascinating thing about the presentations was the tremendous diversity between the research approach out at Lawrence Berkeley Lab and the applied mitigation approach that Art Scott has followed. What information does the mitigation community really need from the research community, and vice versa?

A. Scott: What research needs are there for mitigation? We are limited in mitigation to what we can do at a reasonable cost. What we need from the research community is in essence some kind of easy screening procedure to tell us what is the minimum we can get away with. That is, indeed, the difficult part. I'm running a research program, and from that point of view, I don't care whether it works or not. If it works, we chalk it up as a success. If it doesn't, we again chalk it up as a success. A sub-floor ventilation system is one of the cheapest systems in a retrofit situation, and in our program we find out whether it works by putting one in. The research community could help if they could help answer the questions, "Is it necessary?" "Is it sufficient?" Or do we have to look at other things? Improving our diagnostic capability is an important objective. Moreover, the diagnostics have to be cheap. At the present moment, with our team, we can put in a subfloor system for around $1,000 direct cost, so I would clearly be unwilling to spend $2,000 in diagnostic testing to decide whether to put in a $1,000 system.

R. Sextro: We consider ourselves part of the mitigation community too, because we are finishing research where we have done some mitigation and tested some mitigation systems. Certainly what Arthur said is right. There is a need for some diagnostic development: tools, methodologies, protocols, or whatever. I think Arthur would be the first to agree that in mitigation work, you are sometimes or often surprised that what you think is going to work doesn't work as well as you thought. Or, as in a couple of cases, we've been surprised at how well some things work when in fact we did them as a last desperate resort thinking that we had nothing else left to try. Clearly, you have a situation where houses and soil systems can tend to be very individualistic, and we need to find out more about how those things work.

Can the research community benefit from the mitigation community? Certainly. One area is just the need for additional data. We all run around with anecdotes and numbers written down on scraps of paper about how well certain systems perform and what the soil gas concentrations were, and things like that. What's really needed is some way of trying to put that information on some kind of systematic basis. For example, I would like to get more data on the correlation between measured soil gas concentrations and indoor air concentrations.

R. Tucker: Sylvia, from the standpoint of ground water, do you see any further research questions that need answering?

S. Lowry: There are many things that need to be done. One is to look at the radon daughter buildup inside the GAC bed. We still don't know if Pb210 is retained; we don't know if it is a problem if it is retained. Another thing involves the radon parents, uranium and radium dissolved in water, and how to remove them. If you look at the literature, there is very little information available.

GAC APPROACH FOR AIR

H. Sachs: I have another question for Sylvia Lowry. The GAC approach looks so neat and so quantitative. Why aren't you doing this for radon in air also?

S. Lowry: The idea to use GAC with water came from uranium mines where GAC was used to absorb radon from the air. That is where Dr. Lowry got the initial idea.

R. Sextro: In air there are so many potential poisons for activated carbon and radon is so weakly bound to that system by Van der Waals forces that practically everything you can think of absorbs more strongly on charcoal. You get this process of basically bumping the radon off. It just doesn't stay. We have done some lab experiments, and you can show that with enough granulated carbon, (about 50 kg), you can remove radon in air. But the breakthrough is real quick, and it doesn't stay there long enough to decay.

EFFECT OF WATER TABLE

Question: A recent article suggests that one of the most important driving forces for radon in soil gas is the annual rise and fall of the water table, particularly for areas where the water table is near the surface. Any comments on that?

R. Sextro: That's not something we've been able to measure experimentally. You do get a hydrostatic pumping, and, if the water table is very close to the surface, the effective volume of soil that one deals with is altered by the changing ground water levels. It's fairly clear, at least on theoretical grounds, that that's going to happen. I would like to be able to measure it.

Question: In terms of radon in water, do you see a large variability with shallow well systems?

S. Lowry: As far as I know, the variability is wide.

A. Scott: The trouble in many of these cases is that there is correlation and there is causation. When the rain falls, the water table generally rises, but in fact it has been known for many years that the rise of the water table as indicated by test wells is often higher than the amount of rain that falls. The reason for this is that indeed the pressure in the soil is elevated by the rain coming in, but as the moisture is drawn through the soil by capillary forces, it in fact forces air into the soil ahead of it. So the air is being pushed downwards by an essentially wet surface piston, and, under these circumstances, of course, the flow into buildings will become greater. Whether the rise and fall of the water table has anything to do with the radon levels in houses, except by being correlated, is another matter. It is like the other claim that changes in barometric pressure cause changes in radon concentration in houses. No one has ever shown that; they have correlated changes in barometric pressure to radon changes in houses, but change in barometric pressure is invariably associated with the passage of a front which means the wind speed, direction, and quite often precipitation, change simultaneously. My belief is that the atmospheric pressure has nothing to do with it; that it is the change in these other three that do it, but they are correlated.

AIR CONDITIONING AND SEASONAL RADON LEVELS

Question: Does summertime whole-house central air conditioning produce pressure differences likely to cause influx of soil gas?

A. Scott: Yes, depending on your system. It's probably not so bad in northern states where two-story houses are common. In southern states where single-story slab on grade houses are common, a lot of the air conditioning systems are run outside the house or they are tucked up in the attic. Leakage from the system is out of the house and, therefore, the house becomes depressurized. I've made measurements in Florida where the operation of the air conditioning system has made large changes in the ventilation rate but the radon concentration of the house has stayed about the same because the depressurization has increased the radon supply.

H. Sachs: This question goes back to the question of when you can expect to see the highest values of radon. In the data from Pennsylvania Power & Light, a significant fraction of the houses had their peak values in summer, and some houses had them in spring or fall. We may standardize on making measurements in winter, but we shouldn't do it thinking that that will be the highest season in every house. In particular, you can think of lots of models for centrally air conditioned houses which would indicate why they should have higher values in summer than you see in winter, and you can rationalize the data quite easily. But it doesn't necessarily mean causation.

REMEDIATION BY HOMEOWNERS

Question: Is it prudent to seal the perimeter in the french drain inside the basement, and what kinds of water problems might this lead to?

A. Scott: If it's going to cause water problems, it's probably not prudent to seal it. I am of two minds over this. I conducted a complete program in Canada where the prime mitigation method was, in fact, sealing. But I cannot really recommend that. It's very difficult to do properly, and unless it's done properly, it's not worth doing at all. My preference, if people are looking to do things themselves, would be some kind of active soil ventilation to suck underneath the floor rather than trying to seal perimeters.

Question: You mentioned people doing remediation themselves. For low levels of radon in homes, do you feel that a do-it-yourselfer should attempt a remediation project? How does the panel feel?

A. Scott: In general I would far rather deal with a house at 1 WL (200 pCi/l) than a house that is at 0.03 WL (6 pCi/l). In many cases, the amount of money that has to be spent to reduce the level in a house is independent of the initial radon level. If you are going to stop radon coming in, you have to block up the holes. And it costs just as much to block up the holes or change the pressure distribution regardless of the amount of radon. As to whether there is sufficient information for do-it-yourselfers, I think we can unequivocally say, "No there isn't." There may well be in the future. The EPA is working on at least an initial guide, and the results of my program in the Reading Prong will be published at some point.

R. Sextro: First, I think one of the difficulties is walking this line between apathy and hysteria. The problem is that suppose somebody tells you that you ought to do something in your house if it's above 4 pCi/l. That seems to be the direction the EPA is going, although I'm not sure I agree with it. Suppose your house measures 4.1 pCi/l. First of all, we know that 4.1 pCi/l is probably 4.1 pCi/l plus or minus 3.1 or 2.1 pCi/l. There is a big error on any measurements you do, even if you've integrated over a full year with Track Etch detectors or anything else. The second point is that until you start talking about 20 or 30 or 50 pCi/l, you're talking about a real grey area about what kind of remedial action to take and who does it. That's a case where the homeowner, by just paying more attention to the way he operates the house or supplies outside air for the furnace system, may be able to deal with the problem. And if it's successful, fine. If it's not as successful, well, it may not matter that much. When you get into the 200-300 pCi/l range, I think I agree with Arthur. I'm not sure enough information is known now to give the homeowner a lot of guidance about what he can do. But that's also high enough that ventilation system increases aren't going to do you much good. You can put a whirlwind through your house, increase the air exchanges in your house by a factor of ten, and reduce the radon concentrations by ninety percent. But if you are dealing with a radon level that starts out at 500 pCi/l, that's not going to be enough. The answer to this question, typical of the usual way scientists answer such questions, is a two-handed argument.

H. Sachs: Yes, I'm a one-handed scientist, and I'm going to talk politics. What I want to emphasize is that you have now had three radon mitigation experts from very different backgrounds and interests stand here and say, in cases of houses with fairly high radon concentrations, we would be very reluctant to say we understand how to transfer knowledge or efficiently do mitigation. We are learning, and these gentlemen are making enormous progess now. But I would personally be reluctant to try to teach a course. As Art said, I might like to take one, because we really feel very humble about what we know about cost-effectively or efficiently fixing houses and transferring that knowledge. EPA is planning to teach that course; in fact it may be underway now. They haven't asked me if I want to sign up. I would like to learn, but I want to make the political point that if the agenda of the federal government had been somewhat different in '82 and '83, when the research community was coming to them and saying, "we need much more research in this area," then I believe that many of these questions would have been settled. We would be sitting here today in this meeting transferring much more knowledge and having far fewer questions. There has been a very severe hiatus in the ability of the research community to move forward because of the restrictions imposed by policy decisions that radon research was not something the federal government should be doing. I am angry about it, and I made the point once before. I will promise to try not to make it again until Saturday.

MITIGATION USING VENTILATION

Question: In an average home, is one of the vented under-the-floor pump systems usually enough or would you need several?

A. Scott: We work on a rough rule of thumb that one pipe can cover up to about 15 feet radius. So in a typical house, say 30-by-40, two pipes will do it.

Question: Would having air for combustion for the furnace piped in from the outside help to keep radon levels down?

R. Sextro: We have seen pressure differences in houses go up as soon as the furnace or clothes dryers are turned on. Any source of pumping air out of a building without any replacement makeup air leads to an unbalanced system, whether it's a ventilation system or a combustion system. Yes, supplying outside air to those appliances, a furnace or a clothes dryer, is a good idea in order to offset the extra pressure differential that they develop in the house.

H. Sachs: There is one precautionary note that people hammered into me years ago, when I was thinking of doing this myself. You have to think about whether your particular central furnace is what's called a natural draft, which is true of almost all natural gas furnaces, or a forced draft, like most oil furnaces which have a squirrel cage blower on the front end. If you have a natural draft furnace, you must not use a natural draft outside air supply at a different level from the exhaust. You can get into very severe pulsing and backflow problems. I have no direct data, but the gas association gets very upset about this. I just don't want you to rush out and directly pipe your natural gas furnace to the outside of your house when you've got a chimney that goes up 30 feet. You can get into trouble that way.

MITIGATION ON SLABS WITH AIR DUCTS

Question: Has any work been done on slab constructed homes with forced hot air ducts in the slab?

H. Sachs: I spent a couple of months learning the trade by working on a house which had sub-slab air ducts, and it is difficult to imagine a better system for making sure you get almost all the radon into the house. Particularly if you consider the very high quality systems which use helically-formed air ducts with convectional rectangular risers. The geometric problems in sealing a rectangular riser to a helically-coiled round pipe are such that there is no easy solution. These are basically systems for driving air into the soil and bring it back in through the gaps in the slab. We did mitigation on one with a sub-slab duct system. In fact, we trapped an air system that went around the perimeter of the house, removing about 70-80 percent of the radon. We went the rest of the way with an air-to-air heat exchanger. That one has been written up and published. It was a very difficult challenge, and I think there was some serendipity, some luck that we happened to hit the right places.

MITIGATING BASEMENTS

Question: What effect do wall and floor coverings have on lowering basement radon levels?

H. Sachs: I have no data on whether or not floor coverings make any difference. Considering what the other speakers have said about the difference between diffusion and convection, if you carpet your slab and thereby stop the flow of air through its cracks, you may help. But given Art's extensive experience with trying to seal by any mechanism, I'm not sure I would want to rely on that as being an important part of my radon mitigation strategy.

A. Scott: Almost anything will do as an effective barrier. The only trouble is stopping it coming around the edge. So the question of coverings is to some extent irrelevant; the thing is how does it stop radon from going around the edge.

RADON AND NEW CONSTRUCTION

Question: How do you keep radon out of new houses?

A. Scott: It is easy to give advice. This is one of the pleasant things where the difficulty is not giving the advice, but the difficulty is in following it. The obvious thing to do is: don't leave holes. And in some cases, I think this is not an unreasonable thing to do. I am involved in a project in Florida where they build houses on slabs to demonstrate that we can build houses without significant holes at modest cost. When you get to houses with basements, it becomes more difficult. Possibly the best approach is to build in your mitigation measures ahead of time. For example, Bill Brodhead, a Pennsylvania builder active in radon work, builds houses with gravel underneath the floor slab with some pipes in there led to a riser outside. It doesn't cost much to do when the house is being built, and if it's needed, all you have to do is put the fan on. Apart from that, you are again into questions of what's common in the area. One thing is absolutely certain in the building trade. Something different will cost a lot of money. If they normally build houses on stilts in your area, that would be a low-cost thing; but if they normally build houses on basements, I can tell you that the cost will be astounding.

Section 4

Testing and Measurement

INTRODUCTION AND SUMMARY OF SECTION 4: RADON TESTING AND MEASUREMENT

William J. Makofske*

This section provides an extensive overview of existing instruments, methods and measurement protocols for the indoor detection of radon gas and radon progeny. Because it is the progeny into which radon decays (such as Po-218 and Po-214 which subsequently decay by alpha emission) that actually causes the health risk, some techniques focus on detecting the radon progeny directly. In contrast, other approaches assume that measuring radon gas concentrations will provide an adequate indirect measure of radon progeny and hence risk.

The various approaches or methods for measuring radon levels can be distinguished according to several key characteristics, such as their accuracy, affordability, needed expertise for deployment and reliability. These characteristics can be used to help identify which testing method is best suited to a number of the most common applications. These include the large-scale "screening" for "hot houses" by public agencies, the evaluation or "characterization" of radon levels in homes by the homeowner, "real time" feedback needed for diagnosis and remediation work, certification of radon levels for real estate transfers, and the determination of exposure levels for health studies and risk assessment. Because some of these applications make specific demands, such as differences in the desirable time interval, methods vary in their suitability for different applications. Based upon the papers in this section, it would appear that the current array of instruments and methods are adequate to meet all the varied applications.

In "Instruments and Methods for Measuring Indoor Radon and Radon Progeny Concentrations," Andreas C. George (Physicist with the Environmental Measurements Laboratory, Department of Energy, NYC) provides an extensive overview of existing techniques for measuring radon gas in air, radon progeny in air, and radon sources (from surfaces, in soil, in water). The generic types of testing for radon and progeny in air include grab sampling, continuous sampling and integrated sampling. Some of the applications for which appropriate techniques exist are screening, house

*William J. Makofske is Professor of Physics, Conference Co-Organizer, and Director of the Institute for Environmental Studies at Ramapo College.

characterization (or follow-up), mitigation or diagnostic measures, new construction and special health effects evaluation.

In their paper, "Track Etch Radon Detectors for Long-Term Radon Measurement," Richard A Oswald and Samuel L. Taylor, (Scientists with Terradex Corporation, Walnut Creek, CA), provide details on the theory and operation of a particular type of integrating detector called the alpha track detector ("Track Etch Detector") manufactured by their firm. They emphasize the importance of using integrated measurements over a sufficiently long time period because of the large temporal variations that have been found in radon concentrations in homes. In some cases, variations have been found which span an order of magnitude over time periods of a few days. Since reasonably accurate exposure data is critical for decision-making in every application, collecting integrated data over several months provides the best means of characterizing the radon dose, and, therefore, the risk and health effects to the individual.

Melinda Ronca-Battista, (Health Physicist with the EPA Office of Radiation Programs in Washington, DC), in her paper "Interim Radon and Radon-decay Measurement Protocols"[1,2] describes the EPA's efforts to assure consistency in procedures for data collection as well as for screening and follow-up. Included in the EPA guidances described are recommendations for detector placement, house conditions, time of year, and quality control of instrumentation.

The discussion which followed the presentations underscored some of the uncertainties and issues surrounding radon testing. First, it was generally agreed that there is no substitute for testing individual houses. That is, while it is possible to identify relatively large regions with high radon potential based on geology, soil conditions, or building practices, there is considerable variability in radon concentration in houses in the same neighborhood or area. Furthermore, the lognormal distribution of radon in homes suggests that even in relatively low risk areas, there will be some homes at significantly higher risk. A second conclusion is that screening and follow-up testing for radon instead of radon progeny is a sensible approach. Testing for radon is easier, and the equilibrium factor between progeny and radon gas itself is sufficiently constant (about 0.5 or lower) to justify conversion of pCi/l (a measure of radon gas concentration) to WL (a measure of radiation dose and, therefore, health risk from progeny). Third, it was agreed that screening for radon in water need not be done routinely. If a screening test for air gives sufficiently high values to warrant concern, or if surrounding wells are found to be high, then water should be tested as a possible radon source.

DISCUSSION -- EMERGENT ISSUES REGARDING TESTING

While the EPA has developed protocols for seven different measurement systems, the two instruments that are most commonly used by state agencies, organizations, and homeowners are the activated charcoal canister and the alpha track detector. Both are integrating radon gas monitors which share many of the same advantages in use. They are relatively low cost, completely passive (there is no need for external power), unobtrusive when installed, distributable by mail, and easy for the homeowner to handle and install.

The alpha track detector has an additional advantage in that it can be used over periods typically ranging from 1 month to 1 year. It is the only detector (at reasonable cost) capable of measuring the actual integrated average concentration over these longer periods. Its main disadvantage is that it has a large inherent variability, particularly at low concentrations and when only a small area of the detector is counted. Also, if very high radon concentrations are suspected, the longer time interval for measurement may introduce an undesirable delay in confirming the risk.

In contrast, the charcoal canister integrates over a time period of typically 3-7 days. It, therefore, provides quicker information than the alpha track detector, and it also provides relatively precise results. Although it is sensitive to temperature and humidity, corrections for these variables can be made. Perhaps its major disadvantages are that it is most sensitive to the latter 12-24 hours of exposure, and that the overall duration of the test is relatively short. A critical question is the extent to which this short time period is representative of the actual average concentration of the house. This question becomes relatively more or less important depending on the application.

Thus, perhaps the most important issue involves clarification about when a given testing approach is most suitable. Government agencies concerned with screening studies in order to find "hot spots" and determine those individuals at highest risk have favored the use of the charcoal canister because it is quick, inexpensive and suitable for large sampling groups. However, it is not clear that this is necessarily the best testing method from the perspective of the resident concerned about identifying radon exposure risk in the home, although it is heavily used for this purpose. Despite its precision, the limitations of the charcoal canister are of great concern, given that radon levels can vary dramatically over short time intervals. Short-term results can be misleading since it is the long-term time-averaged concentration which is of interest. To overcome this limitation, charcoal canister measurements can be repeated four times over the year, one during each season. While this approach will identify the average concentration, in practice most actual screening programs stop after only one test. It is argued that if this one test is done according to protocols (closed house conditions, wintertime, basement), it represents a worst case situation, and it is unlikely that a very low reading will result for a very high house. While this is true, it still begs the question for the house where readings over a restricted time period of a few days may come out less than but around 4 pCi/l but where the actual average level may be considerably higher, perhaps between 10 and 20 pCi/l. On the other hand, it may also be argued that waiting 3-6 months for a result for a house with 1000 pCi/l value is not serving the homeowner either. However, given the time delays in setting up government screening programs, this question may be moot.

Once a result is attained, it must be evaluated. According to the draft EPA document[3], the recommended actions and interpretation based on protocol screening measurements are:

Less than 4 pCi/l(0.02 WL) - Relatively low probability of significant health risk from concentrations in general living areas of home; follow-up measurements are not recommended, but are at the discretion of the resident.

Greater than 4 pCi/l(0.02 WL) but less than 20 pCi/l(0.1 WL) -Consider a follow-up measurement over the next 12 months.

Greater than 20 pCi/l(0.1 WL) but less than 200 pCi/l(1 WL) - Consider a follow-up measurement within several months.

Greater than 200 pCi/l(1 WL) - Follow-up measurement and short-term action to reduce levels as soon as possible.

While a homeowner using this guidance can get a good picture of their relative risk and requisite actions, there are some potential sources of confusion. The chart is further discussed in the text of the document, sometimes in an ambiguous fashion. For example, for results less than 4 pCi/l, the guidance does not recommend follow-up "unless the room in which the screening measurement is made is used as a living area." Since every radon test will show some level of radon, does this mean that every test taken in a lived-in area should have follow-up testing, even if it is below 4 pCi/l? Another source of confusion is inherent in any effort to set action levels. Errors in the measurements are likely to be significant, particularly at lower levels where statistics are poor. A homeowner with a 3.9 pCi/l reading might consider no action to be necessary. However, a value of 3.9 pCi/l could actually be 5.1 pCi/l or higher, based on the errors. Furthermore, the health risk, as far as we know, is linear. There is no sharp dividing line where the risk drops significantly. There are essentially no differences in risk between 3.9 and 4.1 pCi/l, between 19 and 21 pCi/l or between 199 and 201 pCi/l. Yet the guidance suggests different responses as determined by dividing points of 4, 20 and 200 pCi/l.

CONCLUSIONS

In summary, we can return to the various applications of testing mentioned earlier:

1) The screening process for finding very high houses. This is a primary concern of government agencies and homeowners in known high radon areas. Either charcoal canisters for a few days or alpha track detectors for a few weeks should suffice to separate out the very high houses from the others. This procedure may not give very good average radon concentrations.

2) The process of house characterization or determining the average radon concentration in a house. The EPA manual refers to this as follow-up measurements. This is or should be the major concern for homeowners who have little reason to suspect high values or who have done a screening measurement resulting in a value that is high enough to warrant some concern. In this case, alpha track detectors over a 3-12 month period, or four charcoal canister tests, one every 3 months over the year, should give a reasonable approximation to the long-term average radon concentration if testing protocols are followed.

3) The process of classifying a house as acceptable or not acceptable in a real estate transaction. Because of the time limitations often imposed in such situations, the usual time-averaged measurement approach is difficult to apply. Often, grab samples and real-time monitoring with relatively sophisticated equipment are used to get quick results. There is often time for charcoal canister screening measurements but house characterization is difficult. In such cases, characterization might not occur until after the transaction is complete. Unfortunately, both the short-term variability of radon levels and the possibility of fraud threaten the meaningfulness of these tests.

4) The diagnostic and mitigative processes of determining the pathways for radon entry in homes with high levels and the subsequent effect of mitigation procedures. This generally refers to the need to have measurements with either short-term or real-time feedback to determine radon source strengths emanating from walls, cracks, sumps, etc. However, because of the very high radon levels under consideration, often charcoal canister or alpha track detectors can give meaningful results in short time intervals as well.

5) The process of determining exposure levels for health studies and risk assessment. In this case, the actual long-term exposure is desired and therefore the use of long-term time-integrated measurement instruments, such as the alpha track detector, is desirable.

6) Post Mitigation Retesting. Finally, in houses that have undergone mitigation, there is a need for post-mitigation measurement to ensure that the house remains at an acceptable level. In most cases, time-integrated measurements similar to those used in house characterization would be most suitable.

The various instruments and methods used for detecting radon in houses must be evaluated carefully in terms of the intended application. Screening measurements involving a single short-term test, such as the charcoal canister, benefits the objectives for large scale radon surveys by government. For suspected or known high radon areas, it may be equally to the benefit of the homeowner. Otherwise, the homeowners' interests are better served by house characterization which provides them with information necessary for decision-making. The confusion of screening and characterization leads to the possibility of both false negatives and positives. Given the risks associated with radon exposure and the costs associated with mitigation, neither of these outcomes is desirable. It, therefore, becomes imperative that screening agencies and testing companies be very clear to the homeowner about the meaning and limitations of various tests.

FOOTNOTES

[1]Ronca-Battista M., P. Magno, S. Windham, E. Sensintaffar, Interim Indoor Radon and Radon Decay Product Measurement Protocols, EPA 520/1-86-04, U.S. Environmental Protection Agency, Office of Radiation Programs, April 1986

[2]Ronca-Battista M., P. Magno, P. Nyberg, Draft Interim Protocols for Screening and Follow-Up Radon and Radon Decay Product Measurements, U.S. Environmental Protection Agency, Office of Radiation Programs, June 1986

[3]Ibid, Table 2, pg. 10.

INSTRUMENTS AND METHODS FOR MEASURING INDOOR RADON AND RADON PROGENY CONCENTRATIONS

Andreas C. George*

INTRODUCTION

A review of the methods currently available for sampling and monitoring radon and progeny shows that several task specific instruments exist that are adequate for the determination of environmental levels of radon and progeny for radiological protection purposes. There is not a single measurement method or type of measurement that will be adequate for all possible applications. To assess the problem of indoor radon, several types of measurements are needed using different protocols and instrumentation:

1. Measurements of the indoor air concentrations or radon and progeny.

2. Measurement of radon sources -- radon emanation from surfaces or the underlying soil, from building materials; radon from water and radium in soil and building materials.

3. Measurements for health effects evaluation -- specialized measurements of radon progeny size distribution, respiratory deposition, airborne particle concentration and the interaction of particles with radon progeny.

The type of method and the instruments to be used depend on the need or application. For screening indoor environments for airborne concentrations of radon and progeny on a large scale, detectors that measure radon or radon progeny or both can be used. Radon measurements, which are easier to perform than those of radon progeny, are generally adequate since they provide information on the upper limit for the potential alpha energy exposure from radon progeny. More intensive and complex measurements are required for in-depth studies of radon transport, radon behavior and its interaction with the indoor environment and

*Andreas C. George is a physicist at the Department of Energy Environmental Measurements Laboratory in New York City.

for the development of control techniques. Measuring sources of radon can be useful in predicting what radon and progeny concentrations will result in new construction and what to do in existing housing to minimize the impact of high concentrations.

The quality of radon and progeny measurement data is determined by the ability to obtain accurate measurements at low concentrations. Therefore, monitoring devices, including the entire system of associated equipment and techniques, should be properly calibrated and be able to stand up against measurement quality, proficiency testing, and evaluation. It should be noted that calibrations provided exclusively by instrument manufacturers are seldom adequate.

MEASUREMENTS OF AIRBORNE CONCENTRATIONS OF RADON AND PROGENY

Instruments and methods for measuring indoor radon and progeny can be divided into several groups depending on the sampling time, on the field application or the type of information needed, on the instrument sensitivity and, to some extent, on cost considerations. Since the inhalation exposure and consequent lung dose from the radon series arises from radon progeny, the appropriate choice is to use devices that measure radon progeny. However, data from many typical residential buildings where both radon and progeny were measured concurrently indicate equilibrium factors between them ranging from 0.3 to 0.7, with an average value usually close to 0.5.[1,2] Using this relationship, the radon progeny, which are the most difficult to measure, can be replaced with measurements of radon and the appropriate corrections made. Radon measurements are generally adequate, providing information on the upper limit for the potential alpha energy exposure from radon progeny. Tables 4.1 and 4.2 list the most commonly used instruments and methods for measuring the airborne concentrations of radon and progeny.

Grab sampling for radon and progeny is useful in some field applications, such as for screening purposes on a small scale, i.e., for indicating whether a building requires detailed study later with more complex instruments. Grab sampling is an attractive and powerful technique in guiding remedial action work. When it is used properly -- by making several measurements in different seasons -- grab sampling is suitable for the assessment of the average indoor air concentration.

Continuous sampling is useful in studies where in-depth measurements of the behavior of radon and progeny are desired and for measuring real-time changes in their concentrations. Peak short-term concentrations can be observed and variations can be correlated with other parameters, i.e., radon source strength and ventilation infiltration. This type of monitoring is not suitable for large scale surveys due to time and cost considerations. It

Table 4.1 - **INSTRUMENTS AND METHODS FOR MEASURING RADON IN AIR**

Instrument and method	Application	Principle of Operation	Sensitivity	Ref.
Scintillation cell	Grab sampling	Scintillation alpha counting	< 0.1-1.0 pCi/liter	3,4
Ionization Chamber	Grab (laboratory only)	Sample transferred into ion chamber. Pulse of current counting.	< 0.05 pCi/liter	5
Active continuous scintillation cell monitor	Continuous	Flow through scintillation cell alpha counted.	<0.1-1.0 pCi/liter	14,15
Passive diffusion electrostatic monitor	Continuous	Radon diffusion into sensitive volume. ^{218}Po collected on scintillation detector electrostatically.	0.5 pCi/liter for 10 minute counting intervals	16
Passive diffusion radon only monitor	Continuous	Radon diffusion into sensitive volume. Radon progeny removed by electret. Count alpha particles from radon only with alpha scintillation counter.	0.1 pCi/liter for 60 minute counting intervals	17
Passive track etch monitor	Integrating	Alpha sensitive film registers tracks when etched in NaOH.	0.2, 0.4 pCi/liter - month depending on size	20-22
Passive activated carbon monitor	Integrating	Radon adsorption on activated carbon. Gamma counting with gamma analyser for ^{214}Pb and ^{214}Bl gamma rays.	0.2 pCi/liter for 100 hour exposure	23,24
Passive electrostatic - thermoluminescence monitor	Integrating	Radon diffusion into sensitive volume. ^{218}Po collects on thermoluminescence detetor electrostatically.	0.03-0.3 pCi/liter depending on size for 170 hr exposure.	25

Table 4.2 - INSTRUMENTS AND METHODS FOR MEASURING RADON PROGENY IN AIR

Instrument and method	Application	Principle of Operation	Sensitivity	Ref.
Kusnetz - Rolle	Grab sampling for WL	Collect sample on filter for 5-10 min. Alpha count.	0.0005 WL	6,7
Tsivoglou and modifications	Grab sampling for individual radon progeny and WL	Collect sample on filter for 5-10 min. Alpha count.	0.1 pCi/liter each of ^{218}Po, ^{214}Pb, ^{214}Bi and 0.0005 WL	8-13
Tsivoglou and modifications	Continuous - Instant radon progeny and WL monitoring	Collect sample on filter for 2-3 min. Alpha and beta counting.	0.1-1.0 pCi/liter 0.001-0.01 WL depending on flow rate	18
Tsivoglou and modifications	Continuous	Collect on filter for 5-10 min. Alpha count. One measurement every 30 min.	0.1-1.0 pCi/liter 0.001-0.01 WL depending on flow rate	19
Thermoluminescence radon progeny integrating sampling unit (RPISU)	Integrating	Collect sample on filter for 1-2 weeks. Detect with thermoluminescence material (CaF_2:DY).	0.0001 WL	26
Thermoluminescence modified WL monitor	Integrating	Collect sample on filter for 1-2 weeks. Detect with thermoluminescence material (LiF).	0.0005 WL	27,28
Surface barrier WL monitor	Integrating	Collect sample on filter continuously. Detect alpha radioactivity with silicon surface barrier detector.	0.00005-0.005 WL depending on flow rate	
Radon/Thoron WL monitor	Integrating	Collect sample on filter continuously. Detect radon and thoron daughter alpha radioactivity with alpha sensitive film.	0.001 WL in 240 hr	

is a necessary method for monitoring radon and radon progeny calibration facilities and for use during mitigation work to reduce indoor concentrations to acceptable levels.

The third group of sampling instruments is the integrating type which yield a single average concentration value for an extended time period from a few days to a week or more. The sampling period depends on the sensitivity of the particular technique being used. Samples usually can be analyzed later at a more convenient time and place. Integrating devices for radon are usually simpler and less expensive than continuous reading devices and can be used to measure concentrations that are too low for grab or continuous sampling methods. Collection of a sample can be accomplished either passively (requiring no power) or actively (requiring power). With the recent development of simpler and smaller integrating devices, a large number of sites can be monitored for both radon and radon progeny. The problem of long-term variations can be partly overcome by performing measurements for 4 to 7 days or more in each season providing a reasonably adequate assessment of the radiation hazard to the occupants of buildings.

Continuous and long-time integrating instruments for radon daughters are used to confirm the results of screening surveys and obtain information for decision making after conditions in a residence have been determined.

Location and Time of Measurement

In single family homes, radon or progeny measurements should be made in the living area. This can be any room on the first or second floor where the occupants are likely to spend most of their time. A second measurement in the basement is recommended because it is very useful in identifying the strength of radon sources and for estimating the maximum potential exposure in the home. In many situations, there is a tendency to convert the basement into a play or family room, and the exposure of the occupants must be assessed accordingly. In multiple dwellings or apartment buildings, measurements should be made in the basement or ground floor only. Radon concentration levels in high rise apartments are usually very low, approaching levels found outdoors.

The best time to make measurements for radon and progeny is during the heating season, at which time the maximum concentration levels are likely to occur. In the northeastern United States, this usually occurs between October 1 through April 30. If this measurement period is not possible, measurements can be made when conditions are maximized for air tightness for at least 12 hours before the measurement begins. An alternate is to make measurements during the winter and summer using short integrating periods of 3 to 30 days per season. If there is no urgent need or concern about measurement results, the average annual exposure of the occupants of buildings can be made using long time integrating instruments using the track etch detector technique.

To facilitate in the selection of the proper instrument and method for a particular application, the following section will present an overview of the principles of detection, sensitivity, availability, cost, and field experience for the commonly used monitoring devices.

Grab Sampling Instruments and Methods for Measuring Radon

Scintillation Cell

The scintillation cell method is one of the oldest and most widely used, both in the laboratory and in the field.[3,4] Scintillation cells range in size from 0.09 to 2.0 liters, and they can be readily constructed by using a container with a transparent bottom, coated internally with silver-activated zinc sulfide phosphor. For counting it is coupled to a photomultiplier tube assembly system. The scintillation cell can be filled with the atmosphere to be measured either by evacuation or by air flow through using a portable pump. Their sensitivities for 1 pCi/l of radon range from 0.5 cpm to 6.0 cpm, respectively. Commercial plastic scintillation cells cost from $50.00 to $100.00 each, and glass and metal range from $200 to $400 each. Scintillation cells can be reused for several years when properly maintained. Measurement errors with the scintillation cell method include improper cell calibration, leaks around the valves or joints and a malfunctioning or improperly calibrated counting system.

In some situations, grab samples for radon are collected in an assortment of metal containers and tedlar bags, impermeable to radon, for shipment to a central laboratory for analysis. The samples are usually transferred into scintillation cells and are handled in the same manner described earlier. In laboratories that maintain calibration facilities which serve as intercalibration and intercomparison centers for radon, the field samples are transferred into pulse or current ionization chambers for counting.[5] Ionization chambers are very specialized and complex laboratory instruments whose results are directly traceable to the U.S. National Bureau of Standards.

Grab Sampling Instruments and Methods for Radon Progeny

A grab sample for radon progeny measurement consists of collecting a known volume of air containing radon progeny on a high efficiency filter for a sampling period of 2 to 10 minutes. The radioactivity of the sample can be counted during and/or after the end of sampling. A number of counting regimes can be used depending on the information desired.

The simplest methods involve a single alpha count made within 90 minutes after sampling, usually at the collection site, and yield results in working levels (WL).* [6,7] The more sophisticated methods also involve short sampling periods (2 to 10 minutes), but utilize several count intervals and yield individual air concentrations of the three radon progeny (^{218}Po, ^{214}Pb, ^{214}Bi), as well as the WL. Counting is performed with alpha scintillation counters[8-10] or alpha spectrometers.[11-13] If alpha spectrometry is used for counting, the volume of air sampled for 2 to 10 minutes is limited because of the nature of the filter paper-detector assembly. Alpha spectrometry is used in special research applications mostly in laboratory settings with limited use in the field. Scintillation type detectors are the most popular because they have the capability to handle large air volume samples, thereby yielding higher sensitivity.

Errors in determining individual radon progeny and WL concentrations can result from inaccurate measurement of the air volume sampled. This is caused by improper calibration of flowmeters, unstable air pumps, faulty filter holders and incorrect timing. Another source of error is the filter medium onto which radon progeny are collected. The filter should be thin, of small pore size, have low air flow resistance, good surface deposition characteristics and high collection efficiency. The counting efficiency of a detector is a function of geometry and energy level. The counting geometry of the counter depends on the distance of the detector from the filter sample and the relative size and shape of the detector and filter. It is essential that alpha counters be calibrated with certified alpha emitting standard sources having the same size and shape as the sample to be counted. Detectors that are energy dependent should be avoided unless they are calibrated with a known radon progeny sample collected in an atmosphere traceable to a National Bureau of Standards reference source.

Continuous Sampling Instruments for Radon

There are basically three types of continuous radon monitors: the continuous scintillation cell, the diffusion electrostatic system, and the diffusion radon-only system. Actually, only the latter can be called a continuous radon monitor, because it detects radon exclusively whereas the others measure radon indirectly by counting radon progeny along with radon. The instruments that measure radon indirectly require a counting time of a few minutes, so in actuality they are semi-continuous.

---

*One WL is the quantity of alpha energy delivered from a mixture of 100 pCi each of ^{218}Po, ^{214}Pb, ^{214}Bi, and ^{214}Po in 1 liter of air = 1.3×10^5 MeV alpha.

Continuous Scintillation Cell

This instrument consists of a scintillation cell[4] in contact with a photomultiplier tube counter.[14,15] The size of the scintillation cell is dictated by the size of the photomultiplier tube and the housing. Air is drawn continuously through the scintillation cell by an air pump so that the counter records the changes in radon concentration. The radon progeny buildup inside the scintillation cell undergoes a lag time that should be taken into consideration during calibration and actual use. To measure varying radon concentrations accurately, corrections for the contribution from the prior counting intervals should be incorporated. For long-term use in the same environment, where radon concentrations do not vary much, a reading taken any time after the first two hours of sampling is adequate to determine the concentration of radon with good accuracy. The cost of a complete continuous scintillation counting system ranges from $3,000 to $4,000. The lower limit of detection with this method for 30-minute counting intervals ranges from 0.01 to 1 pCi/l depending on the size of the scintillation cell.

Diffusion Electrostatic Monitor

This instrument is passive, requiring no air pump. Radon enters the sensitive volume of the instrument by molecular diffusion.[16] The radon progeny resulting from the decay of radon within the sensitive volume are collected electrostatically and counted by a scintillation detection system. The instrument is affected by humidity, but in indoor environments where humidity may range from 30 percent to 50 percent, the error is probably less than 10 percent. There has been an attempt to improve the instrument by reducing its size and improving its sensitivity. For ten-minute counting intervals, the lower limit of detection of this instrument is 0.5 pCi/l. The cost of the monitor with the acquisition data system ranges from $3,000 to $5,000.

Diffusion Radon-Only Monitor

As the name implies, this monitor detects radon only. The radon progeny are removed away from the photomultiplier tube area with an electret[17] so that only the pulses from the radioactive decay of radon are detected. The proper spacing of the electret from the tube and the sensitive volume of the detector housing determine the sensitivity of the monitoring device. The lower limit of detection with this method is 0.1 pCi/l for hourly counting intervals. The cost of the instrument is about $4,000, but it is not commercially available.

Continuous Instruments for Radon Progeny

As with some of the continuous radon monitors, continuous radon progeny monitors are really semi-continuous because the typical operation involves sampling air through a filter for 2 to 5 minutes and counting the collected radioactivity with alpha and beta detectors for counting periods of 2 to 3 minutes.[13] In essence, they measure individual radon progeny and WL every ten minutes. Because of the complexity and high cost of these instruments, they are used only in special studies. The sensitivity of the monitor ranges from 0.1 to 1 pCi/l for individual radon progeny and 0.001 to 0.01 WL depending on the air flow rate.

Another type of instrument that may be considered semi-continuous is the one that uses a variation of the grab sampling method consisting of a 5 to 10 minute air sample counted by using any of the methods that determine individual radon progeny and WL.[8-13] There are commercial monitors[19] that feature an automatic drive that changes the position of the filter for a new grab sample so that a concentration data point can be obtained every 35 minutes. Monitoring devices of this type are complex and require a high volume air pump and a computer for automatic unattended operation. The cost ranges from $15,000 - $30,000. Its use is limited to special studies and as a radon and thoron monitoring device in calibration facilities. The sensitivity is similar to that of the grab sampling methods which is a function of flow rate.

Integrating Sampling Instruments for Radon

The most popular type of monitoring for radon is the integrating sampling, which covers a period of a few days to a week or more. The early development of integrating instruments was exclusively carried out by the Department of Energy and the Environmental Protection Agency and their contractors for use in their respective radiation programs. Usually, trained personnel used them in the field as well as in research laboratories. Recent developments focused on simple monitoring instruments that can be deployed in residential buildings by ordinary citizens. All integrating instruments described here are portable and some are even small enough to be handled readily in the mail.

Track Etch Monitors

This passive device operates on the principle of radiation damage imparted to the detector material (film) by the alpha particles from radon progeny deposited on various surfaces of the sensitive volume of the monitor.[20-22] After exposure, the tracks on the film are revealed by etching with NaOH solution and counted manually or automatically under a microscope. The monitors are available commercially in two configurations depending on the desired application.

For indoor exposures, both the small[21] (pill box size) and the larger one[21-22] (coffee cup size) can be used by leaving them in the test environment for minimum exposures of 2 and 1 months, respectively. Their lower limit of detection are 0.4 and 0.2 pCi/liter/month, respectively. The cost per detector ranges from $20 to $60 depending on the number purchased and on the desired accuracy. Deployment in the field can be handled through the mail. They are usually suited for quarterly measurements when repeated during different seasons of the year. Integrating periods of 1 year are possible if there is no urgent need for the measurement results.

Activated Carbon Monitor

For integrating periods of up to one week, the activated carbon detector is available in different sizes and configurations.[23,24] The device consists of a container filled with activated carbon (usually coconut charcoal). Presently, there are several sizes in use containing 30 to 150 grams of activated carbon.

The activated carbon method uses simple detection principles for gamma rays emitted by the radon progeny adsorbed on the carbon during exposure. Radon diffuses into the monitoring device passively by molecular diffusion. The counting system is usually in a central laboratory where the monitor is returned for analysis. The monitor's small size makes it an attractive device for large surveys where quick information on indoor radon is desired. The lower limit of detection depending on size is 0.1 to 0.2 pCi/l of radon for an exposure period of 4 days. The monitors can be fabricated at a cost of less than $10 and can be reused for several years if maintained properly. The counting system may run from $6,000 to $10,000. The main users of the passive carbon monitors are Federal and State agencies and universities under contract to the Department of Energy and the Environmental Protection Agency. This monitoring device represents the latest development of a passive integrating instrument that can measure indoor radon concentrations with good precision and accuracy.

Electrostatic - Thermoluminescence Monitor

This instrument was developed to monitor sites associated with the uranium fuel cycle. It is passive and is known as the passive environmental radon monitor (PERM). This monitor operates on the principle of electrostatic collection of the radon progeny ions generated in its sensitive volume by the decay of radon which entered by diffusion.[25] Diffusion is through a dessicant bed (10 cm thick) which eliminates the effect of humidity on the instrument's response. The alpha activity that accumulates on the electrode is detected continuously during the exposure by a thin thermoluminescence detector (TLD) (LiF or

CaF_2:Dy). The exposure period usually ranges from 7 to 14 days depending on the humidity conditions. However, by changing the dessicant, the monitor can integrate for periods of one month or longer. After exposure the TLD is returned to a central laboratory for analysis.

The lower limit of detection of the monitor is 0.03 pCi/l in a one-week measurement period. Commercially available instruments cost between $500 and $600. Thermoluminescence detectors cost between $2.00 to $4.00 each and can be reused several times. Thermoluminescence readers range from $5,000 to $10,000.

Integrating Sampling Instruments for Radon Progeny

Time-integrating monitoring instruments and methods for radon progeny were originally developed and applied to measure concentration as WL in buildings suspected to be contaminated with uranium tailings.[26-28] Their use was mostly as area monitors. There are no passive radon progeny monitors, and, since power is needed to move air through a filter to collect radon progeny, these monitors are used mainly by trained personnel. Several commercial instruments (mostly Canadian-made) were built to meet the need for monitoring in and around uranium mills and tailings and in other areas where waste from uranium and radium facilities are stored.

There are basically two types of integrating WL monitors: in one type the WL is measured in a two-step procedure where the detector is exposed in the field and then is returned to a central laboratory for analysis; in the other type, the WL is measured by the detector as an integral part of the monitoring instrument. The latter instruments are more desirable because of the time saved and the considerable reduction in their cost. This added convenience is obtained with more complex instruments.

Thermoluminescence Radon Progeny Integrating Sampling Unit (RPISU)[26]

This radon progeny monitor is an acoustically-shielded low volume air sampler that collects air containing radon progeny on a membrane filter. The filter is located in a detector-filter holder assembly with a CAF_2:Dy thermoluminescence detector (TLD) positioned close to it. The lower limit of detection sampling at 1 liter/minute is ~0.0001 WL. The cost of the RPISU is about $1,000; the TLD reader costs between $5,000 to $10,000.

Thermoluminescence Modified Working Level Monitor (MOD)

The original monitor, designed at the Environmental Measurements Laboratory (EML),[27,28] incorporated a sampling filter-detector head and pump that were developed as a WL dosimeter for uranium mines. For environmental levels of radon progeny, the original monitor was modified to the present unit called the MOD by placing it in a compact, sound proof housing that incorporated a timer. The membrane filter located in the detector-filter head is positioned 2 mm from the detector. The detector is an alpha-sensitive TLD chip (3.2 x 3.2 x 0.38 mm LiF). The lower limit of detection at a flow rate of 0.1 liter/minute is 0.0005 WL for a 7 day measurement period. As in the RPISU, the MOD is an active device requiring power and its deployment in the field requires trained personnel. Commercially available MODs cost between $800 to $1,000 per unit. Thermoluminescence chips cost from $2 to $4 each and they can be reused several times; TLD readers cost between $5,000 and $10,000. The estimated number of MODs and RPSIUs in the United States and Canada totals about 1,000.

Surface Barrier WL Monitor

The most recent development of integrating WL monitors replaces the thermoluminescence detector system entirely with a surface barrier alpha detector.[29,30] By doing so, there is a substantial improvement in sensitivity, time saved and overall cost reduction. The monitor continuously samples air, counts the filtered radon progeny, and displays concentration in terms of WL. The requirements of this instrument are a constant air flow pump and filter-detector geometry such as to eliminate radon progeny plateout upstream from the filter. The integrating period is varied to suit the application. An added feature of this device is that it can be used as a satisfactory continuous WL monitor by using it in conjunction with a recorder or any other electronic device that acquires measurement points on a frequent time basis (every 30 to 60 minutes). The lower limit of detection is 0.00004 WL in a one week measurement period at a flow rate of 0.2 liter/minute. In calibration facilities and during remedial work, it can be used to measure the real-time variation of radon progeny concentrations. Commercially available instruments range in cost from $2,000 to $8,000, depending on data acquisition software. Although this instrument is more expensive than integrating WL monitors that use thermoluminescence detectors, it is favored because of the ease of getting measurement results.

Radon/Thoron Daughter Monitor

This is a small portable instrument designed to measure the integrated WL from both radon and thoron progeny. It consists of a modified aquarium pump and a filter track-etch polycarbonate plastic detector head. The detector is separated from the filter by collimators and alpha particle absorbers for discrimination among the different energies of the radon and thoron progeny. The exposed detectors are analyzed by counting the alpha tracks from the different progeny in an automatic counting system. The sensitivity of the monitor is a function of pump flow rate and exposure time. Using a flow rate of 0.6 liter/minute and an exposure time of 240 hours, the sensitivity is 0.001 WL. Although the monitor currently lacks a timer, one can be incorporated with a minor modification. The instrument is available commercially in Canada and through manufacturer representatives in the United States. Since information on the contribution of the thoron progeny WL to the total WL is lacking, this instrument can be very useful for integrating measurements of both radon and thoron progeny. Simultaneous exposure of this monitor with a passive radon device, such as the activated carbon canister, for 4 to 7 days, can determine the WL ratio:

$$\text{WL ratio} = \frac{\text{WL}}{\text{Radon Concentration (pCi/l)}} \times 100$$

The WL ratio provides very useful information for the accurate estimation of the exposure of the general population to indoor radon progeny.

INSTRUMENTS AND METHODS FOR MEASURING RADON SOURCES

Characterizing the sources of radon indoors requires measurement of the rate at which radon is generated in the source materials and the mode of transport and entry in the indoor environment. Measurements used for diagnostic purposes are radon exhalation from building materials and the underlying soil, infiltration through openings in the foundation and radium and radon concentration in water supplies. Table 4.3 lists commonly used techniques for measuring the sources and transport of radon.

The most direct measurement that determines the source of indoor radon is the exhalation rate of radon from surfaces. This rate is directly dependent upon the fraction of diffusible radon and transport mechanisms in the material. In other words, the emanating power of a material is more critical than its radium content in affecting indoor radon concentrations. Radon flux or areal exhalation rates can be measured by the rapid accumulation or integrating technique.[31,32] Rapid measurement of radon flux usually consists of radon accumulation in a flux can for periods ranging from one to ten hours depending on the size (volume) of the can.[31] As long as the concentration of radon in the can does not exceed ten percent of the concentration of radon in the soil, the method is accurate. Usually an air sample from the can is transferred into a grab sampling device such as a scintillation cell for alpha counting.[4] The method has adequate sensitivity for many field applications with a lower limit of detection of about 10^{-18} aCi cm^{-2} sec^{-1}.

Table 4.3 - INSTRUMENTS AND METHODS FOR MEASURING RADON SOURCES

Instrument and method	Application	Principle of Operation	Ref.
Accumulation can	*In situ* radon exhalation rates from surfaces	Transfer air sample from can to scintillation cell for alpha counting.	31
Activated carbon device	*In situ* radon exhalation rates from surfaces	Radon adsorption on activated carbon Gamma count in NaI (Tl) analyser for ^{214}Pb and ^{214}Bi.	32
Gamma-ray spectroscopy	^{226}Ra content	Measure primary gamma ray flux from ^{214}Pb and ^{214}Bi with high resolution Ge(Li) detector.	33,34
Modified Marinelli beaker	Radon in water	Gamma count water sample in Na(Tl) analyser for ^{214}Pb and ^{214}Bi.	35
Liquid Scintillation vial	Radon in water	Water sample mixed with scintillation fluid. Count on liquid scintillation counter.	36

Integrated measurements of radon flux can be made with activated carbon detectors.[32] By using this method, problems caused by back diffusion into the material is eliminated because the carbon acts as a sink, virtually adsorbing all radon entering the exhalation container. The device used for measuring radon in air passively[23,24] can be used for making radon flux measurements by sealing it against the surface to be measured. The collected radon is determined by gamma counting its decay products. The lower limit of detection with this method is 3×10^{-18} aCi cm^{-2} sec^{-1} for a collection period of one day. The advantage of this technique is that it makes it possible to carry out large scale measurements at a lower cost.

Measurement of the ^{226}Ra content of soil or building materials can be very useful. Areas contaminated with uranium mill tailings, phosphate reclaimed land or areas with known uranium mineralization can be identified by measuring the ^{226}Ra content of 100 to 200 grams of soil, rock, or building material in a laboratory or _in situ_ using gamma spectroscopy.[33,34] Radium concentrations are very variable (0.1 to 10 pCi/g) and may account for some of the differences observed in radon concentration in various buildings.

Radon in water can add significantly to indoor radon concentrations when the source is an underground aquifer and water use is extensive. There are several techniques for measuring the radon content of water, none of which is suitable for field analysis. Water samples collected with care to avoid degassing of the radon gas can be analyzed in the laboratory by two techniques.[35,36] Radon concentrations in surface waters are usually low and need not be measured.

The modified Marinelli Beaker[35] method developed by EML is the more convenient and practical of the two recommended methods. It consists of 1.3- to 2-liter beakers made of polyethelene or lucite, which serve as collectors of water samples directly from the source. The radon progeny in equilibrium with radon in the water sample are gamma-counted in the laboratory. The lower limit of detection depending on the size of the collector is between 5 to 12 pCi of radon per liter of water. Gamma counting equipment used for measuring the radon content of activated carbon detectors can also be used to measure the water content of the beakers.

The alternate method for measuring radon in water is the liquid scintillation technique.[36] This technique is more sensitive than the beaker method, but it is more complicated and requires trained personnel to carry it out. Since radon in water at concentrations of 1,000 pCi/l raises the airborne concentration of radon by about 0.1 pCi/l, the first method (with the lower limit of detection of 5 to 12 pCi/l) is more practical and thus more desirable.

CONCLUSIONS

The concern about unusually high indoor radon concentration levels in some parts of the United States has increased the demand for reliable measurement methods and techniques. Fortunately, the state-of-the-art in the measurement of both radon and progeny has been developed to the point where there are adequate instruments to address particular needs. Looking at the immediate future, there seems to be a strong need for screening indoor environments for possible high concentrations of radon and progeny. Because of the temporal and spatial variations of indoor radon concentration levels, measurements for screening purposes should be averaged over as long a period of time as possible to eliminate short-term perturbations. Instruments that integrate exposures for periods of a few days are becoming popular because they provide information needed in a short time. Grab sampling may be useful in situations where immediate results are desired. When augmented with integrating or continuous sampling methods, grab sampling is very useful.

For research purposes such as real-time changes in radon and progeny concentrations, radon transport, behavior, and interaction with the indoor environment, more complex continuous or semi-continuous monitoring instruments are recommended.

The ability to obtain accurate measurements at low concentrations and the qualifications of the user determines the quality of measurement data. These two criteria, combined with proper instrument calibration, are the essential elements in a successful monitoring program. Since most commercial instruments for radon and progeny are seldom calibrated in field situations, they should be subjected to evaluation in a certified quality assessment program. The entire system of associated equipment such as filters, filter holders, pumps, flowmeters, alpha and gamma counting equipment, thermoluminescence readers, and alpha track counters should be incorporated in the overall evaluation program since any deficiency in any one of these instruments can lead to poor measurement data and potentially erroneous estimates of human exposure.

FOOTNOTES

[1]George, A.C. and A.J. Breslin, "The Distribution of Ambient Radon and Radon Daughters in Residential Buildings in the New Jersey-New York area," Natural Radiation Environment III, Vol. 2, CONF-780422, Technical Information Center, U.S. Department of Energy, 1980, pp. 1272-1307.

[2]Tappan, T.J., "Indoor Radon Engineering Assessment of Residential Structures Located in the Reading Prong Area," Radiological Assessments Department, ARIX, Grand Junction, CO (March) 1985.

[3]Lucas, H.F., "Improved Low-Level Alpha Scintillation Counter for Radon," Rev. Sci. Instrum., 28: 680, 1957.

[4]George, A.C., "Scintillation Flasks for the Determination of Low Level Concentrations of Radon," Proceedings of the Ninth Health Physics Symposium, Denver, CO. (February) 1976.

[5]Fisenne, I.M., and H. Keller, "The EML Pulse Ionization Chamber Systems for ^{222}Rn Measurements," U.S. Department of Energy Report EML-437, New York (March) 1985.

[6]Kusnetz, H.L., "Radon Daughters in Mine Atmospheres. A Field Method for Determining Concentrations," Am. Ind. Hyg. Assoc. J. 17: 85, 1956.

[7]Rolle, R., "Rapid Working Level Monitoring," Health Phys. 22: 233, 1972.

[8]Tsivoglou, E.C., H.E. Ayer and D.A. Holladay, "Occurrence of Nonequilibrium Atmospheric Mixtures of Radon and its Daughters," Nucleonics 1: 40, 1953.

[9]Thomas, J.W., "Measurements of Radon Daughters in Air," Health Phys. 23(6): 783, 1972.

[10]Raabe, O.G., M.E. Wrenn, "Analysis of the Activity of Radon Daughter Samples by Weighted Least Squares," Health Phys. 17 593, 1969.

[11]Martz, D.E., D. F. Holleman, D. E. McCurdy, K.J. Schiager, "Analysis of Atmospheric Concentrations of RaA, RaB and RaC by Alpha Spectroscopy," Health Phys. 17: 131, 1969.

[12]Jonassen, N, E.I. Hayes, "The Measurement of Low-Concentrations of the Short-Lived Radon-222 Daughters in the Air by Alpha Spectrometry," Health Phys. 26: 104, 1974.

[13]Nazaroff, W.W., "Optimizing the Total-Alpha Three Count Technique for Measuring Concentrations of Radon Progeny in Residences," Health Phys. 46: 1984.

[14]Thomas, J.W., and R.J. Countess, "Continuous Radon Monitor," Health Phys. 36: 734, 1979.

[15]Nazaroff, W.W., F.J. Offerman, A.W. Robb, "Automated System for Measuring Air Exchange Rate and Radon Concentration in Houses," Health Phys. 45: 1983.

[16]Spitz, H., M.W. Wrenn, "Design and Application of a Continuous Digital-Output Environmental Radon Measuring Instrument," Radon Workshop, A.J. Breslin (Editor), USERDA Report HASL-325, New York, 1977.

[17]Chittaporn, P., M. Eisenbud, and N. Harley,"A Continuous Monitor for the Measurement of Environmental Radon," Health Phys. 41: 405, 1981.

[18]McDowell, W. B., D.J. Keefe, P.G. Groer, and R.T. Witek, "A Microprocessor Assisted Calibration for a Remote Working Level Monitor," IEEE Trans. Nucl. Sci., NS-24: 1, 1977.

[19]Bigu, J., R. Raz, K. Golden, and P. Dominquez, "A Computer-Based Continuous Monitor for the Determination of the Short-Lived Decay Products of Radon and Thoron," Canada Centre for Mineral and Energy Technologies, Energy, Mines and Resources of Canada, Elliot Lake, Ontario, Division Report MPR/MRL 83 (OP) J, 1983.

[20]Fleischer, R.L., "Dosimetry of Environmental Radon, Methods and Theory for Low-Dose Integrated Measurements," General Electric Research and Development Center, Schenectady, New York, unpublished report, 1980.

[21]Alter. H.W., and R. L. Fleischer, "Passive Integrating Radon Monitor for Environmental Monitoring," Health Phys. 40: 693, 1981.

[22]Urban, M., and E. Piesch, "Low Level Environmental Radon Dosimetry with a Passive Track Etch Device," Radiation Protection Dosimetry 1: 97, 1982.

[23]George, A.C., "Passive Integrated Measurement of Indoor Radon using Activated Carbon." Health Phys. 46: 867, 1984.

[24]Cohen, B.L., and E.S. Cohen, "Theory and Practice of Radon Monitoring with Charcoal Adsorption," Health Phys. 45: 501, 1983.

[25]George, A.C., and A.J. Breslin, "Measurement of Environmental Radon with Integrating Instruments," Workshop on Methods for Measuring Radiation In and Around Uranium Mills, E.D. Harward (Editor), Atomic Industrial Forum, Inc. Vol. 3, No. 9, 1977.

[26]Schiager, K.J., "Integrating Radon Progeny Air Sampler." Am. Ind. Hyg. Assoc. J. 35: 165, 1974.

[27]Breslin, A.J., S.F. Guggenheim, A.C. George, R.T. Graveson, "A Working Level Dosimeter for Uranium Miners," USDOE Report EML-333, New York, 1977.

[28]Guggenheim, F.S., A.C. George, R.T. Graveson, and A.J. Breslin, "A Time-Integrating Environmental Radon Daughter Monitor," Health Phys. 36: 452, 1979.

[29]Latner, N., S. Watnick, and R.T. Graveson, "Integrating Working Level Monitor EML Type TF-11," USDOE Report EML-389, New York, 1981.

[30]Bigu, J., and R. Kaldenbach, "Theory, Operation and Performance of a Time-Integrating Continuous Radon/Thoron Daughter Working Level Monitor," Radiat. Protection Dosimetry 9: 19, 1984.

[31]Wilkening, M. "Measurements of Radon Flux by the Accumulation Method," Workshop on Methods for Measuring Radon In and Around Uranium Mills," E.D. Harward (Editor), Atomic Industrial Forum, Inc. Vol. 3, No. 9, 1977.

[32] Countess, R.J., "Radon Flux Measurement with a Charcoal Canister," Health Phys. 31:455, 1976.

[33]Ingersoll, J.G., "A Survey of Radionuclide Contents and Radon Emanation Rates in Building Materials Used in the United States," Lawrence Berkeley Laboratory Report LBL-11771, (May) 1981.

[34]Beck, H.L., J. DeCampo and C. Gogolak, "In Situ Ge(Li) and NaI(Tl) Gamma-Ray Spectrometry," USAEC Report HASL-258, New York, 1972.

[35]Countess, R.J., "Measurement of Radon-222 in Water," Health Phys. 34: 390, 1978.

[36]Prichard, H.M., and T. Gesell, "Rapid Measurements of Radon-222 Concentrations in Water with a Commercial Liquid Scintillation Counter," Health Phys. 33: 577, 1977.

TRACK ETCH RADON DETECTORS FOR LONG-TERM RADON MEASUREMENT

Richard A. Oswald and Samuel L. Taylor*

INTRODUCTION

Since radon levels vary greatly with time, it is essential to measure average radon concentrations over time periods long enough to provide a representative picture of the overall radon situation in a home.

Track Etch detectors accomplish this by continuously recording radioactive decays at a rate proportional to the radon concentration that exists at any instant in time. These alpha-decays are recorded as alpha-damage tracks. Thus, the total track count provides a measure of the total radon exposure. This can easily be converted to average radon concentration as described later.

RADON VARIATION WITH TIME

Numerous studies[1-5] have documented the fact that indoor radon concentrations vary with time. These studies have shown that numerous factors account for the temporal variation. Influencing factors include meterological conditions, ventilation, season of the year and lifestyle.

It is also true that the different factors influencing the temporal variation contribute for varying lengths of time. For example, the period of variation of a meteorological factor such as barometric pressure is usually on the order of several days. On the other hand, ventilation may vary drastically over a period of several months. Thus, the temporal variability of radon may have a very complex and unpredictable behavior.

A Track Etch radon detector consists of a special dielectric plastic in a cup-like container with a filter over the open end. These devices have been described previously.[6-7] The filter permits entry of radon gas, but excludes radon daughters and other particulate matter. The dielectric plastic records alpha particle tracks which are simply the result of the energetic alpha-particle striking

---

*Both Richard Oswald and Samuel Taylor are scientists associated with Terradex Corporation, 460 N. Wiget Lane, Walnut Creek, California 94598.

the film. The patented[8] configuration for measuring radon in indoor air is shown in Figure 4.1.

When radon diffuses into the interior of the detector, it decays and forms other alpha-emitting progeny. The progeny, of course, also contribute to the tracks recorded on the dielectric plastic; but, because only radon gas itself can enter the interior of the cup, the device measures only radon in the atmosphere being monitored.

After the exposure period, the special dielectric plastic detector is chemically etched, thereby making the alpha particle tracks visible in an optical microscope. Figure 4.2 shows the tracks as they appear, magnified after etching. We count the tracks in a given area and thus determine a track density (tracks per square millimeter). From this track density, the radon exposure in (pCi/l)-days (the product of radon concentration in pCi/l and time) can be computed using the calibration factor. Finally, the average radon concentration in pCi/l can be computed by dividing by the total exposure time in days.

TRACK ETCH DETECTOR CALIBRATION

The Track Etch detectors are calibrated by exposing them in internationally recognized calibration facilities where a known radon concentration can be established and maintained. The detectors are exposed for various times (usually several days to a few weeks) to various radon concentrations (for example, 50-500 pCi/l).

CONCLUSIONS

In almost every radon monitoring application, it is reasonable and prudent to make a long-term integrated radon measurement. Track Etch radon detectors are capable of making the long-term integrated measurements necessary for properly assessing the variable radon levels that exist in homes.

This is because they are capable of recording tracks at a rate proportional to the radon concentration that exists at any particular moment in time.

FIGURE 4.1 -- TRACK ETCH DETECTOR

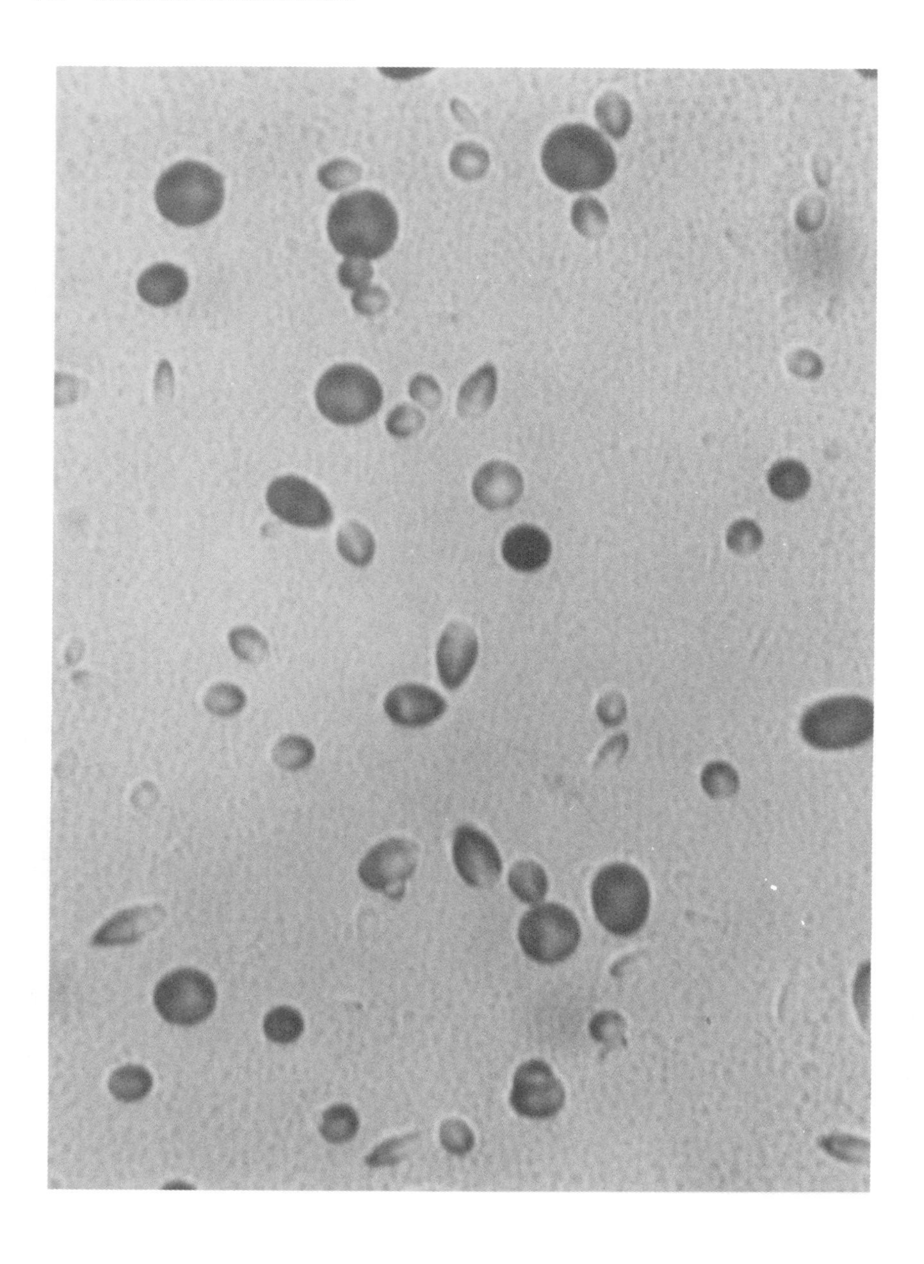

FIGURE 4.2 -- ETCHED ALPHA PARTICLE TRACKS IN DIELECTRIC

FOOTNOTES

[1]Moschandreas, D.J., et al., Radon and Aldehyde Concentrations in the Indoor Environment, Lawrence Berkeley Laboratory, LBL-12590, 1981(April).

[2]Nero, A.V., J.V. Berk, M.L. Boegel, C.O. Hollowell, J.G. Ingersoll, and W.W. Nazaroff, Radon-Daughter Exposures in Energy-Efficient Buildings, LBL-11052, 1981 (October).

[3]Nazaroff, W.W. and S.M. Doyle, Radon Entry Into Houses Having a Crawl Space, Lawrence Berkeley Laboratory, LBL-16627, 1983.

[4]Pearson, J.E., Natural Environmental Radioactivity From Radon-222, U.S. Department of Health, Education, and Welfare, No. 999-RH-26, 1967 (May).

[5]Scott, A., Personal Communication, 1986.

[6]Alter, H.W., and R.L. Fleischer, "Passive Integrating Radon Monitor for Environmental Monitoring," Health Physics, 40, 1981, p. 693.

[7]Alter, H.W., "Passive Integrating Radon Monitor for Environmental Monitoring," Abstract, Second Special Symposium on the Natural Radiation Environment, Bombay, 1981 (January), pp.19-23.

[8]Alter, H.W. and R.A. Oswald, "Compact Detector for Radon and Radon Daughter Products," U.S. Patent, 4,518,860, May 21, 1985.

INTERIM INDOOR RADON AND RADON DECAY PRODUCT MEASUREMENT PROTOCOLS

Melinda Ronca-Battista*

The Environmental Protection Agency (EPA) has issued the first of two radon measurement protocol documents. This document, entitled "Interim Radon and Radon Decay Product Measurement Protocols" (EPA 520/1-86-04), provides guidance for making accurate and reproducible measurements and helps ensure consistency of data among all of the different federal, state, university and private organizations now performing measurements or planning measurement programs. This paper discusses the EPA measurement protocols, as well as EPA guidance on the applications of the measurement methods for different purposes.

There are two opposing interests in measuring radon in homes. First, there is the interest of making the best measurement possible, with state-of-the-art equipment and over a long time period so that the result reflects long term exposure. The other interest is doing it practically, with the equipment that is available and in a way that provides answers quickly. In these protocols, we have balanced these two interests by offering alternatives for different situations and recommending methods that produce accurate and reliable results in a reasonable time frame without requiring extensive measurements in each house.

The objectives of the protocols are three-fold. The first objective is to summarize and make available to the public the extensive knowledge of EPA's laboratories, as well as the others who have helped us with the protocols, in making accurate and reproducible measurements.

The second objective is to assure consistent data from the different groups making the measurements. In other words, we want to make sure that a measurement result of 0.02 WL (4 pCi/l) taken by a private firm in a New Jersey house means the same thing as 0.02 WL measured by a state laboratory in a North Carolina house. To do this, the measurement method must be used the same way, the house conditions during the measurement should be similar, and the criteria that were used to choose the measurement location in the house should be the same.

---

*Melinda Ronca-Battista is a health physicist with the Office of Radiation Programs, United States Environmental Protection Agency, Washington, DC.

The third objective is to assure consistency in the applications of the methods. By this, we mean that measurements made for the same purpose are made in the same way. The most important of these purposes is to determine the need for remedial action. Those decisions that could involve substantial investment by the homeowner should be based on a clearly defined measurement protocol.

The first EPA protocols document provides procedures for measuring radon concentrations with continuous radon monitors, charcoal canisters, alpha-track detectors, and grab radon techniques. The document also provides procedures for measuring radon decay product concentrations with continuous working level monitors, radon progeny integrating sampling units, and grab radon decay product methods. It also provides minimum requirements for quality control, such as how well the results of two instruments placed side-by-side should agree, and how often instruments should be calibrated.

The first protocols document provides guidance for the house conditions that should exist during the measurement. This is an extremely important aspect of making measurements. If measurements are made with the same instrument in a house on two different days, and if the first day the windows in the front of the house are all open, and on the second day all the windows on the back of the house are all open, the results of the two measurements could be completely different. It would be difficult to use only two such results to estimate what the average concentrations in that house are over a longer time period.

A short-term measurement must be made carefully to avoid an erroneous result due to the fact that radon and radon decay product concentrations vary greatly over time. There are many factors that affect the time variations of radon concentrations, but one of the most easily controlled is the amount of ventilation provided by open windows and doors. Our analyses show that radon concentrations are most stable in the winter months when most windows and doors are kept closed. Therefore, we recommend that all short-term measurements less than three months in duration be made during cool months when windows are normally kept closed, or during the summer after carefully instructing the homeowners that windows should be kept closed during the measurement period, and doors should be opened only for a few minutes to get in and out of the house. In addition, external-internal air exchange systems (other than a furnace) such as high-volume attic and window fans should not be operating. These conditions are especially important for 5-minute grab samples and other measurements only hours or days long. The intent of the closed-house conditions is to ensure that measurements are made during the time of highest and most stable conditions.

We are now preparing guidance for how these seven measurement methods should be used for different purposes. The two purposes we have addressed are screening and follow-up measurements. EPA is currently preparing a second protocols document, entitled "Interim Protocols for Screening and Follow-up Radon and Radon Decay Product Measurements." The document addresses where in the home measurements should be made, appropriate sampling times (the duration of the measurement), and how the results should be interpreted.

Screening measurements can be used to measure a large number of homes to quickly find homes that may contain high radon concentrations. They should be relatively fast and inexpensive, providing rough information without wasting time or money on extensive measurements in homes that contain low concentrations. Screening measurements also identify houses that contain very low concentrations and which do not need follow-up measurements. Therefore, the screening measurement alone does not serve as a decision-making tool for remedial action. It serves only for deciding whether additional measurements should be made, and in addition, what types of follow-up measurements are appropriate.

Screening measurements should be made in the lowest livable area in the house, which will in many houses be a basement. Since the highest concentrations of radon or radon decay products will usually be found in the lowest livable area, a basement screening measurement made during closed-house conditions should provide information about the maximum concentrations to which the house occupants may potentially be exposed. A screening measurement made under these conditions makes it more likely that a house with high concentrations in the general living area will be identified. Conversely, if the results of a closed-house basement screening measurement are very low, there is a high probability that the long-term average concentrations in the general living area of that house are also low.

If a house has, as a result of a screening measurement, been identified as having the potential for high exposures, the homeowner should make a follow-up measurement before any permanent actions to reduce the concentrations are taken. A follow-up measurement estimates the long-term average concentrations of radon or radon decay products to which the house occupants are exposed in the living areas, rather than the maximum concentrations in the house. The results of the follow-up measurement can be used to estimate the maximum health risks incurred by the occupants and to compare the measured concentrations to guidance given in terms of annual average concentrations. Follow-up measurements should provide a good estimate of the long-term average concentrations, with a low probability of under-estimation. Our guidance recommends the following criteria for follow-up measurements:

First, follow-up measurements should be made in areas of the house currently used as living areas, such as bedrooms and living rooms. Second, follow-up measurements should not be made in just one living area, but separate measurements should be made on different floors of the house. Whenever possible, one of the measurements should be made in a bedroom, because most people spend more time in their bedrooms than in any other room in the house. The protocols recommend that measurements not be made in kitchens or in bathrooms. The presence of fans or high concentrations of airborne particles caused by cooking can affect the results of radon and radon decay product measurements, especially short-term measurements. After the measurements in the two living areas are made, the results are averaged, and this average result can be used to compare to guidance levels, estimate health risks and decide on the need for remedial action.

The major reason for measuring on two different floors of the house is that radon and radon decay product concentrations tend to be different on different floors of the house. If people do spend time on more than one floor, then they are probably exposed to different radon decay product concentrations. Measuring on only one floor would not, then, be as representative of their actual exposures as measuring on two floors.

The need to make follow-up measurements and the time-frame in which follow-up measurements should be made depends upon the results of the screening measurement. If the screening measurement was made in a lowest livable area of the house, under the closed-house conditions, and its result is less than 4 pCi/l (or 0.02 WL), then it is very unlikely that the annual average concentrations in the general living areas on non-basement floors of the house are greater than 4 pCi/l or 0.02 WL. Therefore, in that case, a follow-up measurement is usually not necessary.

If the result of the screening measurement was less than 20 pCi/l (0.1 WL), but greater than 4 pCi/l (0.02 WL) then EPA recommends that the follow-up measurement consist of a 12-month measurement made in two living areas of the house. Although there is the possibility that the average long-term concentrations in the living areas of the house are significant, they are probably not sufficiently high to warrant immediate action or measurements. Such measurements spanning a year are optimal because the result will incorporate the variations in concentration due to seasonal and lifestyle cycles. This follow-up measurement provides the household with the best measure of the long-term concentrations to which they are actually exposed.

If the screening measurement result was greater than 20 pCi/l or 0.1 WL, then EPA recommends that a follow-up measurement over at least 24 hours be made in two living areas of the house under closed-house conditions. At these levels, an exposure to the radon decay products in the house over one year may cause a significant increase in health risk. The closed-house conditions stabilize the radon and radon decay product concentrations so that the measurement is more precise. Such a short-term follow-up measurement can quickly provide home-owners with a reproducible result that tends to over-estimate, rather than under-estimate, the annual average concentration. However, the increase in reliability of the result outweighs the disadvantage of the over-estimation.

The guidance in the protocols is interim, and we plan to update the documents based on new information as it becomes available. There are plans in several states to gather screening and follow-up measurements, and, as we analyze this data, we may refine the protocols. In the meantime, however, the intent of the protocols is to help both individual homeowners as well as states and other organizations performing measurements to obtain consistent and useful results.

HIGHLIGHTS FROM THE DISCUSSION: TESTING AND MEASUREMENT

COMPARING CHARCOAL CANISTER AND ALPHA TRACK DETECTORS

W. Makofske: For the homeowner choosing a test, what are the meaningful differences between charcoal and track etch detectors?

A. George: It depends on the information that you want. If you worry about your house, or you suspect that it's going to have a high radon level, probably a charcoal canister is the best choice. If it's low, you won't worry about it, and if it's high you can think about other tests.

R. Oswald: I would say the same thing, except I would replace the charcoal canister with Track Etch detector because of the time variation question. If you've been living in your home for years, and you are not in a panic situation where you suspect very high values, you might just as well go ahead and average out the weekly variations and have it measured for a month or so.

W. Makofske: Given the possible ambiguities in the charcoal test due to variations in radon levels, would you recommend a long-term test if the charcoal test shows a reading of around 4 pCi/l? At what levels should you worry about retesting?

M. Ronca-Battista: The protocols say that you measure in the basement under closed house conditions for a screening measurement. That would overestimate the long-term average concentration, and will generally be higher than the levels upstairs. So if you have a 4 pCi/l measurement or greater with a charcoal canister in your basement, we recommend you take follow-up house characterization measurements upstairs. If it is less, we say that it is not needed. It is, of course, at the homeowner's discretion.

G. Nicholls: As a practical matter, to tie these latter two questions and answers together, one of the things we were confronted with in Clinton (New Jersey) was that we had a number of people who had Track Etch detectors in their homes that had been arranged for by the community, and we had a couple of very high houses. We suggested to those people that they reduce the exposure period from three months to one month. You can obtain 4 pCi/l sensitivity with one month albeit with a larger error margin. Of course, what we were concerned with there was just finding the high houses. So the answer to the question of which one is appropriate is based on what you are looking for. If you don't have any reason to suspect a high value, the longest integrating time that you can measure is probably beneficial. If you have some reason to suspect high values, as we did in Clinton, then perhaps the quickest method to get a good result is appropriate. You wouldn't want people to live, as some folks are in Clinton, at 5 WL (1,000 pCi/l) for even one month if you could possibly avoid it. They may have been there for 15 years that way, but one month at 5 WL adds a significant risk no matter how you calculate it.

R. Oswald: I would like to bring up another related aspect to this whole issue. In EPA's guidelines about what to do at levels around 4 or slightly higher, the recommendation is to do follow-up testing in the next 6 to 12 months. There is no point in being in a mad rush to measure 4, 5, or 6 pCi/l in a 2-week period of time. On the other hand, you have to start being concerned about short-term immediate mitigation measures when the levels are much higher. And those much higher levels can in fact be detected with either charcoal or track etch in a short-term measurement, like a week or 2 weeks or 3 weeks. In other words, you have to weigh the level that you are trying to measure as well as consider what action is forthcoming.

APPLICABILITY OF PROTOCOLS

Question: Closed house conditions are not practical for warm-weather situations. Do you expect this feature to come under criticism and hold up adoption of the EPA protocols?

M. Ronca-Battista: I don't expect that it will hold up the issuance of the protocols. I expect that people will try as much as possible to conform to them, and I think that we will find out in fact whether or not people can. If it turns out that closed house conditions are impossible in warm weather situations, then there are two alternatives: either change the condition or tell people that they cannot sample during the summertime. The latter seems to be an untenable conclusion right now, but we may have to change the protocols.

R. Sextro: Melinda, I am really concerned about the closed house in the summertime. People are going to make 3-4 day measurements in a closed house in the summertime, and the worst thing you could get would be a false negative result. In the summertime, I think all bets are off as to whether you are going to get a real measurement.

M. Ronca-Battista: At this point we are just trying to do the best we can. The protocols specify that whenever possible measurements should only be made in the wintertime in northern climates. I haven't seen data on closed house and open house conditions in the south, so that leaves the whole question about what to do about houses in the south open as well. We just have to wait and see. This summer we will get a lot of good data, and we plan to revise the protocols in about 10 to 12 months. For the time being, this is the best we can do.

A. George: Measurements we made in basements at any time of the year usually didn't vary too much because the basements were usually not open to the outside since homeowners wanted them to remain cool. I think that, summertime or wintertime, there is not much variation in the basement. It is only on the first floor that you get these seasonal variations.

R. Sextro: The driving forces are very different in summer and winter, so that can't be right.

A. George: Well all the measurements we have in basements, both summer and wintertime, didn't show much difference. Even the latest data we got a couple of months ago agrees with that.

H. Sachs: Look at the PPL data from '80-'81; it doesn't fit that.

A. George: I only had one season there. But most of the other data that I have usually doesn't vary too much in the basement.

G. Nicholls: I have to agree with Andy, by the way. We just sent Melinda some of the data Jeanette Eng has been collecting from Montclair, Glen Ridge, and West Orange for the past couple of years, and there is an average of perhaps a factor of two difference between summer and winter in a house, but not much more than that. So if you have a bad house (0.1 WL or 20 pCi/l), you are going to find it in the summer. You are not going to get a good representative measurement of the annual average, that's true.

RADON FROM NATURAL GAS

Question: What level of radon is being found in natural gas used in kitchens?

A. George: Well, we made some measurements about 15 years ago in conjunction with Oak Ridge Laboratories. By the time the natural gas traveled 2,000 miles to New York, most of the radon had decayed away. Because of both decay and intermittent use, the radon in natural gas is insignificant. If you are in Alabama or Texas near the source, you might have a small contribution depending on how much gas you use.

FINDING RADON WITH A GEIGER COUNTER

Question: Many people have told me that they have looked for radon with a geiger counter and found nothing, and other people have said that you can't use a geiger counter for looking for radon. Could you qualify when using a geiger counter might tell you something about radon?

G. Nicholls: A geiger counter isn't going to tell you much unless you have a 1 WL (200 pCi/l) house or above. It will tell you if you have a 1 WL house. Put on a polyester suit or a fuzzy sweater and walk around in the house you suspect at 1 WL for about 15 minutes to a half an hour, and then walk outside and run the probe over yourself. You will get a statistically significant number of counts. I was in one of the 5 WL houses in Clinton for about 45 minutes. I didn't have a geiger counter with me, but I had a Ludlum Micro-R meter, and I was reading 40 to 50 micro-R or micro-roentgens per hour off my sweater when I walked out of the house. It is definitely measurable if you have a 1 WL house. If you have something less than that, you will not see it at all. And it depends to a certain extent on what tube you have on the geiger counter. It is definitely not a recommended technique unless you know exactly what you are looking for and you have a real good idea of what the efficiency of the tube is with respect to varying energies.

A BAD HOUSE

Question: Another one for you Gerry. Please define a bad house.

G. Nicholls: A bad house? One with a lot of radon in it. When we think in terms of bad houses, we think of a tenth of a working level (20 pCi/l) and up. If you live in a one-tenth of a working level house, 365 days a year, 24 hours a day, you incur approximately the same risk as if you were to smoke one pack of cigarettes a day. It is extremely subjective, and other people may very well define it in another way, but that is the rule of thumb that we have adopted.

LIQUID SCINTILLATION MEASUREMENTS

Question: Is there a method of measuring radon levels in air, water or soil using a liquid scintillation counter?

A. George: Liquid scintillation counters are more accurate and quite sensitive, but the method requires a transfer and sophisticated counting equipment. I think a modified Marinelli Beaker is probably easier to do and is just as sensitive. People who have sodium iodide or gamma analyzers can do that. So there are two methods, and they differ mainly on how much you are willing to pay.

MEASURING RADON GAS VS. RADON PROGENY

Question: Shouldn't we be measuring radon progeny (Working Levels) instead of radon gas (pCi/l), particularly if we want to look at health effects?

G. Nicholls: If you want to compare your data to the miners' studies, you really do need to know the working level measurement, because that is what was measured in the miners' environment. However, taking time-averaged working level measurements on longer periods is very difficult and expensive. Therefore, most people rely on radon gas measurements. One of the difficulties is that most houses do not have the 50 percent equilibrium factor between radon and its daughters that we mentioned this morning. The 50 percent equilibrium factor is implicit in the assumption that 200 pCi/l is equal to 1 WL. The reality is that there is typically 20 to 40 percent equilibrium so that you are actually over-estimating the working level measurement when you go from picocuries per liter to working levels and use the 50 percent factor. If you want to do an epidemiological study, I would strongly suggest that you use working level measurements. If you want to find out whether you have a bad house, the ease of a radon gas measurement makes this a preferred method. We do many working level measurements using mostly the Kusnetz method, and we are in the process of buying continuous working level monitors. But for a screening technique for the average homeowner, measuring working levels probably is not nearly as convenient or as cost-effective as is a simple radon gas measurement.

R. Oswald: You are probably going to be within a factor of 2 using the 50 percent equilibrium assumption in a house, particularly if you are doing a long-term radon measurement. There is additionally probably uncertainty by a factor of 5 in the risk estimate associated with any given working level measurement. So when you consider how much uncertainty there is in the risks associated with working level months of exposure, then it seems pretty meaningful to go ahead and base risk on a radon gas measurement.

RADON DIAGNOSTICS

Question: If a second-floor kitchen has a reading of 30 pCi/l, what would you expect as a basement-level reading?

A. George: Usually double or triple that found upstairs. It depends on your ventilation system and whether the measurement is done in the wintertime. At that time, the furnace may circulate the air between the basement and upstairs. Even if the measurement is done in the summertime and the heating system is not on, I still expect the basement to be higher. There are some anomalies sometimes that show the second floor higher than the basement, but I personally have never seen one.

G. Nicholls: The only time we have seen an upper level of the home having a higher radon level than a lower level is where there was either a forced air heating system or there was an obvious path in the construction of the house for the gas to rise through it. Sometimes the path is not too obvious. In older construction, with walls that are not insulated and with hot water heat, we have seen higher levels on the second floor of the house than we have seen in the basement. Some people have, in certain homes, cut out a hole in the middle of an upper floor so they can get heat from the woodstove down below. That will very often cause you to see a higher level on the second floor than on the first floor. For the most part, we see the same factor of 2, roughly, between the basement and the first floor that Andy has seen.

MEASURING RADON IN SOIL

Question: There is some question regarding the effect of moisture on the film and restricted air flow when Track Etchs detectors are employed in the soil. Comment please?

R. Oswald: One of the earlier methods of soil gas measurement simply consisted of taping the alpha sensitive detector to the bottom of a cup. In certain environments, condensation could occur on the detector chip itself, which is bad news if you are trying to make a fairly accurate measurement. In exploration studies though, that's not a very critical consideration. However, we have produced a better configuration for making soil gas measurements in which the radon detector itself is now in a second housing. We have attempted, by various means in the lab, to actually produce condensation on the detector. But because of its distance from any surfaces by virtue of the arrangement, there is no possibility of creating condensation. And because the buried cup only creates a pocket or a small chamber in the soil, there is no reason to believe that this is going to particularly impact on the local soil conditions. I don't think there is really an issue about perturbing the thing you are trying to measure.

Question: What kinds of testing ought to be done at a new site to verify that no special precautions ought to be taken because of high radon levels? How closely spaced should the detectors be? Should tests be done after excavation as well?

R. Oswald: In Sweden, there are guidelines that have been promulgated by the Swedish Geologic Survey on potential building sites where they call for soil gas measurements. If the soil gas reading at a 15-inch depth is over 1300 pCi/l, they classify that as a high-risk area and recommend that radon-resistant construction techniques be used. They define a low-risk area as one under 400 pCi/l. At this point in time there is a lot of research that needs to done, and I know that Rich Sextro and others are very interested in identifying and studying what parameters ought to be considered in assessing land from this perspective. The most that this tells you at this stage is whether or not you are about to build a structure on an unusually high level of radon in the soil. However, given all the complicating factors, you could build a house the wrong way on a site where radon is low and still end up with high radon levels in the house...Harvey?

H. Sachs (from the floor): I think a critical aspect was brought out by Jim Otton this morning. The permeability of the soil is at least as important as the soil gas concentration. That is why we have, for the last five years, done as few radon soil gas measurements for that purpose as we could possibly get away with. The other point that was brought out was the Bill Brodhead approach. He is a contractor in eastern Pennsylvania who is very concerned to the point of doing radon measurements for his clients after the houses are built. His method is to put perforated pipe into the foundations in gravel beds before construction and to provide radon-resistant construction of the foundation which I believe involves both poured concrete foundation walls and poured slab. I am not sure of the details, but the point is that this cost is less than it would cost you for the soil tests. There is also an irony involving suburban and rural construction where you are not on any city sewage systems. If you are on soil that is permeable enough that you could pass a perk test and put in a drain field, it is probably permeable enough to accumulate the radon and have a potential radon problem. It may make more sense for builders to put in perforated pipe than to talk to their lawyer about possible liability risks.

A. Scott: Why would anyone be interested in paying for a test that might demonstrate that their land is worthless?

H. Sachs: A number of developers and contractors have come to me with the question of whether or not they need to do the additional work to make houses "radon-proof" (and I am putting that in quotes). From conversations in the last two weeks with a fairly large developer of upper middle class housing, there is a belief that there would be a marketing advantage for such environmentally-sensitive and environmentally-safe houses. That is a partial answer to why someone would want to know, because there is optimism that finding radon need not signal land as worthless. You have only to prevent its entry to buildings.

LIABILITY

Question: What about the liability issue?

A. Scott (from the floor): We know from experience in Florida, where the problem has been identified in the phosphate mining lands for many years, that the usual solution to this in the building trade is to believe that no one will give a damn once the papers stop writing about it, which is probably true, and to put on higher quality brass door knobs and brass lettering and continue to build. The important thing from a liability point of view is in fact not to do any measurements. You say, "Nobody told me about radon." So if you make measurements, then you are admitting you know something about it. The most powerful argument, from a legal point of view, is to say, "I'm just a simple building contractor. If it was a leaking roof, I could be expected to be sued over it. But such items as your scientific stuff, that's quite beyond my ken. Nobody ever told me about it. If you'd asked, I'd have done it. If it was in the law, I'd have complied. But why should I go looking for trouble?"

Question: How do you prevent homeowners who are selling their houses from skewing test results?

R. Oswald: You can't. Harvey Sachs will explain tomorrow why, and what some of the alternatives are.*

CONTINUOUS RADON MONITOR

Question: How much does a computerized continuous printout radon monitoring device cost?

A. George: $5,000-$10,000. I'm giving a range because some of the Canadian instruments are more sophisticated than American instruments, and they go up to $10,000. These are just for radon gas, not for radon daughters.

EPA DELAYS

Question: When will the public find out the results of the EPA Radon Measurement Proficiency Round?

M. Ronca-Battista: Any day now. We are waiting for the results to come back from the companies themselves, so we have to wait for them to figure out what they want to tell us. Then the results will be published.

---

*The reader is referred to section 7 for more on the liability issue.

Question: Why is the EPA taking so long to come out with brochures for homeowners when you know the damage and health hazards from radon conditions?

M. Ronca-Battista: I know that the second protocols document is delayed until the release of the EPA guidance because it does have policy implications in it. The recommendations for what to do for a result of a screening measurement must be consistent with the EPA guidance document. This question is probably better addressed on Saturday when the governmental agency review process will be discussed.

EFFECT OF PARTICLE SIZES

Question: The radon daughters are absorbed onto particles or aerosols in the home. Do you have any information on the effect of the variation in particle size distribution on the measurement of radon or daughter concentrations obtained by different sampling techniques?

M. Ronca-Battista: I don't have information on that. We are starting a project with DOE and specifically some people at Oak Ridge Laboratories to look at the effects of changing particle size on working levels and other factors relating to unattached fractions.

A. George: I think particle size really is no problem. All the filters that you use to measure the working level are 100 percent efficient for all particles. So all you have to do is make sure your filter assembly does not cause any plate out of radon daughters before it gets onto the filter, and most of the instruments usually take that precaution.

Section 5
Radon and Health

INTRODUCTION AND SUMMARY OF SECTION 5: RADON AND HEALTH

Robert A. Michaels*

This section addresses three issue areas relating to radon and health, only the first of which is central to the science of environmental toxicology:

Risk Assessment, or the measure of how much risk is associated with given exposures,

Risk Management, the attempt to control the degree of risk, and

Acceptable Risk, or how much risk does society choose to tolerate.

Each of these can be summarized in turn.

RISK ASSESSMENT

Risk assessment encompasses assessment of exposure and health risks. In "The Association between Radon Daughter Exposure and Lung Cancer: Evidence and Implications from Studies of Miners," Geoffrey Howe (Professor of Preventive Medicine, University of Toronto) and Kristan L'Abbe review available data relating radon exposure to its principal health effect, lung cancer, as well as evidence of incremental stomach cancer risk. In addition, they present findings of Howe's own research into the lung cancer issue. In all, some six epidemiological investigations have contributed to elucidating a dose-response relationship describing lung cancer risk following radon exposure. Each investigation focused upon a significantly large population of hard-rock miners (ores included uranium, fluorspar, and iron), who were occupationally exposed to elevated levels of radon gas--the principal radioactive hazard even in uranium mines.

Lung cancer rates among the mining populations were compared with rates in comparable non-mining populations, yielding relative risk ratios of from 1.9 to 6.7; that is, the mining populations exhibited lung cancer rates of from 1.9 to 6.7 times those of control populations selected for

*Robert A. Michaels, Co-Chairman of the Radon and Health Session, is a principal of Ramtrac, a consulting firm dealing with environmental toxicology and risk assessment.

comparison. The relative risk ratios were found to be related to independently calculated levels of radon exposure, measured in working level months (WLM), where one working level is roughly equivalent to a radon radioactivity level of 200 picocuries per liter (pCi/l) of air inhaled.

Lung cancer risks associated with residential radon exposure, when quantified by extrapolating the dose-response curve to lower radon exposure levels, are overestimated. In "Updating Radon Daughter Risk Projections," Naomi Harley, (Professor of Environmental Medicine, New York University), documents this overestimation, and attempts to explain it. She concludes that children appear not to constitute a particularly sensitive population group, and that lung cancer risk appears to decline gradually once radon exposure is discontinued.

RISK MANAGEMENT

In "Comments on the Heath Presentations," Judith Klotz, (Coordinator of Environmental Epidemiology, New Jersey Department of Health), notes the significance of Harley's conclusions with respect to the issue of mitigation. Mitigation of excessive radon levels will not only decrease future lung cancer risk in proportion to reduction of _future_ radon exposure. As a bonus, mitigation of future radon exposure is also likely to decrease the lung cancer risk incurred by _past_ exposure to radon. In this context, it is unfortunate that, as Howe observed, radon and smoking appear to contribute to lung cancer risk in an additive rather than a multiplicative (synergistic) manner. Were the contributions multiplicative, reduction of either risk would accomplish a disproportionately large reduction in lung cancer risk posed by the other.

ACCEPTABLE RISK

Acceptable risk is not decided by scientific principles, but by regulatory agency decisions, which should reflect each agency's role in promoting the values embraced by our society. In "Overview of Federal Planning and Guidance," Christie Eheman, (Health Physicist, Centers for Disease Control, Atlanta), discusses the basis for federal agencies recommending mitigation of residential radon levels exceeding 4 pCi/l. Radon at this level is acknowledged to pose a worst-case incremental lifetime lung cancer risk in the range of about 1 to 3 percent (10,000 to 30,000 per million), assuming continuous exposure over 70 years. The federal recommendations were clearly based upon cost-benefit considerations, in particular, the high cost and diminishing returns of mitigating residential radon levels which are intially as low as 4 pCi/l or lower.

Robert Michaels (Consultant in Environmental Toxicology and Risk Assessment, Long Island City, NY) notes in the discussion that confusion has resulted from the failure of health agencies to separate concepts of risk assessment, acceptable risk, and economic factors, all of which may limit risk management. For example, one company which analyzes radon samples of residential air interprets results for clients as follows: "The maximum safe level for radon in homes tentatively recommended by EPA is approximately 4 pCi/l" (emphasis added). In fact, the worst-case incremental cancer risks associated with this radon level appear to be some 10,000 to 30,000 times the one-per-million level that, in other circumstances, would trigger EPA action to reduce them.

A second effect of the federal recommendations may be to inhibit investment of venture capital toward development technologies to improve the economics of radon mitigation. Such investment requires unequivocal federal agency commitment to achieving lower levels as they become technically feasible. Michaels suggests that federal agencies can drive radon mitigation technology by establishing mitigation targets, just as automobile engine efficiency is driven by miles-per-gallon targets which become more demanding over the years. Until action is taken to reduce average residential radon exposure instead of just extreme levels, which exert little effect upon average exposure, Klotz notes, no appreciable reduction of the radon contribution to rising lung cancer rates in the U.S. can be anticipated.

THE ASSOCIATION BETWEEN RADON DAUGHTER EXPOSURE AND LUNG CANCER: EVIDENCE AND IMPLICATIONS FROM STUDIES OF MINERS

Geoffrey R. Howe and Kristan A. L'Abbe*

Smoking is by far the most important single cause of lung cancer in North America. Many studies have shown that smoking accounts for 80 percent or more of lung cancer cases which occur in males, and a smaller though still substantial fraction of those cases occurring in females. However, lung cancer is a very common disease. In Canada, some 7,000 cases of lung cancer occur annually in males and more than 2,000 cases in females. Since the population of the United States is approximately ten times that of Canada, by extrapolation the corresponding estimated numbers in the U.S. are 70,000 males and 20,000 females. Thus, if 20 percent of lung cancer is unaccounted for by smoking in males, and as much as 50 percent in females, causes other than smoking could be contributing some 14,000 cases in males and close to that number in females.

This latter fact has focused attention in the past several decades upon the possible contribution of exposure to radon daughters to lung cancer risk from sources other than smoking. Exposure to the radioactive daughter products of radon-222, in particular polonium-218, is known to cause lung cancer in several animal species.[1] As discussed below, there is also considerable evidence from human studies that these compounds can cause cancer particularly in the proximal bronchi. Radon is a naturally occurring gas which is found in varying concentrations throughout the United States and Canada, and in some areas and in certain situations the exposure of the general population can be non-negligible. However, at the typical levels of exposure found, it is very difficult from a statistical point of view to detect excess cases of lung cancer occurring above those expected from smoking and other sources in the general population. In order to quantify the problem of radon daughter exposure, it is necessary to quantify risk estimates, i.e., to be able to define how much extra risk is conferred per unit of exposure in the general population,

*Geoffrey R. Howe is Professor of Preventive Medicine at the University of Toronto. Both he and Kristan A. L'Abbe are associated with the National Cancer Institute of Canada, Epidemiology Unit, Faculty of Medicine at the University of Toronto, Ontario, Canada.

and to decide how these risks are modified by other factors such as smoking, age at exposure, or age at risk. This is a classical epidemiologic problem and has a corresponding classical epidemiologic solution. Studies are carried out on groups who have received a much wider range of exposures than the general population, including individuals with substantial exposures. By using mathematical models, it is then possible to extrapolate the risk per dose relationship to the levels at which the general population is exposed and thus to be able to estimate risk parameters for these levels of exposure, and also to determine the nature of any interactions of radon daughters with other factors.

Hard-rock miners in recent decades have been exposed to substantial amounts of radon daughters, and this is the group that has provided us with the best current evidence relating to risk parameters for radon daughters. In this paper, five major cohort studies of such miners are very briefly described, together with more detailed results of a recently completed cohort study of uranium miners at the Beaverlodge Mine in northern Saskatchewan. The implications of these results for occupational exposure and for the general population are then discussed.

COHORT STUDIES OF HARD-ROCK MINERS

The first observation of excess lung cancer occurring among miners was made in Bohemia in the fifteenth century. Kunz and others reported a follow-up study of 2,300 uranium miners from the Czechoslovakian uranium mines[2] (Table 5.1).

Number of Miners	2,530
Exposure Level (mean)	313 WLM
Observed	212
Expected	42.5
Relative Risk	5.0

Table 5.1. Lung Cancer Deaths, Czechoslovakia Uranium Miners (1948-75).

During an average follow-up period of 26 years, they observed 212 cases of lung cancer using the national cancer registry together with industry health records, as compared to an expected number of 42.5, based on national age specific and year specific lung cancer rates. This leads to an estimate of the relative risk of 5.0. The average exposure of these miners was 313 working level months (WLM) although most of these miners were long-term employees and the actual levels of exposure to radon daughters were substantially lower than those in the U.S. mines discussed next.

A second cohort study has looked at the mortality experience of 3,356 uranium miners in the Colorado Plateau region.[3] Miners from a large number of mines have been followed from 1950 to 1974 (Table 5.2).

Number of Miners	3,356
Exposure Level (mean)	1180 WLM
Observed	159
Expected	25.2
Relative Risk	6.3

Table 5.2. Lung Cancer Deaths, United States Uranium Miners (1950-74).

A total of 159 lung cancer cases has been reported, compared to an expected number 25.2, based on national age and year specific lung cancer rates. The relative risk of 6.3 is comparable to that of the Czechoslovakian miners, although the average level of exposure in this group was much higher. A much larger series of 15,000 uranium miners employed in the Province of Ontario, Canada, have been followed since 1959, with deaths being monitored through the national Canadian mortality data base using computerized record linkage techniques [4] (Table 5.3).

Number of Miners	15,000
Exposure Level (mean)	36 WLM
Observed	62
Expected	25.5
Relative Risk	2.4

Table 5.3. Lung Cancer Deaths, Ontario Uranium Miners (1959-77).

Again, there is a two-fold excess of lung cancer despite the fact that the average exposure of these miners was lower by an order of magnitude than the Czechoslovakian experience. Table 5.4 shows the experience of another cohort of Canadian miners, that of fluorspar miners in Newfoundland.[5]

Number of Miners	2,414
Exposure Level (mean)	246 WLM
Observed	65
Expected	9.7
Relative Risk	6.7

Table 5.4. Lung Cancer Deaths, Newfoundland Fluorspar Miners (1951-71).

Exposures to radon daughters in this cohort were comparable to the Czechoslovakian miners having an average value of 246 WLM, and again, there is a six-fold increase in risk of lung cancer. The fifth major study, namely that of Swedish iron mine workers in two mines north of the Arctic Circle, has recently been reported by Radford and St. Clair Renard (see Table 5.5).[6]

Number of Miners	1,294
Exposure Level (mean)	93.7 WLM
Observed	50
Expected	14.6
Relative Risk	3.4

Table 5.5. Lung Cancer Deaths, Sweden Iron Miners (1951-76).

These miners were employed for long periods of time at levels of exposure close to the currently accepted occupational standards. The three-fold excess of lung cancer observed has given further rise to concern as to the adequacy of those occupational standards. This study was of particular importance since smoking histories for the miners were available and the data indicate that radon daughters add to the risk from smoking, rather than multiplying it, the two usual alternative models considered for extrapolating risk throughout life.

THE BEAVERLODGE STUDY

Howe, et al.[7] have recently reported the mortality experience of 8,487 uranium miners employed at the Eldorado Resources Ltd. Beaverlodge Mine in northern Saskatchewan since the early 1950's. Mortality in the group has been determined through to the end of 1980, again using computerized record linkage to the Canadian national mortality data base. Again, an excess of lung cancer is observed in the entire group (Table 5.6), despite the fact that the average exposure was only 16 working level months.

Number of Miners	8,487
Exposure Level (mean)	16.6 WLM
Observed	65
Expected	34.2
Relative Risk	1.9

Table 5.6. Lung Cancer Deaths, Beaverlodge, Saskatchewan Uranium Miners (1948-80).

When this excess is examined by dividing the cohort into those who were essentially unexposed (less than 5 WLM cumulative experience)(Table 5.7) and those receiving more than 5 WLM (Table 5.8), there is no excess of lung cancer among the non-exposed, with all the excess being limited to the exposed group.

Cause	Observed	Expected	Relative Risk
Lung Cancer	19	18.4	1.0
All other cancers	42	45.6	0.9

Table 5.7. Beaverlodge, Saskatchewan Study, Unexposed.

Cause	Observed	Expected	Relative Risk
Lung Cancer	46	15.9	2.9
All other cancers	24	36.7	0.7

Table 5.8. Beaverlodge, Saskatchewan Study, Exposed.

There is also no evidence of any excess of any other form of cancer in either group. This latter point is in contrast to the excess of stomach cancer seen in both the Newfoundland and Swedish cohorts (Table 5.9).

Study	Observed	Expected	Relative Risk
Newfoundland	23	11.7	2.0
Sweden	28	15.1	1.9

Table 5.9. Stomach Cancer Deaths.

When the data from Beaverlodge are examined for a dose-response relationship (Table 5.10 and Figure 5.1), there is an obvious and increasing risk corresponding to increasing exposure.

Exposure (WLM)	Observed	Expected	Relative Risk
0-4	14	14.5	1.0
5-24	12	6.5	1.9
25-49	5	2.6	1.9
50-99	6	2.5	2.4
100-149	7	1.2	6.0
150-249	6	0.8	7.9
250+	4	0.3	14.2

Table 5.10. Dose-Response Relationship in the Beaverlodge, Saskatchewan Study.

Figure 5.1 - BEAVERLODGE: LUNG CANCER DOSE-RESPONSE

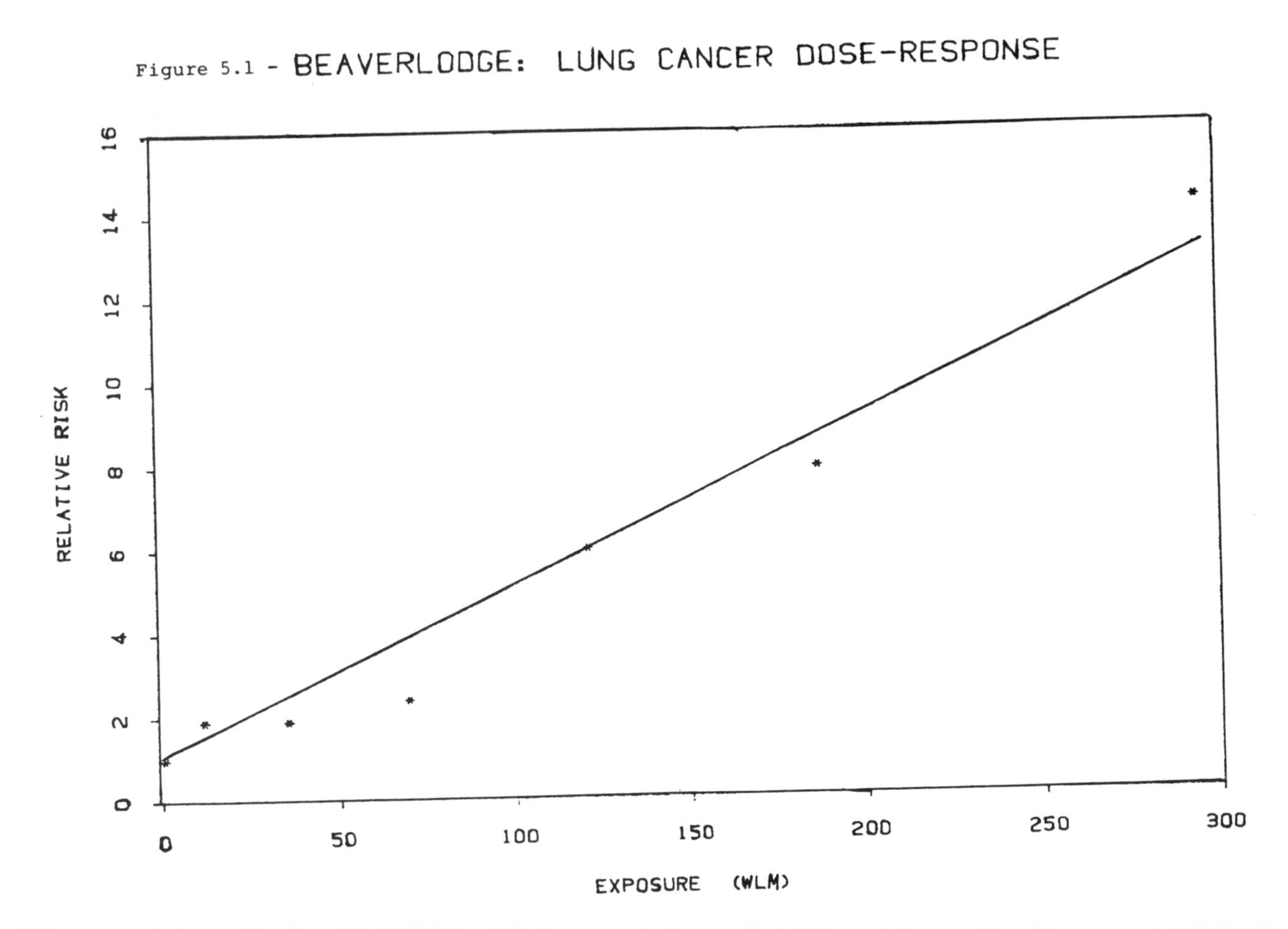

The data are well fitted by a straight line whose slope gives rise to a value of 20.8 excess lung cancers per 10^6 person units per WLM (additive scale) or 3.28 percent per WLM (multiplicative scale). These values are similar to the Czechoslovakian data and to the Swedish data in particular, and are in turn substantially greater than those reported from the U.S. study. The major difference between these studies is the very much higher exposure rates in the U.S. data, and it is possible that this may partly account for the differences, together with differences in dosimetry. A further result of interest from the Beaverlodge study is that the excess appears to be correlated with age of first exposure, with those exposed before age 30 having significantly lower risks than those exposed after this age. A similar result has been reported in the Czechoslovakian data. The excess risk first manifests itself five years after first exposure, though general risk estimates are made from this as from other studies allowing for a ten-year latent period. Unfortunately, smoking data are not available for the Beaverlodge cohort at this point, so the question of the appropriateness of the additive or multiplicative model could not be addressed by consideration of smoking data. There is some indication in these data, however, that the relative risk does drop off with age at risk, and these data are in fact consistent with a model which is intermediate between the relative and additive models.

IMPLICATIONS FOR OCCUPATIONAL STANDARDS AND THE GENERAL POPULATION

Applying the results from the Beaverlodge study to miners exposed to the current occupational maximum (i.e., 4 WLM a year, for a period of five years starting at age 30), gives the data shown in Table 5.11.

Unexposed	30.6/1000
Exposed (*)	
Model 1	50.0/1000
Model 2	40.8/1000

(*) Exposed to 4 WLM/yr for 5 years from age 30

Table 5.11. Beaverlodge, Saskatchewan Study. Risk of dying from Lung Cancer by age 70 (males).

An individual who is not exposed to these radon daughters would be expected to have a risk of 30 per 1,000 of dying of lung cancer by age 70. In contrast, a miner exposed at this level would have an expectation of 50 per thousand using the relative risk model (Model 1) and 40 per thousand using the additive model (Model 2). Caution must be exercised in extrapolating the results from studies of miners to the general population. Dose rates are much lower, and the atmosphere in homes is very different from that in mines, with respect to dust and other factors. It is also possible that miners may breathe more deeply in view of their physical exertion, and this in turn could carry the radon daughter products past the sensitive proximal bronchi into the lung itself. Given these caveats, it is possible to use the mining data to consider the possible effect of radon daughter exposure in the general population. A number of studies have estimated the geometric average working level exposures in homes in various states and countries (Table 5.12).

Location (date)	Mean WL	WLM/yr[1]
Great Britain (1980)	0.003	0.10
New York/New Jersey (1980)	0.004	0.14
Ontario (1980)	0.004	0.14
Colorado (1975)	0.006	0.22
Tennessee (1980)	0.008	0.30
Sweden (1981)	0.020	0.72

[1]Assumption: 17 hours/day in home

Table 5.12. Radon daughter levels in homes.

If one assumes that exposure to radon daughters confers an increased risk of approximately 20 cases of lung cancer per million persons per year per working level month, for the male population of the United States this would lead to an estimate of some 2,000 extra cases of lung cancer per year per working level month. This of course assumes that the risk applies uniformly to all males and makes no allowance for a latent period or a varying risk for age at exposure. Consideration of the levels of exposure seen in Table 5.12 suggests, however, that based on these risk estimates, a considerable fraction of the lung cancer cases unexplained by smoking could be attributable to general radon daughter exposure. More complex specific and detailed models have been developed, as discussed in other papers in this series, but nonetheless it is obvious that it is essential to continue monitoring the mortality and cancer incidence experience of the mining cohorts since these still provide our best estimates of risk, and thus are critical to any subsequent risk benefit analyses.

FOOTNOTES

[1]Kennedy, A. and J.B. Little, "Radiation Carcinogenesis in the Respiratory Tract," Pathogenesis and Therapy of Lung Cancer, Ed. C.C. Harris, New York: Marcel Dekker Inc., 1978, pp. 189-261.

[2]Kunz, E., J. Sevc, V. Placek, and J. Horacek, "Lung Cancer in Man in Relation to Different Time Distribution of Radiation Exposure," Health Physics, 36 (1979), pp. 699-706.

[3]Archer, V.E., J.K. Wagoner, and F.E. Lundin, "Lung Cancer among Uranium Miners in the United States," Health Physics, 25 (1973), pp. 351-71.

[4]Muller, J., W.C. Wheeler, J.F. Gentleman, G. Suranyi, and R.A. Kusiak, Study of Mortality of Ontario Miners 1955-77, Part I. Toronto: Ontario Ministry of Labour, Ontario Worker's Compensation Board, Atomic Energy Control Board of Canada, 1983.

[5]de Villiers, A.J., and D.T. Wigle, Mortality Experience of Fluorspar Miners Exposed to Radon and Radon Daughter Products. (Unpublished manuscript).

[6]Radford, E.P. K.G. St. Clair Renard, "Lung Cancer in Swedish Iron Miners Exposed to Low Doses of Radon Daughters," New Engl. J. of Med. 310 (1984), pp. 1481-4.

[7]Howe, G.R., R.C. Nair, H.B. Newcombe, A.B. Miller, and J.D. Abbatt,"Lung Cancer Mortality (1950-1980) in relation to Radon Daughter Exposure in a Cohort of Workers at the Eldorado Beaverlodge Uranium Mine," Forthcoming in JNCI.

UPDATING RADON DAUGHTER RISK PROJECTIONS

Naomi H. Harley*

INTRODUCTION

There are presently four case control studies in the U.S., Canada, and Sweden which are attempting to determine whether an effect from environmental levels of radon daughters can be detected. However, the individual risk of lung cancer from environmental radon daughter exposure is small. Given the problems of population mobility and recreating personal exposure history, which is perhaps more difficult than those attempted in underground mines, there is slight chance of a significant lung cancer excess being detected by these studies. Thus, the underground mining data provides the best source for devising a risk projection model to estimate the lifetime risk of lung cancer from radon daughter exposure in environmental situations.

In this work the relative risk model is used to calculate lifetime risk as a function of age at first exposure and exposure duration. A range of relative risk coefficients of 1 to 4 percent per WLM has been adopted by the USEPA. These values are used as a basis for calculation to show that risk coefficients this large applied to a standard relative risk calculation produces results that are inconsistent with the observed mining studies. Reducing the risk coefficient exponentially with an effective half-time of ten years produces values which are plausible. These values compare with the National Council on Radiation Protection and Measurements (NCRP) modified absolute risk model projections[1].

UNDERGROUND MINING STUDIES

There are four large epidemiological studies of underground miners which provide the basic data for risk projection. These are the U.S. uranium miner study[2,3] of 3,362 miners followed from 1950 to the present with exposures ranging from 60 to 7,000 WLM with an average exposure of 800 WLM. In the Canadian uranium miners study[4], 15,984 miners have been followed since 1955, with exposure ranging from 5 to 510 WLM with an average exposure

*Naomi H. Harley is Professor of Environmental Medicine at New York University, New York.

of 74 WLM. 2,400 Czechoslovakian uranium miners[5] were followed since 1948, with exposures ranging from 72 to 716 WLM with an average of 200 WLM, and 1,415 Swedish iron miners[6] were followed since 1951, with exposures ranging from 27 to 218 WLM with an average exposure of 80 WLM.

The total lung cancer experience of these four mining cohorts to the end of reported follow-up is shown in Table 5.13.

Table 5.13–The Number of Miners, Average Radon Daughter Exposure and Lung Cancer Mortality in the 4 Major Underground Miner Follow-Up Studies

	Number of Miners	Average Exposure (WLM)	Lung Cancer Mortality Observed/Expected	Lung Cancers Million Persons (WLM)
USA	3,362	800	185/38	50
Canada	15,984	33-74	119/66	60
Czechoslovakia	2,400	200	212/55	320
Sweden	1,415	80	50/15	310

It is from these data and the indications of the temporal patterns of lung cancer that inferences must be made concerning environmental exposures.

RISK PROJECTION MODELS

One currently popular form of risk projection is the use of the relative risk model. In this model the natural disease mortality is increased by a constant fraction for each unit (WLM) exposure. The natural or baseline mortality is different for groups such as white males, smoking males, women, etc. In this study two baselines will be used. One for lung cancer in white males in 1979[7] along with the 1980 life table[8] to correct for competing causes of death. The other is the American Cancer Society (ACS) age specific lung cancer base for nonsmokers corrected by the smoker/nonsmoker mortality ratio[9] to yield a baseline for smokers. This is used with the ACS life table for male smokers to correct for their competing causes of death[10].

The relative risk projection model used here assumes a minimum latent interval of five years from exposure to the first expression of lung cancer mortality. The baseline rate is then increased by a multiple of either 0.01 or 0.04 for each 1 WLM/year exposure and the summations done for different ages at first exposure and exposure durations. These lifetime lung cancer risk projections are shown in Tables 5.14A and 5.14B.

Table 5.14 A—Calculated Lifetime Lung Cancer Mortality per Million Persons from Radon Daughter Exposure of 1 WLM/yr as a Function of Age at First Exposure and Exposure Duration Relative Risk Calculation Based on 0.01 Increased Mortality per WLM and 5 yr Minimum Latent Interval

Exposure	..................Age at First Exposure..................				
Duration	Birth	10	20	30	50
1 yr	670	680	680	700	650
10 yr	6,800	6,800	6,900	7,000	6,000
20 yr	14,000	14,000	14,000	14,000	10,000
30 yr	20,000	21,000	21,000	20,000	12,000
Lifetime	46,000	39,000	32,000	25,000	12,000

Note: Calculated values for white males, 1979 age specific lung cancer mortality, 1980 life table.

Table 5.14 B—Calculated Lifetime Lung Cancer Mortality per Million Persons from Radon Daughter Exposure of 1 WLM/yr as a Function of Age at First Exposure and Exposure Duration Relative Risk Calculation Based on 0.04 Increased Mortality per WLM and 5 yr Minimum Latent Interval

Exposure	..................Age at First Exposure..................				
Duration	Birth	10	20	30	50
1 yr	2,700	2,700	2,700	2,800	2,600
10 yr	27,000	27,000	28,000	28,000	24,000
20 yr	56,000	56,000	56,000	56,000	40,000
30 yr	80,000	84,000	84,000	80,000	48,000
Lifetime	184,000	156,000	128,000	100,000	48,000

Note: Calculated values for white males, 1979 age specific lung cancer mortality, 1980 life table.

The U.S. uranium miners, from Table 5.13, have an average exposure duration of 10 years and an average exposure of 800 WLM. This would yield an exposure of 80 WLM/year. Examination of Tables 5.14A and 5.14B show that for a group of 3,362 miners the predicted lifetime lung cancers would be 1,800 for a 1 percent relative risk and more than 100 percent of the miners for the 4 percent relative risk.

Clearly this is an absurdity given that only 150 excess cancers have been seen in this population.

The same calculation is done in Tables 5.15A and 5.15B using the estimated smokers' lung cancer age specific mortality and the life table for male smokers.[11]

Table 5.15 A—Calculated Lifetime Lung Cancer Mortality per Million Persons from Radon Daughter Exposure of 1 WLM/yr as a Function of Age at First Exposure and Exposure Duration Relative Risk Calculation Based on 0.01 Increased Mortality per WLM and 5 yr Minimum Latent Interval

Exposure	Age at First Exposure				
Duration	Birth	10	20	30	50
1 yr	540	540	540	540	490
10 yr	5,400	5,400	5,400	5,200	4,300
20 yr	11,000	11,000	11,000	10,000	7,100
30 yr	16,000	16,000	16,000	14,000	8,100
Lifetime	34,000	29,000	24,000	18,000	8,000

Note: Calculated values for male smokers. Age specific lung cancer mortality and lifetable from Garfinkel (6a86). American Cancer Society Study.

Table 5.15 B—Calculated Lifetime Lung Cancer Mortality per Million Persons from Radon Dauthter Exposure of 1 WLM/yr as a Function of Age at First Exposure and Exposure Duration Relative Risk Calculation Based on 0.04 Increased Mortality per WLM and 5 yr Minimum Latent Interval

Exposure	Age at First Exposure				
Duration	Birth	10	20	30	50
1 yr	2,200	2,200	2,200	2,200	2,000
10 yr	22,000	22,000	22,000	21,000	17,000
20 yr	44,000	44,000	44,000	40,000	28,000
30 yr	64,000	64,000	64,000	56,000	32,000
Lifetime	136,000	116,000	96,000	72,000	32,000

Note: Calculated values for male smokers. Age specific lung cancer mortality and lifetable from Garfinkel (6a86). American Cancer Society Study.

The lifetime radon daughter-related mortality for the same 80 WLM/year average exposure for 10 years is actually somewhat lower than that for the white male population. This is due to the effect of the life table correction for competing causes of death which is considerably larger for smoking males. However, the number of lung cancer fatalities in the U.S. miners that would be calculated in Tables 5.15A and 5.15B is still untenable.

What is the problem with a relative risk projection model? Hornung and Meinhardt[12] reported an analysis of this U.S. mining cohort and find that miners who have the same total exposure but who have not mined for 10 years have 1/2 the risk as currently mining miners. It appears that there is a reduction in lung cancer risk with time post exposure which is now emerging in the well-studied populations. This effect is most probably related to the same phenomenon which appears as a higher risk of lung cancer for older ages at first exposure which is seen in the Czech underground uranium miners.

Applying an exponential correction following exposure with a half-time of 10 years to reduce the relative risk coefficients of 0.01 and 0.04 in Tables 5.14A and 5.14B yields the values in Tables 5.14C and 5.14D.

Table 5.14 C—Calculated Lifetime Lung Cancer Mortality per Million Persons from Radon Daughter Exposure of 1 WLM/yr as a Function of Age at First Exposure and Exposure Duration Relative Risk Calculation Based on 0.01 Increased Mortality per WLM, 5 yr Minimum Latent Interval and Risk Coefficient Reduction with 10 yr Half-Time

Exposure	Age at First Exposure				
Duration	Birth	10	20	30	50
1 yr	8	17	34	69	200
10 yr	120	240	480	940	2,100
20 yr	360	720	1,400	2,500	4,000
30 yr	840	1,700	3,000	4,700	5,000
Lifetime	7,200	7,100	6,800	7,500	5,000

Note: Calculated values for white males. 1979 age specific lung cancer mortality. 1980 life table.

Table 5.14 D—Calculated Lifetime Lung Cancer Mortality per Million Persons from Radon Daughter Exposure of 1 WLM/yr as a Function of Age at First Exposure and Exposure Duration Relative Risk Calculation Based on 0.04 Increased Mortality per WLM, 5 Year Minimum Latent Interval and Risk Coefficient Reduction with 10 yr Half-Time

Exposure	Age at First Exposure				
Duration	Birth	10	20	30	50
1 yr	32	68	140	280	800
10 yr	480	960	1,900	3,800	8,400
20 yr	1,400	2,900	5,600	10,000	16,000
30 yr	3,400	6,800	12,000	19,000	20,000
Lifetime	29,000	28,000	27,000	30,000	20,000

Note: Calculated values for white males, 1979 age specific lung cancer mortality, 1980 life table.

It can be seen that these projected deaths from lung cancer for the miners whose exposures are given in Table 5.14 are now more consistent with those observed. The same exponential decrease applied to the American Cancer Society data yields the lifetime risks in Tables 5.15C and 5.15D.

Table 5.15 C—Calculated Lifetime Lung Cancer Mortality per Million Persons from Radon Daughter Exposure of 1 WLM/yr as a Function of Age at First Exposure and Exposure Duration Relative Risk Calculation Based on 0.01 Increased Mortality per WLM 5 Year Minimum Latent Interval and Risk Coefficient Reduction with 10 Year Half-Time

Exposure	Age at First Exposure				
Duration	Birth	10	20	30	50
1 yr	9	17	34	69	160
10 yr	120	240	480	810	1,600
20 yr	360	720	1,300	2,000	3,000
30 yr	840	1,500	2,500	3,700	3,600
Lifetime	6,500	6,400	6,100	5,600	3,600

Note: Calculated values for male smokers. Age specific lung cancer mortality and lifetable from Garfinkel (6a86). American Cancer Society Study.

Table 5.15 D—Calculated Lifetime Lung Cancer Mortality per Million Persons from Radon Daughter Exposure of 1 WLM/yr as a Function of Age at First Exposure and Exposure Duration Relative Risk Calculation Based on 0.04 Increased Mortality per WLM 5 Year Minimum Latent Interval and Risk Coefficient Reduction with 10 year Half-Time

Exposure	Age at First Exposure				
Duration	Birth	10	20	30	50
1 yr	36	68	140	280	640
10 yr	480	960	1,900	3,200	6,400
20 yr	1,400	2,900	5,200	8,000	12,000
30 yr	3,400	6,000	10,000	15,000	14,000
Lifetime	26,000	26,000	24,000	22,000	14,000

Note: Calculated values for male smokers. Age specific lung cancer mortality and lifetable from Garfinkel (6a86). American Cancer Society Study.

The use of a reduction in the risk coefficient was utilized in the NCRP[13] risk projection model. This model used a modified absolute approach and a baseline risk of 10 million persons per year per WLM, decreasing the risk coefficient exponentially after exposure by a 20-year half-time. A five-year minimum latent interval was used but never expressing risk until after age 40 regardless of the age at exposure since this is the normal age at which lung cancer appears in the population. The NCRP risk projections are shown for comparison in Table 5.16 using the same life table utilized in the calculations for Table 5.14A to 5.14D (1980 white males).

Table 5.16—Calculated Lifetime Lung Cancer Mortality per Million Persons from Radon Daughter Exposure of 1 WLM/yr as a Function of Age at First Exposure and Exposure Duration NCRP Modified Absolute Risk Model. Risk Coefficient Reduction with 20 yr Half-Time Lung Cancer Not Expressed Before Age 40 Regardless of Age at First Exposure. 5 Year Minimum Latent Interval

Exposure	Age at First Exposure				
Duration	Birth	10	20	30	50
1 yr	43	62	89	130	120
10 yr	510	730	1,000	1,500	990
20 yr	1,200	1,800	2,500	2,800	1,600
30 yr	2,300	3,200	3,800	3,800	1,900
Lifetime	7,000	6,500	5,700	4,700	1,900

Note: Calculated values for white males. 1980 life table. Risk coefficient 10/million persons-year-WLM.

The effect of radon daughters on nonsmokers is not discussed here since there is insufficient data to describe their response. Samet et al.[14] show that the relative risk coefficient is higher for nonsmoking Navajos than that for the white miner smoking cohort. Radford and Renard[15] report a relative risk coefficient much higher for nonsmokers than for smokers in the Swedish underground iron miners. These studies suggest that the effect of smoking is not multiplicative with radon daughter exposure but perhaps lower than a multiplicative effect. Muller[16] has also reported this to be true for the Canadian uranium miners. The risk projection for nonsmokers awaits further data from the mining cohorts.

SUMMARY

It can be shown that a relative risk model with risk coefficients of 0.01 to 0.04 increase in lung cancer mortality per WLM radon daughter exposure produces results that are grossly incompatible with the observed underground mining data. Reanalysis of the U.S. uranium miner data suggests a reduction in risk coefficient with time since exposure when a relative risk was examined. If this effect is factored into the calculations, both the NCRP modified absolute model which uses this approach and the corrected relative risk models are in accord with the observed lung cancer mortality in the mining studies. While the two types of models produce numerically comparable results, they are based on widely different assumptions. Their validity can only be shown when the complete temporal pattern of miner lung cancer becomes available.

ACKNOWLEDGEMENT

The author acknowledges Dr. Lawrence Garfinkel for allowing us to use his unpublished data and John H. Harley for his helpful suggestions. This work was supported by USDOE Grant DE AC02 EV10374 and in part by Center Grants from NIEHS (ES00260) AND NCI (CA13343). All support is gratefully acknowledged.

FOOTNOTES

[1]National Council on Radiation Protection and Measurements, Evaluation of Occupational and Environmental Exposures to Radon and Radon Daughters in the United States, Report Number 78. Bethesda, MD, 1984.

[2]National Academy of Sciences, The Effect on Populations of Exposure to Low Levels of Ionizing Radiation, (BEIR III Report) Washington, D.C.: National Academy Press, 1980.

[3]Waxweiler, R.J., "Updated Mortality Analysis of U.S. Uranium Miner Study Group," Proc. Int. Conf. on Radiation Hazards in Mining: Control, Measurements and Medical Aspects, Golden CO: New York Society of Mining Engineers, 1981.

[4]Muller, J., W.C. Wheeler, J.F. Gentleman, G. Suranyi, and R.A. Kusiak, "Study of Mortality of Ontario Miners," Proc. International Conference on Radiation Safety in Mining, Toronto. Ontario, Canada: October, 1984.

[5]Sevc, J., E. Kunz, and V. Placek, "Lung Cancer in Uranium Miners and Long Term Exposure to Radon Daughter Products," Health Physics, 30 (1976), pp.433.

[6]Radford, E.P., and K.G. Renard, "Lung Cancer in Swedish Iron Miners Exposed to Low Doses of Radon Daughters," New England Journal of Medicine, 310 (1984), pp.1495-99.

[7]National Center for Health Statistics, Vital Statistics of the United States-1979 Vol.II Mortality Part B, Hyattsville, MD: U.S. Dept. of Health and Human Services, 1984.

[8]National Center for Health Statistics, Vital Statistics of the United States-1980 Vol. II Life Tables Section 6, Hyattsville, MD: U.S. Dept. of Health and Human Services, 1984.

[9]Hammond,C.E., L. Garfinkel, and E.A Lew, "Longevity, Selective Mortality and Competitive Risks in Relation to Chemical Carcinogens," Env. Res., (1978) pp. 153-173.

[10]Garfinkel, L., Private Communication, 1986.

[11]Ibid.

[12]Hornung, R.W., and T.J. Meinhardt, "Quantitative Risk Assessment of Lung Cancer in U.S. Uranium Miners," Cincinnati, OH: NIOSH, Centers for Disease Control, 1986.

[13]National Council on Radiation Protection, Evaluation of Occupational and Environmental Exposures to Radon and Radon Daughters in the U.S., op. cit.

[14]Samet, J.M., D.M. Kutvirt, R.J. Waxweiler, and C.R. Key, "Uranium Mining and Lung Cancer in Navajo Men," N. Eng. J. Med., 310 (1984), pp. 1481-1484.

[15]Radford and Renard, op. cit.

[16]Muller, Wheeler, Gentleman, Suranyi, and Kusiak, op. cit.

OVERVIEW OF FEDERAL RADON PLANNING AND GUIDANCE

Christie R. Eheman*

Radon is an inert gas that results from the radioactive decay of radium. Both radium and radon are products of the decay of uranium to lead and occur wherever natural uranium deposits exist. Elevated rates of lung cancer and other respiratory diseases among underground miners exposed to radon and radon daughters were documented early in this century. Lung cancer caused 30 to 70 percent of the deaths among miners in the mountains of Central Europe where ores high in uranium have been mined for centuries. An increased risk of lung cancer has been demonstrated in studies of fluorspar, iron, lead, zinc, and uranium miners.

Radon diffuses into the air of a mine from the surrounding rock and soil and decays into radioactive atoms of polonium, bismuth, and lead. If these decay products, called radon daughters, are inhaled, they can deposit along the trachea and within the lungs. Only a small percentage of the radiation dose to the bronchial epithelium arises from radon itself; almost 95 percent of the dose comes from the daughters deposited in the bronchi. Radon daughters are measured as working levels, a measure of radon daughter concentrations in air. Exposure to radon daughters is expressed in working level months (WLM) and is a function of the time of exposure and level of radon daughters in working levels.

Radon was viewed primarily as an occupational hazard until the mid 1970's when elevated levels were found inside homes constructed using the waste from uranium processing -- a sandy material called mill tailings. In the same period, houses in central Florida were being tested and elevated levels were found in houses built over land reclaimed from phosphate mining. Radon can enter a building directly from building materials contaminated with radon sources such as uranium mill tailings, or (if the source of radon is underneath the structure) through cracks, such as those in a basement floor, and through openings around pipes and wiring. Radon occurs in all buildings at low concentrations because of natural radium and uranium in soil and rock. However, the recent trend to reduce heat loss by weatherization processes, such as the use of storm windows

---

*Christie R. Eheman is a health physicist with the Centers for Disease Control, Atlanta, GA.

and weatherstripping, reduces ventilation in buildings and may increase indoor radon levels and the exposure to residents.

In the case of homes contaminated with uranium mill tailings or built on reclaimed phosphate mining land, the U.S. Environmental Protection Agency (EPA) established an annual average level of 0.02 WL (4 pCi/l) as the goal for remediation. The 0.02 WL was based on the availability and cost of the technology required to lower the radon daughter concentrations in homes below 0.02 WL rather than on acceptable risk. The risk of dying from lung cancer from lifetime exposure to radon at EPA's recommended level for action is 1.6 to 3.3 percent. The normal risk of dying from lung cancer for a nonsmoker is about 1 percent, so that at this level an individual's risk may be doubled or tripled. Since the remedial action level was not based solely on risk, this level is not defined as safe. Current knowledge does not support a cutoff point where radiation exposure is safe or without risk.

The EPA has estimated that approximately one to two million homes in the U.S. exceed the guideline for remedial action of 0.02 WL. Current data indicate that from 5,000 to 30,000 lung cancer deaths may be due to radon each year, with the most likely estimates between 10,000 and 15,000.

The examples of elevated indoor radon given above are all associated with enhanced or artificial sources such as mill tailings. However, the discovery of extremely high concentrations of radon in homes in the Reading Prong has focused attention on natural uranium deposits as sources of radon. While indoor radon measurements slightly above 0.02 WL have been reported for years in various parts of the country, no one suspected that levels above 1.0 WL (200 pCi/l) could be due to natural sources alone or that so many houses could exist above this level.

The indoor radon problem is unique because the source of exposure is natural rather than the result of industrial practices so that there is no individual or company to blame. Also, radon is the only example of environmental radiation where exposure to the public can approach or exceed the limits for occupational workers. This is particularly important since the risks from radon are orders of magnitude higher than the risks predicted for most chemical environmental contaminants. Because of the large risk from radon, the discovery of extremely high levels of indoor radon in Pennsylvania and New Jersey has stimulated activity on radon in state and federal agencies. On the federal level, the EPA has taken the lead role, but other agencies such as the Centers for Disease Control, the Department of Energy, and the Department of Housing and Urban Development have also been active.

The Centers for Disease Control has been involved in various aspects of the radon issue for some time. Within the Centers for Disease Control, the focus of the Center for Environmental Health's (CEH) activities has been on environmental exposures to the public from natural or enhanced sources of radon. The Centers for Disease Control, CEH has a number of functions and activities that define our role in the area of radon. CDC hopes to:

1. Alert states to the radon problem and serve as a source of information for state and local health officials.

2. Serve as a liaison or coordinator for information exchange between health officials and other federal agencies.

3. Assist state and federal agencies in explaining risks from radon to members of the public.

4. Disseminate information to the public through the press and through individual contact.

Since the federal program is non-regulatory, the heaviest responsibility rests with the individual states. State agencies have to evaluate the particular geology in their region and decide if a measurement program is warranted and, if so, how to undertake such a task. Also, since state representatives are the most accessible, they are often the ones who explain the nature of the radon problem to the public, advise individuals on how to best remediate a home and deal with the hard issues on a daily basis.

Compared to our knowledge about many chemical carcinogens found in the environment, the scientific community knows a great deal about the risk from radon exposure, principally because of the data from underground miners. However, miners were usually exposed to higher levels of radon than members of the public so that the risk has to be extrapolated from occupational levels to typical indoor concentrations. Some houses, though, are at levels above those permitted in mines. In the Reading Prong, residents in about 10 percent of the homes measured could receive exposures similar to underground miners. Also, we have no real information from studies of miners about the effects on children or women since the studies have included adult males only.

The primary cause of lung cancer in this country is smoking. In 1981, there were 80,000 to 90,000 lung cancer deaths attributed to smoking and about 10,000 to 20,000 deaths potentially due to radon. The interaction of tobacco smoke and radon is difficult to determine from mining studies, but the risks appear to be additive. However, in a particular community, increases in lung cancer due to radon will be masked by the large number of deaths caused by smoking.

There are no federal legal limits on the level of indoor radon from natural sources so that remediation becomes a matter of choice for the homeowner, hopefully an informed choice. A guidance booklet for members of the public which explains the health risks from radon is being developed by the EPA and CDC with input from other federal agencies and the states, but that document is not yet available to the public. In some cases, the national and local media are the primary source of information for members of the public.

Indoor radon poses unique problems for both public health and for environmental protection agencies. An effective program for reducing exposure to the public requires interaction and cooperation between these two groups on a state and federal level. There has already been an encouraging level of communication between agencies on this problem. We should continue to improve that interaction so that all of the agencies can actively deal with the radon problem.

COMMENT ON THE HEALTH PRESENTATIONS

Judith B. Klotz*

From the foregoing papers, some important themes emerge which are particularly pertinent to the seriousness of the radon problem and the remaining informational needs for directing our efforts most effectively to prevent lung cancer from radon exposure.

First, the epidemiological data from the underground mining studies reviewed today are extremely convincing in their consistency, strength of their association, dose-response relationship, specificity of effect, and biological plausibility, i.e., they satisfy all of the classic epidemiological principles for demonstrating causation.

Second, some dwellings in this region have been shown to have radon progeny levels which result in cumulative exposure to the residents surpassing many of the occupational exposures which have been unequivocally associated with excess lung cancer risk.

However, in light of the current paucity of data on excess lung cancer associated with radon in the residential setting, the major unresolved issues of extrapolation of occupational to residential populations take on particular importance.

Among the themes which were explicitly or implicitly described in this session were:

The risks for population subgroups such as children, the interaction of smoking with radon exposure, the effect of breathing rate of miners engaged in strenuous work, possible contribution of other environmental factors in mining to the lung cancer risk (which could result in different risk in residents), and possible departure from linearity of the dose-response relationship at very high or very low doses.

The data presented by Howe illustrate that for practical purposes there does not appear to be any threshold of exposure below which there is no increase in risk, but there may be cumulative exposures above which the increase in risk per unit dose begins to plateau. The suggestions emerging from new data analysis that cessation of exposure may result in decreased risk is both important and exciting;

---

*Judith B. Klotz is Coordinator of Environmental Epidemiology, New Jersey Department of Health.

it implies greater benefit from identifying and remediating high residential exposures.

The conclusions that we can draw from the existing information seem clear, despite the remaining issues. Even if the actual risks of lung cancer to the general population were a hundred-fold less than simple application of risk coefficients derived from hard-rock miners, those estimated risks are enormous compared to those which are generally used to set guidelines and standards for exposure to environmental contaminants (even considering the protective assumptions which we appropriately use in generating these estimates). Therefore, the energetic programs which state and federal agencies have designed to identify and mitigate elevated radon exposures are highly appropriate. These include (1) the discovery and remediation of the most highly exposed dwellings and (2) the decrease of average radon levels in new buildings.

HIGHLIGHTS FROM THE DISCUSSION: RADON AND HEALTH

CHILDREN AND RADON

J. Klotz: I want to ask the panel to deal with a set of specific and related questions. Whenever we talk about the risk coefficients, the easiest way for us to deal with them is to consider some kind of excess of relative risk or excess of attributable risk per working level month. But when we do that, we are making an assumption that the excess--whichever model we are using--is linear along the entire length of exposure, from the average that would be accrued in this country to the extremes that mining populations used to be exposed to and that some residents are as well. This raises several questions about the possibility of departures from linearity. First, what are the effects on people who have been exposed to extremely high exposures over a long period of time? Second, what are the implications for the elderly? Given the latency period for cancer development and the possible reduction of risk if exposure is reduced, there may be little efficacy for the elderly on fixed incomes to spend their time and money on doing remediation to their homes. Third, how might exceptions to linearity apply to children?

N. Harley: In the modeling, of course, the exponential correction diminishes the estimation of risk at childhood. There is an odd effect in the dosimetry that for children about age 10, the actual radiation dose received per unit exposure is somewhat higher by about a factor of 2 than the adult. The effect of this on the overall lifetime risk is not clear. The only data set that is in existence is a study in China where children were exposed in tin mines at early ages--7 to 10. Because of their size they were sent down into these very small mine tunnels. Radon exposure levels in the mines were quite high up in the many working level values. Since these exposures took place in the 1940's, the children are now attaining the age -- 40 or 50 -- when you would expect lung cancers. However, the preliminary data coming out of China does not indicate that there is any higher risk.

G. Howe: With respect to the question of the age effect, as Dr. Harley has just pointed out, in fact there is virtually no data, other than possibly the Chinese data, directly on the effect of radon daughter exposures in the young. However, if we are willing to accept the analogy with what is called low LET--Linear Energy Transfer--radiation, i.e., that alpha irradiation has the same type of mechanism as X- or Gamma irradiation in the way that it impacts the various stages of carcinogenesis, then we can use findings from the atomic bomb survivors which are consistent with an increasing effect with increasing age of exposure. However, while we don't have any evidence on which to suggest that children are particularly prone to be high risk, I would be loathe to say that children are at a lower risk.

SMOKING AND RADON

J. Klotz: Dr. Howe, I am interested in your statement that smoking seems to be additive. What is the evidence for that conclusion?

G. Howe: We don't as yet have smoking data in our study. In fact, I am in the process of following up those individuals who have died of lung cancer and a random control group to ascertain smoking histories from next of kin. I was referring primarily to the Swedish data which seems to suggest that the pure relative risk model might be inappropriate and certainly additivity seems to explain their data reasonably well. This is consistent with the concept of a relative risk model applying and then there being a decay in the relative risk after exposure ceased. There is also currently a joint analysis going on of all the data sets, with the exception of the Czech data, which shows a decrease in the relative risk following the cessation of exposure. So, in fact, the point Dr. Harley made from the U.S. data is supported by the other data sets too.

J. Klotz: This has very far-reaching implications for risk management and for inhabitants of houses with high levels. The data indicate that not only are they not increasing their risk by avoiding further exposure by remediation, but they are also possibly decreasing the risk that they have already accrued. This is, of course, a model that does work for smoking although we don't get back to the baseline. At least in the New Jersey data, we certainly have seen a decrease of risk after a certain number of years of cessation of smoking exposure. I'm very glad this is probably going to be shown to be true as well for radon exposure. I think it is going to make a tremendous amount of difference in the hope that we can give people as to the efficacy of what they can do from the time of discovery onward.

N. Harley: There is a recent article in the journal of the National Cancer Institute within the last two months on the effect of stopping smoking. This is a study of Jonathan Samet's in New Mexico. Apparently the exponential decrease is in vogue, because if you stop smoking, their study showed that the lung cancer mortality risk decreased with a 5-year halftime.

THE FEDERAL GUIDANCE OF 4 PCI/L (0.02 WL)

T. Sall: Could any of you tell us how the recommended guidance of 4 pCi/l came about? How was that established?

C. Eheman: I've said about as much as I know on the subject. Based on talking with EPA, and in looking at the Federal Register where they made that decision, it seems that they tried to estimate how much money you would have to put in to a house that was close to 0.02 WL (4 pCi/l) to significantly lower the radon levels, and they found that it was a lot. For example, if you had a house at 1 WL (200 pCi/l), you could put in X dollars to get it to almost background. Even at 0.02 WL you could put 2-times X dollars and lower it to almost background. So the level was based on the technology they had at that time. If that technology changes then maybe that level will change. I don't want to give too much emphasis to that level, because there is not a big difference between 0.02 and 0.03 or 0.19 WL. While it is a legal limit for phosphate lands and for uranium mill tailings, for the rest of us it's a recommendation, just a point of reference that we can use to try to get down below. Or, should we get significantly above that level, maybe we should do something.

R. Michaels: Christie, why isn't it the role of radon mitigation companies and not the EPA or CDC to make mitigation recommendations like that based upon cost effectiveness? Why not just deal with the health issue and set a level?

C. Eheman: Because if you set a level for radon like you did for chemical exposures, the level would be below background. The reasons EPA first got involved in this in terms of remedial efforts is because at mill tailing sites and for phosphate mining lands they were responsible for figuring out how to do the remediation as well as doing the risk assessment, so they were forced to put those two together to come up with some reasonable number.

T. Sall: This level (4 pCi/l)would yield a possible death rate of approximately one percent, is that not so, one out of a hundred?

C. Eheman: Yes, over a lifetime.

T. Sall: Using your rationale, a pharmaceutical company might say, "Gee we have to spend X number of millions of dollars to insure that an implantable device is sterile (that is, would not be sterile out of one chance in a million), but we could save a million dollars if it was only one chance out of a hundred." Isn't there some fallacy in that argument?

C. Eheman: You can look at it two different ways. Is it reasonable to spend 93 million dollars (which I've seen at some sites) to prevent a risk of dying for an individual of one in a million, and then look at the other hand and say, well, one in two hundred or one in a hundred is too high a risk? I don't think that anybody is arguing that 0.02 WL is an acceptable risk unless they believe in an exposure threshold. I think the argument is maybe in terms of practicality. That's as low as we can go. I don't think EPA would tell anybody below 0.02 WL (4 pCi/l) not to remediate. I think the point is that they found as they were doing the remediation that they were putting in a lot of money to save a very small risk, and also that the largest risk to anyone who smokes is smoking, unless you're in the 0.1 percent of the houses that are above occupational levels. There are different ways of looking at it, and I don't think anybody is trying to say 0.02 WL is safe, or good, or anything like that. But to get to one in a million, you'd have to lower background, and that's a pretty difficult thing to do in millions and millions of houses.

H. Sachs: First I think that the superfund reference made by Christie Eheman is inexact. Superfund addresses both public health and environmental protection. So it may not be fair to charge a superfund expenditure of 93 million dollars on a one-in-a-million risk to public health alone. Of course, protecting the environment may pay off in long term protection of human health.

That aside, I wanted to raise some major reservations about what we have been told this morning about how the EPA set 0.02 WL (4 pCi/l) as a limit. We've been told that twice now, and I think it's very important that its meaning be put into a metaphor which is perhaps more accessible to us. We are told that this level has been set on the basis of the EPA's estimates from work in a research environment of the incremental cost of risk reduction at each level. It's not based on an evaluation of the public health threat. Even if you succeed at going to 0.02 WL (4 pCi/l), which is where my own house is, you still have a one-in-a-hundred chance of dying of lung cancer, or at least of contracting lung cancer. That is still a very high risk by public health standards. The point I want to make is that this guidance wasn't set from a risk estimation standpoint, but from an incremental cost standpoint. I think we would be dubious if we were told today that based on some preliminary research findings in a couple of experimental reactors and some extrapolations from a few consultants, we think that the risks involved in running a reactor are fairly small, and consequently at those risks it's not cost effective or societally beneficial to add containment structures or remove the reactors from populated areas. However, in the radon case, this is exactly analogous to a set of calculations based on research findings without significant field experience, the kind of experience you get from doing a few thousand houses in least cost-mitigation strategies.

If Bernie Cohen were here he'd be tearing his hair because he's very concerned about consistency in risk assessments for radioactivity. We're using a very different set of requirements for risk estimation in superfund sites, reactor sites, and in things that involve our own houses. The only way you can logically interpret that is to assume that radiation from natural causes is obviously less dangerous than radiation from superfund sites because we're willing to spend so much less as a society to take care of it. I understand the public policy constraints the EPA was under. But it leaves us with a logically inconsistent set of standards which to me, based as it is on such thin mitigation evidence of what actual cost will be, is hardly a way to make public health policy.

C. Eheman: I agree with just about everything you said. When I talked about superfund sites and money spent, I was not implying that it was unacceptable. I was just trying to give different views. Even if there were no deaths from a superfund site, we are obligated to clean up what we messed up. I hope that no one has gotten the idea that 0.02 WL (4 pCi/l) was CDC's doing. I'm not here to defend EPA, but I agree--that their decision was not based on acceptable risk. If my house were at 0.02 WL or even a little below that, and I had some easy way to fix it, fixing some holes in the basement, or ventilating the crawl space, I would do it. I hope that the message is not going out that 0.02 WL is safe or that CDC or EPA don't want people to remediate below that level. If that idea has gone out in any form, I regret that. Most of the deaths are at levels below 0.02 WL. We have to think of indoor air quality in general and really start concentrating, as members of the public, on the quality of air inside our homes.

J. Klotz: Although EPA may have been one of the agencies that has been setting guidances for a long time, it is not the only one. To my knowledge there is not a single guidance below 0.02 WL (4 pCi/l) for remediation any place in the world, and there have been a number of countries that have set guidances above that. New Jersey has adopted, for the time being, the 0.02 WL guidance, which Pennsylvania previously adopted, and which CDC has promoted in the past. Again, we are not doing it implying safety. We are doing it as a logical place for us to encourage people to remediate down to. The state is going to be participating in that process. When we talk about added environmental risks that have been caused by either negligence or ignorant actions by humans, I think we have to separate those we can avoid from those which happen to be there because this is the planet we are living on. We can't do anything about cosmic rays and nobody talks about not flying in planes because that may increase radiation doses. There are certain risks that are inherent in being a human being or an animal on this planet. Unavoidable risks that are natural may have to at times be on a slightly different scale from avoidable risks which are put there by human beings by negligent or ignorant activity and which can be regulated against in a prospective sense. On that same note, some countries that are dealing with radon now are looking at two different criteria for acceptability of exposure. One is in new housing where new construction codes can be geared to preventing radon accumulation in houses, and the other is in old houses where the current residents would have to spend a great deal of time and money to correct them. One can argue whether or not that is appropriate, but certainly we are talking about two different risk levels under those circumstances.

G. Howe: I just had one related point not directed at the general population environmental exposure, but going back to working standards. To keep this in perspective, one could point out, in fact, that even in uranium miners, within the last two decades, more deaths (excess deaths) were attributable to accidents, cirrhosis, suicide, and homicide than were ever induced by lung cancer. I'm not using this as an argument that we shouldn't extrapolate from mines, I'm simply pointing out that one has to keep a perspective on all this.

N. Harley: There is also the other side of the coin from other countries and their guidelines. For example, I understand Canada has just suggested 0.1 WL (20 pCi/l) as a guideline. Sweden is even higher. In the United States, NCRP, for example, suggested 2 working level months per year which translates to 0.04 WL (8 pCi/l). Meanwhile, EPA steps back and has taken a non-regulatory stance. People use the 0.02 WL (4 pCi/l) from the mill tailings cases, but it is not really any sort of official EPA guideline for environmental situations. I agree with what Christie said about risk because the guideline means that you fix your home and spend money. But the people up at these high levels are few, so you're not saving people from lung cancer when you work with the high end of the distribution. To save people from lung cancer you really need to work with the average level. That is the public health issue, and the guideline is a monetary issue--how much money you are willing to spend.

R. Michaels: It is my personal opinion that public health protection agencies should stick to public health protection, and I think that the message is being delivered that levels at the 0.02 WL or 4 pCi/l level are acceptable from a public health standpoint. That is the message that I got in attending two conferences, this one and a previous one. I think that the agencies should not abandon their societal responsibility to set forth levels that are regarded as acceptable and leave it to the venture capitalists to implement technologies that will become capable of realizing those levels. If the message gets out that levels which we now regard as minimum are also acceptable from a public health standpoint, this will inhibit expenditures of venture capital money, government research dollars and other sources of innovation which could change our perception of what is reality for tomorrow.

STUDIES OF NEW JERSEY RADIUM SITES

Audience: Any particular comment on the health study of the people in Montclair, West Orange and Glen Ridge who lived or are currently living in homes built on the radium waste sites?

N. Harley: I have heard that study presented, but I don't recollect the details offhand. There is a problem with studying small groups with poor exposure history. One would not expect to find any effects, and none are really found. Is that correct?

J. Klotz: To my knowledge, that study is still underway, and no data results that I know of have been released. We are still in the process of finding the people who lived there and following them as far as their vital status and cause of death.

REPRODUCTIVE RISK AND RADON

Audience: Are there any health studies or any comments about the reproductive risks of individuals who have been exposed to radon for 30 years or so?

N. Harley: To my knowledge there is no other effect than lung cancer from radon daughter exposure with the exception of the gastric cancer data that Dr. Howe presented. No one has ever shown any excess tumor or genetic effects.

G. Howe: In fact if we go by analogy with the actual radiation data, I don't think anybody expects, by extrapolation, a measurable genetic effect.

N. Harley: It seems to revert always back to the dosimetry. Given the short-lived daughters, the lung is the target and material cleared from the lung goes to the stomach, perhaps implicating other substances in the mining environment. Doses to other organs would be really very, very small. There are some published data on this, by the way.

CANCER PREVENTION

Audience: In a recent newspaper article, it was stated that "the U.S. is losing its fight on cancer." Yet according to Dr. Garfinkel of the American Cancer Society, "If you prevent lung cancer, you have a 13 percent decrease in cancer deaths instead of an 8 percent increase." The point is that cancer prevention is more promising than cancer treatment. Shouldn't this be the cause for prompt attention to the radon crisis?

C. Eheman: Certainly, I would agree that prevention is better than trying to treat a cancer, both on an individual level as well as the cost to the population. Certainly, at CDC prevention is the reason we are trying to address the radon issue. In terms of public health, you try to think in numbers of cancers prevented. Unfortunately, the vast majority of deaths are not at 1 WL (200 pCi/l) or 10 WL's (2000 pCi/l) or even at 0.1 WL (20 pCi/l). Statistically, the largest number of deaths are around 0.02 WL (4 pCi/l) or below. In terms of public health, that has some important implications. I wouldn't desert people above 1 WL because there aren't a large number of them dying. I think that those levels have to be addressed immediately, and that is where most of the attention has been focused. In the long term, however, I think we are going to have to think about indoor air quality in general. Maybe we should make some changes in the way we build our homes to lower background levels of radon and levels of formaldehyde and other materials in the indoor environment? I think that is going to be the answer in the long term.

AIR FILTRATION

Audience: What are your views on hepa-filtration with regards to alleviating the radon problem in a home?

N. Harley: I've been very interested in air cleaning as a means for reducing radon daughter exposure, and I had recently done some dosimetry looking at the effects of air cleaning. People talk about the famous unattached fraction of radon daughters; that is, the Polonium-218 or the Radium A that stays in the atmosphere and really does not combine with the ambient aerosol. This is a small fraction of the total activity, but it deposits in the upper airways very efficiently--100 percent deposition. So the question has always been, if you clean up the air you have a larger fraction of this unattached daughter, and perhaps the dose is actually higher. In looking at some preliminary calculations, it turned out that the dose per unit exposure (if you make certain assumptions about this unattached fraction), is about twice what you have if you don't clean the air. That means that if you clean the air of radon daughters by more than a factor of 2, you have an overall net gain. So, I'm sort of in favor of air cleaning. I know many groups such as Lawrence Berkeley Laboratory are working on air cleaning as a method. Also, because of the gastric cancers in mines, for example, other substances might be implicated. I think air cleaning in mines should be a very sound way to go.

Audience: Whether you use electron precipitation or hepa-filter methods, does it make any difference as long as the air is clean?

N. Harley: No.

Section 6

Perception of Risk and Psychosocial Impacts of Radon Exposure

INTRODUCTION AND SUMMARY OF SECTION 6: PERCEPTION OF RISK AND PSYCHOSOCIAL IMPACTS OF RADON EXPOSURE

Michael R. Edelstein*

In planning the conference Radon and the Environment, I asked Richard Guimond of the EPA what we might include that would address an area about which little was currently understood regarding radon. Without hesitation, he responded that the real challenge was to learn how to get people to respond effectively to the threat of radon without panicking. In numerous other conversations with officials addressing the radon issue, this communications theme repeatedly emerged.

The question of risk communcation is of particular importance in responding to radon gas exposure, as Caron Chess, session chairperson, noted in her introduction to the conference session and various authors here reiterate. This importance stems from the very scope of the radon problem and the fact that it materializes within private dwellings. As a result, a basic assumption made by government is that the homeowner, rather than government, must take the primary action to identify and correct the threat. The homeowners' perception of the risks, therefore, becomes central to the determination of whether they will act effectively to do this.

A second area of inquiry addressed in the session stems from the area of research spearheaded by Margaret Gibbs and myself into the psychosocial impacts of toxic exposure. In numerous communities across the United States, we have had the opportunity to examine the consequences for individuals, families and neighborhoods of the discovery of either some toxic contaminant or the proposal to site a facility that might in the future pollute the area. While Dr. Gibbs has documented the psychological damage from stress in such situations, I have profiled the resulting changes in the ways that people live and think as they attempt to cope with the realization that they and their family have been exposed to an environmental threat. Accordingly, any response to the radon issue should proceed with an awareness of what it means to discover that one's home has "high" radon levels, the resulting adjustments in lifestyle and the impacts on how one views the home, the family and the future.

---

*Michael Edelstein is Associate Professor of Psychology, Ramapo College, and a conference co-organizer.

The first paper in this section, "Rating the Risks: The Structure of Expert and Lay Perceptions" by Paul Slovic and his colleagues Baruch Fischhoff and Sarah Lichtenstein, provides a basic overview of the growing literature on risk perception, to which they are prime contributors. The paper amplifies the points made by Dr. Slovic in his opening presentation during the conference session. Slovic and his colleagues present an argument based upon two foundations: that there are cognitive limitations in our ability to comprehend complex issues and that we have a psychological need for predictability in our lives. The result of both these characteristics is that we easily deny uncertainty, distort risks and become overconfident in certain statements of fact. We deal with complexity by using various "heuristics" or judgmental biases which serve as shortcuts in our thinking. These cause us frequently to misperceive risks. For example, the "availability heuristic" involves our using whatever images we recall in judging a risk. We may apply an inappropriate image which exaggerates a risk needlessly; if no image is available to us, we may overlook the risk altogether. Interestingly, experts engage in heuristics much as do lay people. The major difference between lay and expert perceptions of risk appears to be the use of qualitative risk criteria by lay people which is absent from expert projections. Two factors summarize these qualitative criteria: the extent to which a risk is "unknown" (new, poorly understood, involuntary and delayed) and "dreaded" (it will lead to catastrophic outcomes for large numbers of people). Given these basic rules about risk perception, it is not surprising that controversy and disagreement often occur in determining what level of risk is acceptable.

While this overview of risk perception was not directed to radon, per se, Slovic is currently collaborating on a study of radon funded by the New Jersey Department of Environmental Protection. His two colleagues in this work, Neil Weinstein and Peter Sandman, have both contributed papers based upon their conference presentations which make a direct link between risk perception and the radon issue.

Weinstein, in "Response of the New Jersey Public to the Risk from Radon," and Sandman, in "Communicating Radon Risk: Alerting the Apathetic and Reassuring the Hysterical," independently argue in their contributions that panic over radon is not evident. In fact, their major concern is that the public is apathetic to the radon risk. This is not because the public is uninformed about radon, as Weinstein reports, but rather because many of the things that they believe about radon cause them to underestimate its risk. Sandman argues that the reason for this underestimation can be found in the characteristics of radon: the fact that radon involves no enemy to blame, is naturally occurring, is widely dispersed, is remediable, is an individual problem under individual control, is invisible, affects a secure place -- the home, invokes no mental image, is a complication for home sellers, and it is associated with a long latency disease.

Citing evidence from their work, they show that only 18 percent of the New Jersey public have tested or plan to. Half have never even thought about it. The intention to test appears to be related to factors such as the presence of radon in the neighborhood and the level of concern among the neighbors. This raises the question of how government should communicate about the issue in order to increase the active concerns of citizens.

In her contribution to this section, "Considering the Public's Perception of Risk in the Development of a Radon Communications Program - The Case of the New Jersey Department of Environmental Protection," Caron Chess of the New Jersey Department of Environmental Protection describes that agency's novel approach to designing a risk communication strategy for radon. Using Sandman and Weinstein as consultants, the DEP convened a symposium of risk experts to devise a plan. The communication goals elucidated by this panel stressed the need to provide the public with an accurate understanding and healthy degree of concern. This would stem from an active public campaign on the part of the state to fill the information vacuum and establish and keep trust and credibility. Of special importance were attempts to help the homeowner feel in control of the situation and to make certain that homeowner concerns are addressed. This outcome could only be achieved, the panel argued, if communications are carefully planned.

A major contributor to the public's radon understanding of course, is the press. An analysis of newspaper articles presented by Edelstein and Boyle in "Media and the Perception of Radon Risk," reveals basic inconsistency, contradiction and inaccuracy in dealing with the radon issue. Furthermore, when college undergraduates were presented one of two articles varied according to the level of threat attributed to radon, it became clear that threat perception can be strongly influenced by media accounts. Beyond the overall perception of risk, the media can influence specific beliefs which may be inaccurate. As Weinstein also reported, these beliefs, once ingrained in the public mind, are hard to erase. The clearest example involves the "Myth of the Reading Prong," the continually reinforced but inaccurate belief that the radon problem in Pennsylvania, New Jersey and New York is confined to the geologic formation which now publicly is associated with that label.

A further exploration of psychological issues associated with radon is found in Margaret Gibbs' collaboration with Edelstein and Susan Belford, "Psychological Impacts of Radon Gas Exposure: Preliminary Findings." They have been conducting a study which attempts to identify the characteristics of people who undertake radon testing as contrasted with nontesters, and which compares people in perceived high risk locations (the Reading Prong) with those in areas perceived to be safer. The study also seeks to achieve pre-post comparisons, rare in the study of disasters, by contrasting perceptions of radon by those testing their homes before and then after their results are received. Based upon preliminary data, the authors conclude that their comparison groups have a similar concern with environmental hazards in general, but the group which elected to test is more concerned with radon specifically. Thus, general environmental concern may not be predictive of testing for a specific hazard. Why was the test group more concerned with radon? It would appear that all respondents were heavily influenced by media coverage in their views of radon. Testers saw themselves as more vulnerable to harm and thus took action.

Very different from the public's perception of radon is the perception of families who have already discovered severe radon levels in their homes. This is underscored by "Impacts of Living in High Level Radon Homes," the contribution from Kay Jones and Kathy Varady, the leaders of the organization "People Against Radon" (PAR). As residents of the Boyertown area in Pennsylvania and neighbors of Stanley Watras, whose residential radon level launched the radon issue, Jones and Varady were swept up in the response to this identified "hot spot." Their own families were subject to levels of radon more than 100 times the EPA guideline for exposure. They spoke in moving terms of living in a neighborhood where the equivalent of a Three Mile Island occurs every week without the same level of government assistance. Kathy Varady discussed her anger as a parent in discovering that her four childrens' bedrooms contained 0.3 Working Levels of radon (60 pCi/l), a level at which OSHA would shut down a uranium mine. "It was okay for my children to be there," she noted, "just don't hire anyone to work there because there are unacceptable occupational levels!"

Both Jones and Varady had their homes selected for the EPA's remedial demonstration project, subjecting their families to a lengthy process of experimentation and disruption. They discuss the weight placed upon them by the concern for their families' health and the particular devastation for their husbands because of their perceived failure to protect their families. The invasion of their homes, first by radon, then by government officials and contractors and finally by the radon issue, has fundamentally changed their views of life. Becoming activists, they organized PAR and have placed themselves at the center of radon controversy and policy formulation ever since. From their perspective as victims, they have come to question heavily the government response to the radon issue.

The discussion following the conference presentations established a number of additional points about the radon issue. Acting as a responder, Edelstein noted the importance of distinguishing between radon victims and those "merely flirting with the possibility of exposure." The latter are coming to grips with accepting that the radon risk is real while the former are coping with a risk that is no longer abstract but highly personal. It is hard for non-victims as "outsiders" to understand what those affected by high levels may be going through. This difference may influence how risk communications are received, with there being a distinction between hearing information that may not apply to you and hearing information that helps to define your degree of risk from an existing condition. These differences also help to explain why experts are likely to view radon differently than victims. Edelstein also pointed out the effect of public disinformation on the belief systems of people trying to decide whether to take the radon threat seriously.

Edelstein noted the paucity of organized public concern about radon since the ease of testing has satisfied the desire of those concerned about radon to define their threat. The lag in the availability of mitigation, however, is likely to intensify concerns among those who discover significant exposures. These people will serve as the constituency for an organized public reaction to radon. The radon issue is too diffuse and universal to create the conditions for community organizing, except in cases such as Boyertown. There the problem was bounded and localized, and the lag in government assistance intensified concern within the immediate community. These are the conditions most likely to produce organized community response.

Responding to the frequent analogies of radon levels to cigarette smoking, Sandman further noted the possibility of distortion involved in comparing risks from different sources. There was also discussion of the effects of people's expectations about the ease of mitigation. Sandman argued for the importance of addressing in risk communication three different groups: those not having as yet tested, those finding low but significant radon levels, and those facing major exposures. He received support from fellow panelists for his goal of encouraging universal testing in northern New Jersey and other areas where radon is known to exist.

RATING THE RISKS: THE STRUCTURE OF EXPERT AND LAY PERCEPTIONS*

Paul Slovic,** Baruch Fischhoff, and Sarah Lichtenstein

People respond to the hazards they perceive. If their perceptions are faulty, efforts at public and environmental protection are likely to be misdirected. In order to improve hazard management, a risk assessment industry has developed over the last decade which combines the efforts of physical, biological, and social scientists in an attempt to identify hazards and measure the frequency and magnitude of their consequences.[1]

For some hazards extensive statistical data are readily available; for example, the frequency and severity of motor vehicle accidents are well documented. For other familiar activities, such as the use of alcohol and tobacco, the hazardous effects are less readily discernible and their assessment requires complex epidemiological and experimental studies. But in either case, the hard facts go only so far and then human judgment is needed to interpret the findings and determine their relevance for the future.

Other hazards, such as those associated with recombinant DNA research or nuclear power, are so new that risk assessment must be based on theoretical analyses such as fault trees (see Figure 6.1), rather than on direct experience. While sophisticated, these analyses, too, include a large component of human judgment. Someone, relying on educated intuition, must determine the structure of the problem, the consequences to be considered, and the importance of the various branches of the fault tree.

Once performed, the analyses must be communicated to the various people, including industrialists, environmentalists, regulators, legislators, and voters, who are actually responsible for dealing with the hazards. If these people do not see, understand, or believe these risk

---

*Reprinted by permission from Environment, Vol. 21 (April, 1979).

**Paul Slovic presented an overview of risk perception as part of the session on perception of risk and psychosocial impact due to radon. He is a principal of the firm Decision Research, located in Eugene, Oregon. This paper is a an overview of risk perception work written by Slovic and his colleagues.

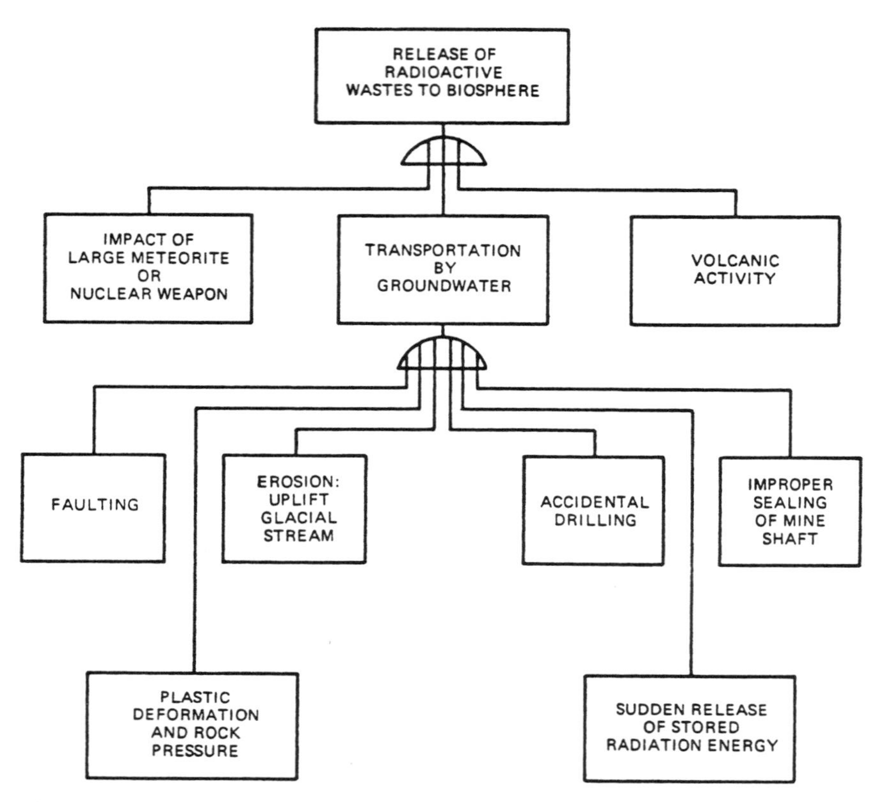

Figure 6.1 - Illustration of a Fault Tree. Fault trees are used most often to characterize hazards for which direct experience is not available. The tree shown here indicates the various ways in which radioactive material might accidentally be released from nuclear wastes buried within a salt deposit. To read this tree, start with the bottom row of possible initiating events, each of which can lead to the transportation of radioactivity by groundwater. This transport can in turn release radioactivity to the biosphere. As indicated by the second level of boxes, release of radioactivity can also be produced directly (without the help of groundwater) through the impact of a large meteorite, a nuclear weapon, or a volcanic eruption. Fault trees may be used to map all relevant possibilities and to determine the probability of the final outcome. To accomplish this latter goal, the probabilities of all component stages, as well as their logical connections, must be completely specified. (Source: P.E. McGrath, "Radioactive Waste Management," Report EURFNR 1204, Karlsruhe, Germany, 1974.) Reprinted by permission.

statistics, then distrust, conflict, and ineffective hazard management may result.

JUDGMENTAL BIASES

When lay people are asked to evaluate risks, they seldom have statistical evidence on hand. In most cases they must rely on inferences based on what they remember hearing or observing about the risk in question. Recent psychological research has identified a number of general inferential rules that people seem to use in such situations.[2] These judgmental rules, known technically as heuristics, are employed to reduce difficult mental tasks to simpler ones. Although valid in some circumstances, in others they can lead to large and persistent biases with serious implications for risk assessment.

Availability

One heuristic that has special relevance for risk perception is known as "availability."[3] People who use this heuristic judge an event as likely or frequent if instances of it are easy to imagine or recall. Frequently occurring events are generally easier to imagine and recall than rare events. Thus, availability is often an appropriate cue. Availability, however, is also affected by numerous factors unrelated to frequency of occurrence. For example, a recent disaster or a vivid film such as The China Syndrome may seriously distort risk judgments.

Availability-induced errors are illustrated by several recent studies in which we asked college students and members of the League of Women Voters to judge the frequency of various causes of death, such as smallpox, tornadoes, and heart disease.[4] In one study, these people were told the annual death toll (50,000) for motor vehicle accidents in the United States; they were then asked to estimate the frequency of forty other causes of death. In another study, participants were given two causes of death and asked to judge which of the two is more frequent. Both studies showed people's judgments to be moderately accurate in a global sense; that is, people usually knew which were the most and least frequent lethal events. Within this global picture, however, there was evidence that people made serious misjudgments, many of which seemed to reflect availability bias.

Figure 6.2 compares the judged number of deaths per year with the actual number according to public health statistics. If the frequency judgments were accurate, they would equal the actual death rates, and all data points would fall on the straight line making a 45-degree angle with the axes of the graph. In fact, the points are scattered about a curved line that sometimes lies above and sometimes below the line of accurate judgment. In general, rare causes of death were overestimated and common causes of

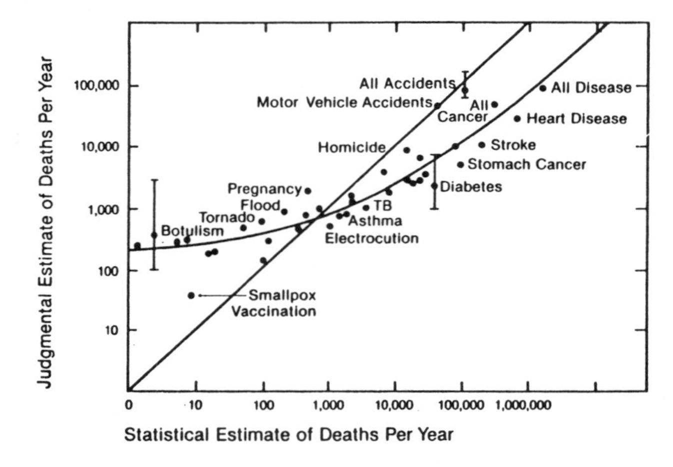

Figure 6.2 - Relationship between Judged Frequency of Death and the Actual Number of Deaths Per Year for 41 Causes of Death. If judged and actual frequencies were equal, the data would fall on the straight line. The points, and the curved line fitted to them, represent the averaged responses of a large number of lay people. While people were approximately accurate, their judgments were systematically distorted. As described in the text, both the compression of the scale and the scatter of the results indicate this. To give an idea of the degree of agreement among subjects, vertical bars are drawn to depict the 25th and 75th percentile of individual judgment for botulism, diabetes, and all accidents. Fifty percent of all judgments fall between these limits. The range of responses for the other 37 causes of death was similar. From Lichtenstein et al.[4] Copyright 1978 by the American Psychological Association. Reprinted by permission.

death were underestimated. As a result, while the actual death toll varied over a range of one million, average frequency judgments varied over a range of only a thousand.

In addition to this general bias, many important specific biases were evident. For example, accidents were judged to cause as many deaths as diseases, whereas diseases actually take about fifteen times as many lives. Homicides were incorrectly judged to be more frequent than diabetes and stomach cancer. Homicides were also judged to be about as frequent as stroke, although the latter actually claims about eleven times as many lives. Frequencies of death from botulism, tornadoes, and pregnancy (including childbirth and abortion) were also greatly overestimated.

Table 6.1 lists the lethal events whose frequencies were most poorly judged in our studies.

Most Overestimated	Most Underestimated
All accidents	Smallpox vaccinations
Motor vehicle accidents	Diabetes
Pregnancy, childbirth, and abortion	Stomach cancer
Tornadoes	Lightning
Flood	Stroke
Botulism	Tuberculosis
All cancer	Asthma
Fire and flames	Emphysema
Venomous bite or sting	
Homocide	

Table 6.1 Bias in Judged Frequency of Death

In keeping with availability considerations, overestimated items were dramatic and sensational, whereas underestimated items tended to be unspectacular events which claim one victim at a time and are common in nonfatal form.

In the public arena the availability heuristic may have many effects. For example, the biasing effects of memorability and imaginability may pose a barrier to open, objective discussions of risk. Consider an engineer's demonstrating the safety of subterranean nuclear waste disposal by pointing out the improbability of each branch of the fault tree in Figure 6.1 Rather than reassuring the audience, the presentation might lead individuals to feel that "I didn't realize there were so many things that could go wrong." The very discussion of any low-probability hazard may increase the judged probability of that hazard regardless of what the evidence indicates.

In some situations, failure to appreciate the limits of "available" data may lull people into complacency. For example, we asked people to evaluate the completeness of a fault tree showing the problems that could cause a car not to start when the ignition was turned.[5] Respondents' judgments of completeness were about the same when looking at the full tree as when looking at a tree in which half of the causes of starting failure were deleted. In keeping with the availability heuristic, what was out of sight was also out of mind.

Overconfidence

A particularly pernicious aspect of heuristics is that people are typically very confident about judgments based on them. For example, in a follow-up to the study on causes of death, participants were asked to indicate the odds that they were correct in their judgment about which of two lethal events was more frequent.[6] Odds of 100:1 or greater were given often (25 percent of the time). However, about one out of every eight answers associated with such extreme confidence was wrong (fewer than 1 in 100 would have been wrong if the odds had been appropriate). About 30 percent of the judges gave odds greater than 50:1 to the incorrect assertion that homicides are more frequent than suicides. The psychological basis for this unwarranted certainty seems to be people's insensitivity to the tenuousness of the assumptions upon which their judgments are based (in this case, the validity of the availability heuristic). Such overconfidence is dangerous. It indicates that we often do not realize how little we know and how much additional information we need about the various problems and risks we face.

Overconfidence manifests itself in other ways as well. A typical task in estimating failure rates or other uncertain quantities is to set upper and lower bounds so that there is a 98 percent chance that the true value lies between them. Experiments with diverse groups of people making many different kinds of judgments have shown that, rather than 2 percent of true values falling outside the 98 percent confidence bounds, 20 percent to 50 percent do so.[7] People think that they can estimate such values with much greater precision than is actually the case.

Unfortunately, experts seem as prone to overconfidence as lay people. When the fault tree study described above was repeated with a group of professional automobile mechanics, they, too, were insensitive to how much had been deleted from the tree. Hynes and Vanmarcke[8] asked seven "internationally known" geotechnical engineers to predict the height of an embankment that would cause a clay foundation to fail and to specify confidence bounds around this estimate that were wide enough to have a 50 percent chance of enclosing the true failure height. None of the bounds specified by these experts actually did enclose the true failure height. The multi-million dollar Reactor Safety Study (the "Rasmussen Report"),[9] in assessing the probability of a core meltdown in a nuclear reactor, used a procedure for setting confidence bounds that has been found in experiments to produce a high degree of overconfidence. Related problems led a review committee, chaired by H.W. Lewis of the University of California, Santa Barbara, to conclude that the Reactor Safety Study greatly overestimated the precision with which the probability of a core meltdown could be assessed.[10]

Another case in point is the 1976 collapse of the Teton Dam. The Committee on Government Operations has attributed this disaster to the unwarranted confidence of engineers who were absolutely certain they had solved the many serious problems that arose during construction.[11] Indeed, in routine practice, failure probabilities are not even calculated for new dams even though about 1 in 300 fails when the reservoir is first filled. Further anecdotal evidence of overconfidence may be found in many other technical risk assessments. Some common ways in which experts may overlook or misjudge pathways to disaster include:

1. Failure to consider the ways in which human errors can affect technological systems. Example: Due to inadequate training and control room design, operators at Three Mile Island repeatedly misdiagnosed the problems of the reactor and took inappropriate actions.

2. Overconfidence in current scientific knowledge. Example: Use of DDT came into widespread and uncontrolled use before scientists had even considered the possibility of the side effects that today make it look like a mixed and irreversible blessing.

3. Insensitivity to how a technological system functions as a whole. Example: Though the respiratory risk of fossil-fueled power plants has been recognized for some time, the related effects of acid rains on ecosystems were largely missed until very recently.

4. Failure to anticipate human response to safety measures. Example: The partial protection offered by dams and levees gives people a false sense of security and promotes development of the flood plain. When a rare flood does exceed the capacity of the dam, the damage may be considerably greater than if the flood plain had been unprotected. Similarly, "better" highways, while decreasing the death toll per vehicle mile, may increase the total number of deaths because they increase the number of miles driven.

Desire for Certainty

Every technology is a gamble of sorts, and, like other gambles, its attractiveness depends on the probability and size of its possible gains and losses. Both scientific experiments and casual observation show that people have difficulty thinking about and resolving the risk/benefit conflicts even in simple gambles. One way to reduce the anxiety generated by confronting uncertainty is to deny that uncertainty. The denial resulting from this anxiety-reducing search for certainty thus represents an additional source of overconfidence. This type of denial is illustrated by the case of people faced with natural hazards, who often view their world as either perfectly safe or as predictable enough to preclude worry. Thus, some flood victims interviewed by Kates[12] flatly denied that floods could ever recur in their areas. Some thought (incorrectly) that new dams and reservoirs in the area would contain all potential floods, while others attributed previous floods to freak combinations of circumstances unlikely to recur. Denial, of course, has its limits. Many people feel that they cannot ignore the risks of nuclear power. For these people, the search for certainty is best satisfied by outlawing the risk.

Scientists and policy-makers who point out the gambles involved in societal decisions are often resented for the anxiety they provoke. Borch[13] noted how annoyed corporate managers get with consultants who give them the probabilities of possible events instead of telling them exactly what will happen. Just before a blue-ribbon panel of scientists reported that they were 95 percent certain that cyclamates do not cause cancer, Food and Drug Administration Commissioner, Alexander Schmidt, said, "I'm looking for a clean bill of health, not a wishy-washy, iffy answer on cyclamates."[14] Senator Edmund Muskie has called for "one-armed" scientists who do not respond "on the one hand, the evidence is so, but on the other hand..." when asked about the health effects of pollutants.[15] Such demands may tempt scientists to issue "certain" answers which, however convenient for regulators, are unsupportable by science.

The search for certainty is legitimate if it is done consciously, if the remaining uncertainties are acknowledged rather than ignored, and if people realize the costs. If a very high level of certainty is sought, those costs are likely to be high. Eliminating the uncertainty may mean eliminating the inevitable. Efforts to eliminate it may only alter its form. We must choose, for example, between the vicissitudes of nature on an unprotected flood plain and the less probable, but potentially more catastrophic, hazards associated with dams and levees.

ANALYZING JUDGMENTS OF RISK

In order to be of assistance in the hazard management process, a theory of perceived risk must explain people's extreme aversion to some hazards, their indifference to others, and the discrepancies between these reactions and

	Group 1 LOWV	Group 2 Coll. Stud.	Group 3 Active Club	Group 4 Experts
Nuclear power	1	1	8	20
Motor vehicles	2	5	3	1
Handguns	3	2	1	4
Smoking	4	3	4	2
Motorcycles	5	6	2	6
Alcoholic beverages	6	7	5	3
General (private) aviation	7	15	11	12
Police work	8	8	7	17
Pesticides	9	4	15	8
Surgery	10	11	9	5
Fire fighting	11	10	6	18
Large construction	12	14	13	13
Hunting	13	18	10	23
Spray cans	14	13	23	26
Mountain climbing	15	22	12	29
Bicycles	16	24	14	15
Commercial aviation	17	16	18	16
Electric power	18	19	19	9
Swimming	19	30	17	10
Contraceptives	20	9	22	11
Skiing	21	25	16	30
X-rays	22	17	24	7
High school & college football	23	26	21	27
Railroads	24	23	20	19
Food preservatives	25	12	28	14
Food coloring	26	20	30	21
Power mowers	27	28	25	28
Prescription antibiotics	28	21	26	24
Home appliances	29	27	27	22
Vaccinations	30	29	29	25

[a]The ordering is based on the geometric mean risk ratings within each group. Rank 1 represents the most risky activity or technology.

Table 6.2 - Ordering of Perceived Risk for 30 Activities and Technologies.

experts' recommendations. Why, for example, do some communities react vigorously against locating a liquid natural gas terminal in their vicinity despite the assurances of experts that it is safe? Why do other communities situated on flood plains and earthquake faults or below great dams show little concern for the experts' warnings? Such behavior is doubtless related to how people assess the quantitative characteristics of the hazards they face. The preceding discussion of judgmental processes was designed to illuminate this aspect of perceived risk. The studies reported below broaden the discussion to include more qualitative components of perceived risk. They ask, when people judge the risk inherent in a technology, are they referring only to the (possibly misjudged) number of people it could kill or also to other, more qualitative, features of the risk it entails?

Quantifying Perceived Risk

In our first studies, we asked four different groups of people to rate thirty different activities and technologies according to the present risk of death from each.[16] Three of these groups were from Eugene, Oregon; they included 30 college students, 40 members of the League of Women voters (LOWV), and 25 business and professional members of the "Active Club." The fourth group was composed of 15 persons selected nation-wide for their professional involvement in risk assessment. This "expert" group included a geographer, an environmental policy analyst, an economist, a lawyer, a biologist, a biochemist, and a government regulator of hazardous materials.

All these people were asked, for each of the thirty items, "to consider the risk of dying (across all U.S. Society as a whole) as a consequence of this activity or technology." In order to make the evaluation task easier, each activity appeared on a 3-by-5-inch card. Respondents were told first to study the items individually, thinking of all the possible ways someone might die from each (e.g., fatalities from non-nuclear electricity were to include deaths resulting from the mining of coal and other energy production activities as well as electrocution; motor vehicle fatalities were to include collisions with bicycles and pedestrians). Next, they were to order the items from least to most risky and then assign numerical risk values by giving a rating of 10 to the least risky item and making the other ratings accordingly. They were also given additional suggestions, clarifications, and encouragement to do as accurate a job as possible. For example, they were told "A rating of 200 means that the item is 20 times as risky as the least risky item, to which you assigned a 10...." They were urged to cross-check and adjust their numbers until they believed they were right.

Table 6.2 shows how the various groups ranked the relative riskiness of these 30 activities and technologies.

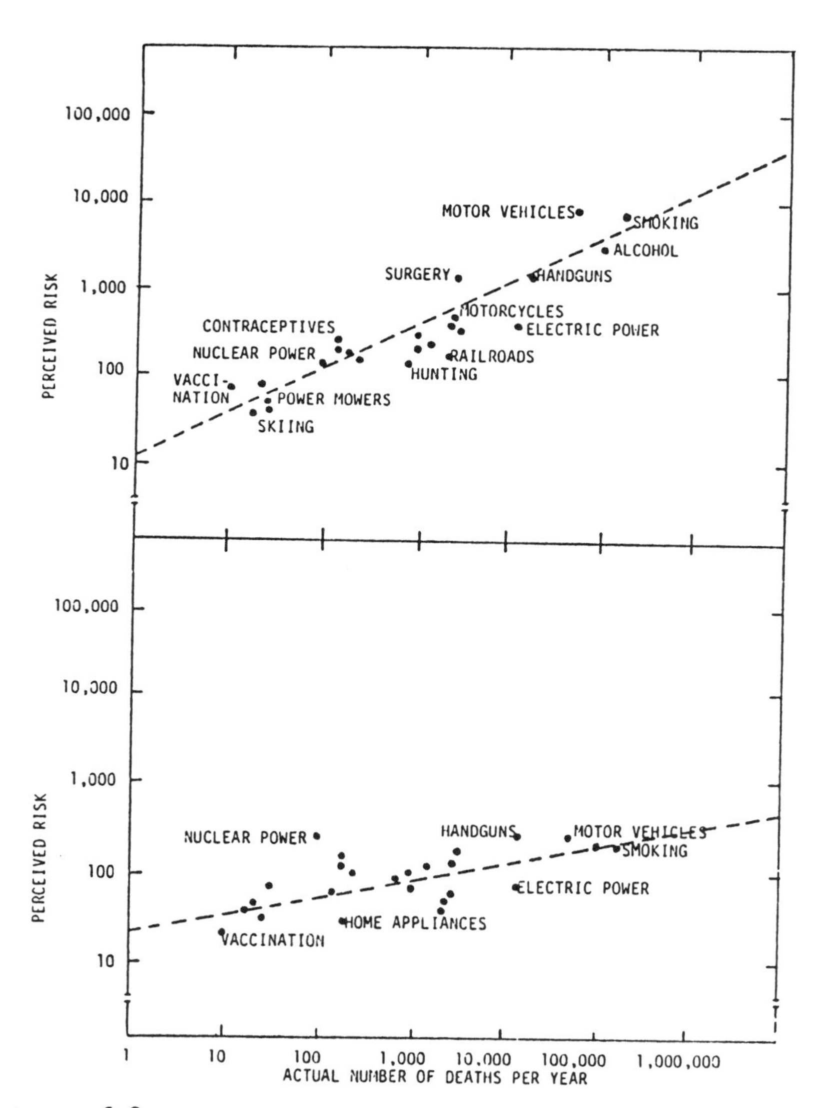

Figure 6-3 – Judgments of perceived risk for experts (top) and lay people (bottom) plotted against the best technical estimates of annual fatalities for 25 technologies and activities. Each point represents the average responses of the participants. The dashed lines are the straight lines that best fit the points. The experts' risk judgments are seen to be more closely associated with annual fatality rates than are the lay judgments.

There were many similarities between the three groups of lay persons. For example, each group believed that motorcycles, other motor vehicles, and handguns were highly risky, and that vaccinations, home appliances, power mowers, and football were relatively safe. However, there were strong differences as well. Active Club members viewed pesticides and spray cans as relatively much safer than did the other groups. Nuclear power was rated as highest in risk by the LOWV and student groups, but only eighth by the Active Club. The students viewed contraceptives and food preservatives as riskier and swimming and mountain climbing as safer than did the other lay groups. Experts' judgments of risk differed markedly from the judgments of lay persons. The experts viewed electric power, surgery, swimming, and X-rays as more risky than the other groups, and they judged nuclear power, police work, and mountain climbing to be much less risky.

What Determines Risk Perception?

What do people mean when they say that a particular technology is quite risky? A series of additional studies was conducted to answer this question.

<u>Perceived Risk Compared To Frequency Of Death</u>

When people judge risk, as in the previous study, are they simply estimating frequency of death? To answer this question, we collected the best available technical estimates of the annual number of deaths from each of the thirty activities included in our study. For some cases, such as commercial aviation and handguns, there is good statistical evidence based on counts of known victims. For other cases, such as the lethal potential of nuclear or fossil-fuel power plants, available estimates are based on uncertain inferences about incompletely understood processes. For still others, such as food coloring, we could find no estimates of annual fatalities.

For the 25 cases for which we found technical estimates for annual frequency of death, we compared these estimates with perceived risk. Results for experts and the LOWV sample are shown in Figure 6.3 (the results for the other lay groups were quite similar to those from the LOWV sample). The experts' mean judgments were so closely related to the statistical or calculated frequencies that it seems reasonable to conclude that they viewed the risk of an activity or technology as synonymous with its annual fatalities. The risk judgments of lay people, however, showed only a moderate relationship to the annual frequencies of death[17], raising the possibility that, for them, risk may not be synonymous with fatalities. In particular, the perceived risk from nuclear power was disproportionately high compared to its estimated number of fatalities.

Activity or Technology	Technical Fatality Estimates	Geometric Mean Fatality Estimates Average Year		Geometric Mean Multiplier Disastrous Year	
		LOWV	Students	LOWV	Students
Smoking	150,000	6,900	2,400	1.9	2.0
Alcoholic beverages	100,000	12,000	2,600	1.9	1.4
Motor vehicles	50,000	28,000	10,500	1.6	1.8
Handguns	17,000	3,000	1,900	2.6	2.0
Electric power	14,000	660	500	1.9	2.4
Motorcycles	3,000	1,600	1,600	1.8	1.6
Swimming	3,000	930	370	1.6	1.7
Surgery	2,800	2,500	900	1.5	1.6
X-rays	2,300	90	40	2.7	1.6
Railroads	1,950	190	210	3.2	1.6
General (private) aviation	1,300	550	650	2.8	2.0
Large construction	1,000	400	370	2.1	1.4
Bicycles	1,000	910	420	1.8	1.4
Hunting	800	380	410	1.8	1.7
Home appliances	200	200	240	1.6	1.3
Fire fighting	195	220	390	2.3	2.2
Police work	160	460	390	2.1	1.9
Contraceptives	150	180	120	2.1	1.4
Commercial aviation	130	280	650	3.0	1.8
Nuclear power	100[a]	20	27	107.1	87.6
Mountain climbing	30	50	70	1.9	1.4
Power mowers	24	40	33	1.6	1.3
High school & college football	23	39	40	1.9	1.4
Skiing	18	55	72	1.9	1.6
Vaccinations	10	65	52	2.1	1.6
Food coloring	--[b]	38	33	3.5	1.4
Food preservatives	--[b]	61	63	3.9	1.7
Pesticides	--[b]	140	84	9.3	2.4
Prescription antibiotics	--[b]	160	290	2.3	1.6
Spray cans	--[b]	56	38	3.7	2.4

[a] Technical estimates for nuclear power were found to range between 16 and 600 annual fatalities. The geometric mean of these estimates was used here.

[b] Estimates were unavailable.

Table 6.3 - Fatality Estimates and Disaster Multipliers for 30 Activities and Technologies.

Lay Fatality Estimates

Perhaps lay people based their risk judgments on annual fatalities but estimated their numbers inaccurately. To test this hypothesis, we asked additional groups of students and LOWV members "to estimate how many people are likely to die in the U.S. in the next year (if the next year is an average year) as a consequence of these thirty activities and technologies." We asked our student and LOWV samples to consider all sources of death associated with these activities.

The mean fatality estimates of LOWV members and students are shown in columns 2 and 3 of Table 6.3. If lay people really equate risk with annual fatalities, one would expect that their own estimates of annual fatalities, no matter how inaccurate, would be very similar to their judgments of risk. But this was not so. There was a moderate agreement between their annual fatality estimates and their risk judgments, but there were important exceptions. Most notably, nuclear power had the lowest fatality estimate and the highest perceived risk for both LOWV members and students. Overall, lay people's perceptions were no more closely related to their own fatality estimates than they were to the technical estimates (Figure 6.3).

These results lead us to reject the idea that lay people wanted to equate risk with annual fatality estimates but were inaccurate in doing so. Instead, we are led to believe that lay people incorporate other considerations besides annual fatalities into their concept of risk.

Some other aspects of lay people's fatality estimates are of interest. One is that they were moderately accurate. The relationship between the LOWV members' fatality estimates and the best technical estimates is plotted in Figure 6.4. The lay estimates showed the same overestimation of those items that cause few fatalities that was apparent in Figure 6.2 for a different collection of hazards. Also, as in Figure 6.2, the moderate overall relationship between lay and technical estimates was marred by specific biases (e.g., the underestimation of fatalities associated with railroads, X-rays, electric power, and smoking).

Disaster Potential

The fact that the LOWV members and students assigned very high risk values to nuclear power along with very low estimates of its annual fatality rates is an apparent contradiction. One possible explanation is that LOWV members expected nuclear power to have a low death rate in an average year but considered it to be a high-risk technology because of its potential for disaster.

In order to understand the role played by expectations of disaster in determining lay people's risk judgments, we

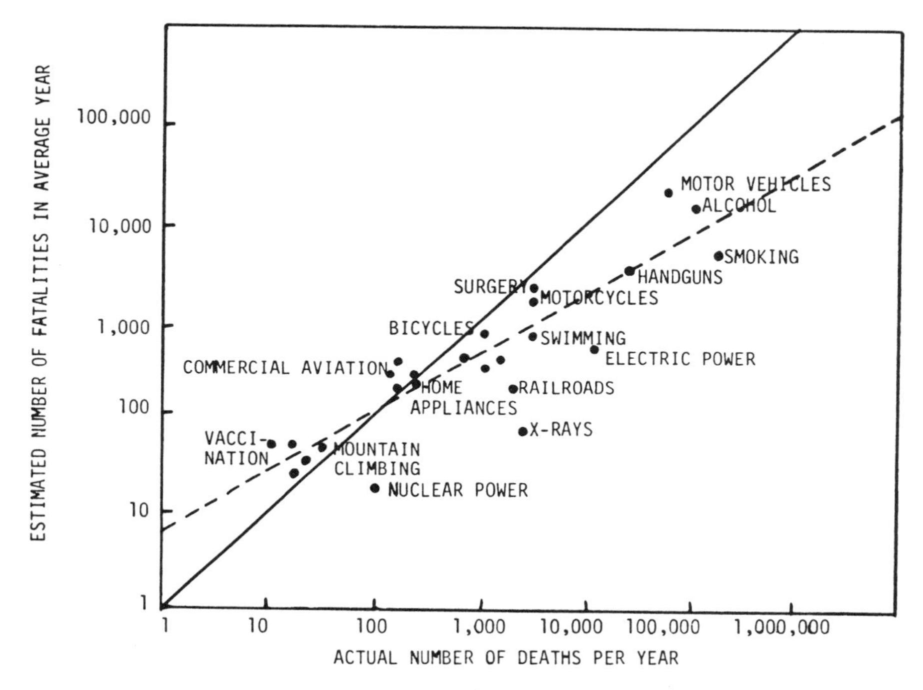

Figure 6.4. Lay People's Judgments of the Number of Fatalities in an Average Year Plotted Against the Best Estimate of Annual Fatalities for 25 Activities and Technologies. The solid line indicates accurate judgment, while the dashed line best fits the data points. These results have much the same character as those shown in Figure 6.2 for a different collection of hazards. Low frequencies were overestimated and high ones were underestimated. The overall relationship is marred by specific biases (e.g., the underestimation of fatalities associated with railroads, x-rays, electric power, and smoking).

asked these same respondents to give for each activity and technology a number indicating how many times more deaths would occur if next year were "particularly disastrous" rather than average. The averages of these multipliers are shown in Table 6.3. For most activities, people saw little potential for disaster. For the LOWV sample all but five of the multipliers were less than 3, and for the student sample all but six were less than 2. The striking exception in both cases is nuclear power, with a geometric mean disaster multiplier in the neighborhood of 100.

For any individual, an estimate of the expected number of fatalities in a disastrous year could be obtained by applying the disaster multiplier to the estimated fatalities for an average year. When this was done for nuclear power, almost 40 percent of the respondents expected more than 10,000 fatalities if next year were a disastrous year. More than 25 percent expected 100,000 or more fatalities. These extreme estimates can be contrasted with the Reactor Safety Study's conclusion that the maximum credible nuclear accident, coincident with the most unfavorable combination of weather and population density, would cause only 3,300 prompt fatalities.[18] Furthermore, that study estimated the odds against an accident of this magnitude occurring during the next year (assuming 100 operating reactors) to be about 2,000,000:1.

Apparently, disaster potential explains much or all of the discrepancy between the perceived risk and frequency of death values for nuclear power. Yet, because disaster plays only a small role in most of the thirty activities and technologies we have studied, it provides only a partial explanation of the perceived risk data.

Qualitative Characteristics

Are there other determinants of risk perceptions besides frequency estimates? We asked experts, students, LOWV members, and Active Club members to rate the thirty technologies and activities on nine qualitative characteristics that have been hypothesized to be important.[19] These ratings scales are described in Table 6.4.

Mean ratings were quite similar for all four groups. Particularly interesting was the characterization of nuclear power, which had the dubious distinction of scoring at or near the extreme on all of the undesirable characteristics. Its risks were seen as involuntary, delayed, unknown, uncontrollable, unfamiliar, catastrophic, dreaded, and fatal. This contrasted sharply with the characterizations of non-nuclear electric power and another radiation technology, X-rays. Electric power and X-rays were both judged more voluntary, less certain to be fatal, less catastrophic, less dreaded, more familiar, and less risky than nuclear power (see Figure 6.5).

Voluntariness of risk
Do people face this risk voluntarily? If some of the risks are voluntarily undertaken and some are not, mark an appropriate spot towards the center of the scale.

risk assumed voluntarily	1 2 3 4 5 6 7	*risk assumed involuntarily*

Immediacy of effect
To what extent is the risk of death immediate--or is death likely to occur at some later time?

effect immediate	1 2 3 4 5 6 7	*effect delayed*

Knowledge about risk
To what extent are the risks known precisely by the persons who are exposed to those risks?

risk level known precisely	1 2 3 4 5 6 7	*risk level not known*

To what extent are the risks known to science?

risk level known precisely	1 2 3 4 5 6 7	*risk level not known*

Control over risk
If you are exposed to the risk, to what extent can you, by personal skill or diligence, avoid death?

personal risk can't be controlled	1 2 3 4 5 6 7	*personal risk can be controlled*

Newness
Is this risk new and novel or old and familiar?

new	1 2 3 4 5 6 7	*old*

Chronic/Catastrophic
Is this a risk that people have learned to live with and can think about reasonably calmly, or is it one that people have great dread for--on the level of a gut reaction?

common	1 2 3 4 5 6 7	*dread*

Severity of consequences
When the risk from the activity is realized in the form of a mishap or illness, how likely is it that the consequence will be fatal?

certain not to be fatal	1 2 3 4 5 6 7	*certain to be fatal*

Table 6.4 - Risk Characteristics Rated by LOWV Members, Active Club Members, Students, and Experts.

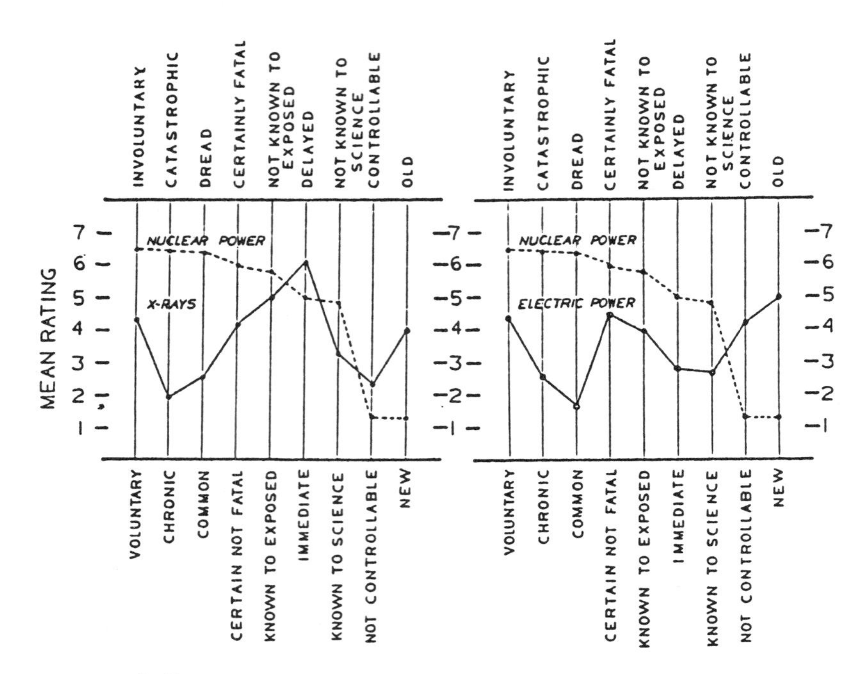

Figure 6.5 - Qualitative characteristics of perceived risk for nuclear power and related technologies. In the right-hand diagram, risk profiles for nuclear power and non-nuclear electric power are compared. In the left-hand diagram, nuclear power and X-rays are compared. Each profile consists of nine dimensions rated on a seven-point scale. The instructions that elicited these responses are reproduced in Table 6.4. The perceived qualities of nuclear power are dramatically different from the comparison technologies. The source of this data is the LOWV sample studied by Fischhoff et al.[16]

Across all 30 hazards, ratings of dread and of the likelihood of a mishap's being fatal were closely related to lay judgments of risk. In fact, the risk judgments of the LOWV and student groups could be predicted almost perfectly from ratings of dread and lethality and the subjective fatality estimates for normal and disastrous years.[20] Experts' judgments of risk were not related to any of the nine risk characteristics.[21]

Many pairs of risk characteristics tended to be correlated with each other across the 30 activities and technologies. For example, risks faced voluntarily were typically judged as well known and controllable. These interrelations were sufficiently high to suggest that all the ratings could be explained in terms of a few basic dimensions of risk. In order to identify such dimensions, we conducted a factor analysis of the correlations from each group (principal components analysis with varimax rotation to simple structure). We found that the nine characteristics could be represented by two underlying factors which appeared to be the same for each group. Figure 6.6 illustrates the factor scores for each hazard within the common space. Hazards at the high end of the vertical dimension or factor (e.g., food coloring, pesticides) tended to be new, unknown, involuntary, and delayed in their effects. Hazards at the other extreme of this factor (e.g., mountain climbing, swimming) had the opposite characteristics. High (right-hand) scores on the horizontal factor (e.g., nuclear power, commercial aviation) were associated with events whose consequences were seen as certain to be fatal, often for large numbers of people, should something go wrong. Hazards low on this factor (e.g., power mowers, football) were seen as causing injuries, rather than fatalities, to single individuals. We have labeled the vertical factor as "Unknown Risk" and the horizontal factor as "Dread Risk." In sum, even though the four groups had somewhat different perceptions of the riskiness of the various hazards (Table 6.2), they tended to characterize these hazards similarly.

Judged Seriousness of Death

In a further attempt to improve our understanding of perceived risk, we examined the hypothesis that some hazards are feared more than others because the deaths they produce are much "worse" than deaths from other activities. We thought, for example, that deaths from risks imposed involuntarily, from risks not under one's control, or from hazards that are particularly dreaded might be given greater weight in determining people's perceptions of risk.

When we asked students and LOWV members to judge the relative "seriousness" of a death from each of the thirty activities and technologies, however, the differences were slight. The most serious forms of death (from nuclear power and handguns) were judged to be only about two to four times

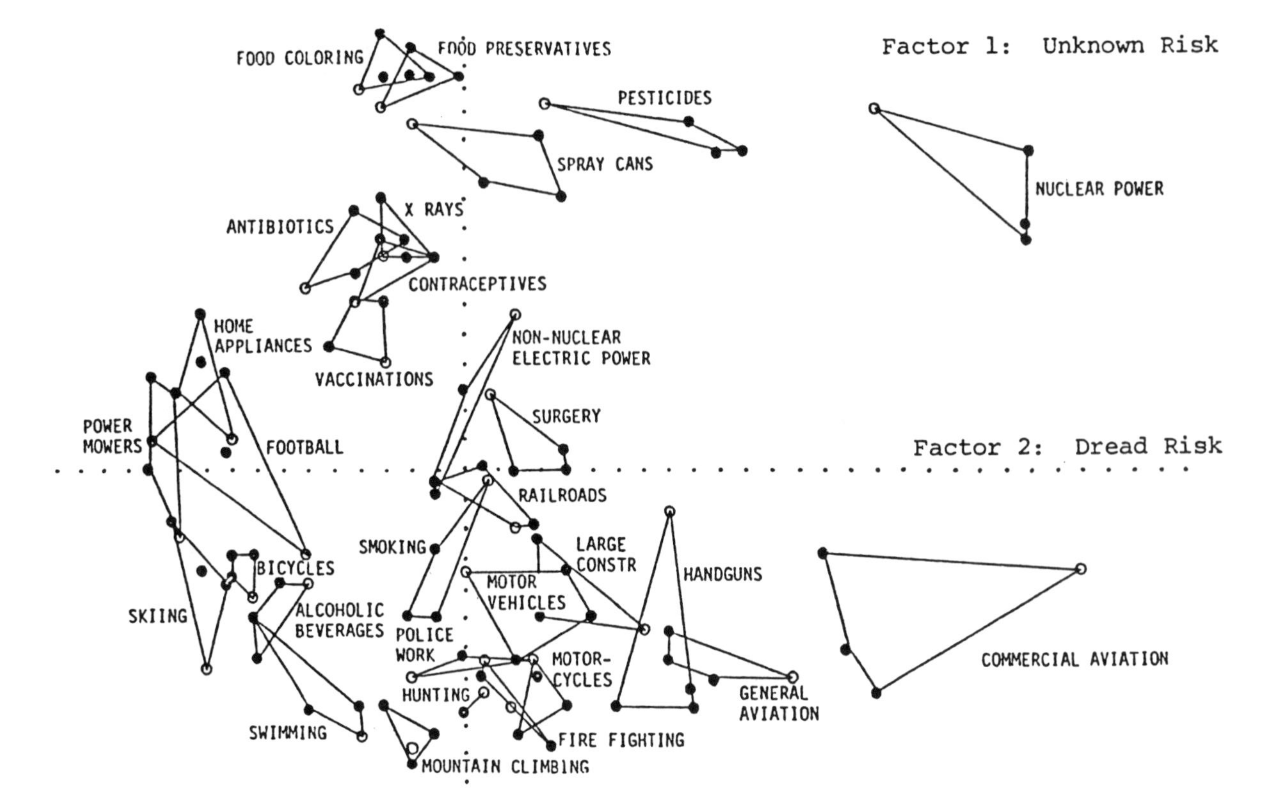

Figure 6.6 – Location of 30 hazards within the two-factor space obtained from LOWV, student, Active Club and expert groups. Connected lines join or enclose the loci of four group points for each hazard. Open circles represent data from the expert group. Unattached points represent groups that fall within the triangle created by the other three groups.

worse than the least serious forms of death (from alcoholic beverages and smoking). Furthermore, across all thirty activities, judged seriousness of death was not closely related to perceived risk of death.

Reconciling Divergent Opinions

Our data show that experts and lay people have quite different perceptions about how risky certain technologies are. It would be comforting to believe that these divergent risk judgments would be responsive to new evidence so that, as information accumulates, perceptions would converge towards one "appropriate" view. Unfortunately, this is not likely to be the case. As noted earlier in our discussion of availability, risk perception is derived in part from fundamental modes of thought that lead people to rely on fallible indicators such as memorability and imaginability.

Furthermore, a great deal of research indicates that people's beliefs change slowly and are extraordinarily persistent in the face of contrary evidence.[22] Once formed, initial impressions tend to structure the way that subsequent evidence is interpreted. New evidence appears reliable and informative if it is consistent with one's initial belief; contrary evidence is dismissed as unreliable, erroneous, or unrepresentative. Thus, depending on one's predispositions, intense effort to reduce a hazard may be interpreted to mean either that the risks are great or that the technologists are responsive to the public's concerns. Likewise, opponents of a technology may view minor mishaps as near catastrophes and dismiss the contrary opinions of experts as biased by vested interests.

From a statistical standpoint, convincing people that the catastrophe they fear is extremely unlikely is difficult under the best conditions. Any mishap could be seen as proof of high risk, whereas demonstrating safety would require a massive amount of evidence.[23] Nelkin's case history of a nuclear siting controversy[24] provides a good example of the inability of technical arguments to change opinions. In that debate each side capitalized on technical ambiguities in ways that reinforced its own position.

THE FALLIBILITY OF JUDGMENT

Our examination of risk perception leads us to the following conclusions:

1. Cognitive limitations, coupled with the anxieties generated by facing life as a gamble, cause uncertainty to be denied, risks to be distorted, and statements of fact to be believed with unwarranted confidence.

2. Perceived risk is influenced (and sometimes biased) by the imaginability and memorability of the hazard.

People may, therefore, not have valid perceptions even for familiar risks.

3. Our experts' risk perceptions correspond closely to statistical frequencies of death. Lay people's risk perceptions were based in part upon frequencies of death, but there were some striking discrepancies. It appears that for lay people, the concept of risk includes qualitative aspects such as dread and the likelihood of a mishap's being fatal. Lay people's risk perceptions were also affected by catastrophic potential.

Disagreements about risk should not be expected to evaporate in the presence of "evidence." Definitive evidence, particularly about rare hazards, is difficult to obtain. Weaker information is likely to be interpreted in a way that reinforces existing beliefs.

The significance of these results hinges upon one's acceptance of our assumption that subjective judgments are central to the hazard management process. Our conclusions mean little if one can assume that there are analytical tools which can be used to assess most risks in a mechanical fashion and that all decision makers have perfect information and the know-how to use it properly. These results gain in importance to the extent that one believes, as we do, that expertise involves a large component of judgment, that the facts are not all in (or obtainable) regarding many important hazards, that people are often poorly informed or misinformed, and that they respond not just to numbers but also to qualitative aspects of hazards.

Whatever role judgment plays, its products should be treated with caution. Research not only demonstrates that judgment is fallible, but it shows that the degree of fallibility is often surprisingly great and that faulty beliefs may be held with great confidence.

When it can be shown that even well-informed lay people have difficulty judging risks accurately, it is tempting to conclude that the public should be removed from the hazard-management process. The political ramifications of such a transfer of power to a technical elite are obvious. Indeed, it seems doubtful that such a massive disenfranchisement is feasible in any democratic society.

Furthermore, this transfer of decision-making would seem to be misguided. For one thing, we have no assurance that experts' judgments are immune to biases once they are forced to go beyond their precise knowledge and rely upon their judgment. Although judgmental biases have most often been demonstrated with lay people, there is evidence that the cognitive functioning of experts is basically like that of everyone else.

In addition, in many if not most cases, effective hazard management requires the cooperation of a large body of lay people. These people must agree to do without some things and accept substitutes for others; they must vote sensibly on ballot measures and for legislators who will serve them as surrogate hazard managers; they must obey safety rules and use the legal system responsibly. Even if the experts were much better judges of risk than lay people, giving experts an exclusive franchise on hazard management would involve substituting short-term efficiency for the long-term effort needed to create an informed citizenry.

For those of us who are not experts, these findings pose an important series of challenges: to be better informed, to rely less on unexamined or unsupported judgments, to be aware of the qualitative aspects that strongly condition risk judgments, and to be open to new evidence that may alter our current risk perceptions.

For the experts, our findings pose what may be a more difficult challenge: to recognize their own cognitive limitations, to temper their assessments of risk with the important qualitative aspects of risk that influence the responses of lay people, and somehow to create ways in which these considerations can find expression in hazard management without, in the process, creating more heat than light.

ACKNOWLEDGMENT

The authors wish to express their appreciation to Christoph Hohenemser, Roger Kasperson, and Robert Kates for their many helpful comments and suggestions. This work was supported by the National Science Foundation under Grant ENV77-15332 to Perceptronics, Inc. Any opinions, findings and conclusions, or recommendations expressed herein are those of the authors and do not necessarily reflect the views of the National Science Foundation.

FOOTNOTES

[1]We have not attempted here to review all the important research in this area. Interested readers should see C.H. Green, "Risk: Attitudes and Beliefs," in Behavior in Fires, ed. D.V. Canter (Wiley, New York, 1980).; R.W. Kates, Risk Assessment of Environmental Hazard, New York: Wiley, 1978; and H.J. Otway, D. Maurer, and K. Thomas, "Nuclear Power: The Question of Public Acceptance," Futures, 10 (April) 1978, pp. 109-118.

[2]Tversky, A. and D. Kahneman, Judgment under Uncertainty: Heuristics and Biases," Science, 185, 1974, pp. 1124-1131.

[3]Tversky, A. and D. Kahneman, Availability: A Heuristic for Judging Frequency and Probability," Cognitive Psychology, 4, 1973, pp. 207-232.

[4]Lichtenstein, S., P. Slovic, B. Fischhoff, M. Layman, and B. Combs, "Judged Frequency of Lethal Events." Journal of

Experimental Psychology: Human Learning and Memory, 4, 1978, pp. 551-578.

[5]Fischhoff, B., P. Slovic, and S. Lichtenstein, "Fault Trees: Sensitivity of Estimated Failure Probabilities to Problem Representation," Journal of Experimental Psychology: Human Perception and Performance, 4, 1978, pp. 342-355.

[6]Fischhoff, B., P. Slovic, and S. Lichtenstein, "Knowing with Certainty: The Appropriateness of Extreme Confidence," Journal of Experimental Psychology: Human Perception and Performance, 3, 1977, pp. 552-564.

[7]Lichtenstein, S. B. Fischhoff, and L.D. Phillips, "Calibration of Probabilities: The State of the Art," Decision Making and Change in Human Affairs, ed. H. Jungermann and G. de Zeeuw (D. Reidel, Dordrecht, the Netherlands, 1977).

[8]Hynes, M. and E. Vanmarcke, "Reliability of Embankment Performance Predictions," Proceedings of the ASCE Engineering Mechanics Division Specialty Conference. Waterloo, Ontario: University of Waterloo Press, 1976.

[9]U.S. Regulatory Commission, Reactor Safety Study: An Assessment of Accident Risks in U.S. Commercial Nuclear Power Plants (WASH-1400, NUREG-75/014), Washington, D.C.: Nuclear Regulatory Commission, 1975.

[10]U.S. Regulatory Commission, Risk Assessment Review Group, Risk Assessment Review Group Report to the U.S. Nuclear Regulatory Commission (NUREG/CR-0400), Washington, D.C.: Nuclear Regulatory Commission, 1978.

[11]U.S. Congress, House, Committee on Government Operations, Subcommittee on Conservation, Energy, and Natural Resources, Teton Dam Disaster: Hearings..., 94th Congress, 2d Session, Washington, D.C.: Government Printing Office, 1976.

[12]Kates, R.W., Hazard and Choice Perception in Flood Plain Management, Research Paper, 78, Chicago: University of Chicago, Department of Geography, 1962.

[13]Borch, K., The Economics of Uncertainty, Princeton, NJ: Princeton University Press, 1968.

[14]"Doubts Linger on Cyclamate Risks," Eugene Register-Guard, 14 (January) 1976.

[15]David, E.E., "One-Armed Scientists?," Science, 189, 1975, p.891.

[16]Fischhoff, B., P. Slovic, S. Lichtenstein, S. Read, and B. Combs, "How Safe is Safe Enough? A Psychometric Study of

Attitudes towards Technological Risks and Benefits," Policy Sciences, 8, 1978, pp. 127-152; Slovic, P., B. Fischhoff, and S. Lichtenstein, "Expressed Preferences," Unpublished Manuscript (Decision Research, Eugene, Oregon, 1978).

[17]The correlations between perceived risk and the annual frequencies of death were .92 for the experts and .62 , .50, and .56 for the League of Women Voters, students, and Active Club samples, respectively.

[18]U.S. Nuclear Regulatory Commission, 1975, op. cit.

[19]Lowrance, W. Of Acceptable Risk, Los Altos, California: William Kaufman, 1976.

[20]The multiple correlation between the risk judgments of the LOWV members and students and a linear combination of their fatality estimates, disaster multipliers, dread ratings, and severity ratings was .95.

[21]A secondary finding was that both experts and lay persons believed that the risks from most of the activities were better known to science than to the individuals at risk. The experts believed that the discrepancy in knowledge was particularly great for vaccinations, X-rays, antibiotics, alcohol, and home appliances. The only activities whose risks were judged better known to those exposed were mountain climbing, fire fighting, hunting, skiing, and police work.

[22]Ross, L., "The Intuitive Psychologist and His Shortcomings," Advances in Social Psychology, ed. L. Berkowitz, New York: Academic Press, 1977.

[23]Green, A.E., and A.J. Bourne, Reliability Technology, New York: Wiley Interscience, 1972.

[24]Nelkin, D., "The Role of Experts on a Nuclear Siting Controversy," Bulletin of the Atomic Scientists, 30, 1974, pp. 29-36.

RESPONSE OF THE NEW JERSEY PUBLIC TO THE RISK FROM RADON

Neil D. Weinstein*

The public health problems presented by radon are different from those posed by most other environmental hazards. Government regulations and enforcement can reduce many environmental threats, but reducing the substantial health risk from radon requires individual citizens to act. Legislatures are no more likely to pass laws forcing homeowners to reduce the radon levels in their houses than to pass laws forbidding cigarette smokers from lighting up in their homes. Nor is it likely that government agencies will go door-to-door offering to measure radon levels or to provide free radon mitigation. No matter how ambitious a program is developed by local, state, and federal governments, citizens will still have to act to protect themselves.

What kind of a public response would we like? Ideally, everyone should realize that radon can be a very serious hazard. We would like people to attend to information indicating where radon could be a problem, and we would want them to monitor their homes wherever significant radon concentrations are a possibility. We would like the public to understand the results of these tests, feeling reassured if their levels are quite low and taking the risk seriously if they are not. We would want people to make decisions about home mitigation that accurately reflect the risk they face, choosing mitigation strategies that are appropriate for their level of risk and their type of home construction. Finally, they should realize that reducing radon still involves some trial and error, and retesting is needed to see whether mitigation efforts have succeeded.

This ideal of healthy concern and informed decision-making must be contrasted with two other plausible patterns of response: panic and apathy. In a full-scale panic, people will be terribly frightened whatever their radon levels. They will demand the immediate and total elimination of radon in their homes, be vulnerable to scare tactics of companies that offer radon reduction measures of dubious value, and remain frightened when they find that

---

*Departments of Human Ecology and Psychology, Rutgers -- The State University of New Jersey. He is a collaborator in radon research sponsored by the New Jersey Department of Environmental Protection.

detectable levels of radon persist in their homes. Their frustration and fear will become hostility directed toward government agencies that can not satisfy their impossible expectations. If the predominant reaction is apathy, on the other hand, no one tests and no one cares. Relations between the public and government agencies are not strained, but the serious health risk continues unabated.

Because radon is so different from other hazards, the public's actual response is difficult to predict a priori. Below are some of the factors in the radon situation that support predictions of both panic and apathy.

Panic Producing

*The threat is cancer, and a particularly nasty form of cancer at that.

*People tend to be afraid of radiation.

*Radon in your home can be harming you even though you cannot see it.

*Radon permeates the home, normally our haven of safety.

*The risk applies to all family members, including children.

*The actual levels of risk can be quite high.

*Mitigation can be difficult and expensive.

*The value of one's home may be seriously decreased by high radon levels.

Apathy Producing

*Radon is naturally occurring; there is no villain to arouse us with anger.

*Negative health effects are only a slim possibility for most people.

*Radon is invisible; there are no sensory reminders that it is present.

*There are no short-term symptoms indicating dangerous levels; it would be years before any health effects might appear.

*Obtaining a home measurement requires considerable individual initiative, so like other chores, it is easier to put off until tomorrow.

*Radon reduction measures may be costly and complicated; it may be easier to avoid the whole issue.

In meetings I have attended, in conversations about radon, and even in recent news articles (e.g., New York Times, September 2, 1986, pp. C1, C7), it appears that many scientists and government officials believe panic over radon is very likely and that government actions must be carefully planned to avoid panic. Yet, the individual citizens who reach officials and scientists with concerns and questions about radon, particularly at this early stage of the issue, are highly atypical; they come primarily from the segment of the population that is most alarmed. The impression created by such contacts with the public may be highly misleading.

Recognizing that its radon program needs accurate information about public reactions, the New Jersey Department of Environmental Protection decided to support an ambitious public survey. My colleague Peter Sandman and I set out to discover what people think they know about radon, what they really know, where they are getting their information, whom they trust, what emotions they are feeling, whether they are planning to obtain a home radon measurement, and many related issues. Using a random sample derived from telephone directories, we contacted people in April, 1986 who live in the Reading Prong and in bordering areas of New Jersey.

Because most media coverage has focused on the radon problem in the Reading Prong, our results indicate the response of people who reside in an area where radon has already received considerable attention. The general public is surely less informed and less concerned. This cautionary note receives support from data obtained in a statewide poll carried out by the Center for Public Interest Polling at Rutgers University in July, 1986. Their results showed that 84 percent of the population in the Prong region had heard of radon, 87 percent of those living in our "border" region had heard of radon, but only 62 percent of those living in other parts of northern New Jersey, where radon is also likely to be a problem, had heard of this hazard. A few of our findings are reported below. They are based on responses to a mailed questionnaire sent to owners of single family homes. We succeeded in obtaining data from 61.7 percent of those we contacted, with a total sample of 650.

WHAT PEOPLE KNOW AND DON'T KNOW

On questions concerning matters of fact, people tend to give "don't know" answers rather than wrong answers. These are favorable conditions for a public education campaign because it is easier to provide information to people who realize that they are uninformed than to try to convince people that they are mistaken.

Most people (80% or more) know what radon is, where it comes from, that it varies from house to house, and how it enters houses. There is less widespread understanding (60-79% correct) of the fact that most radon is not industrial and that it can cause lung cancer. There is poor understanding of most health aspects of radon: that it does not cause skin cancer (28% correct), that it does not irritate the eyes or throat (20% correct), that there is a risk from prior exposure even when radon is eliminated (53% correct), that there would not be symptoms to warn of a dangerous radon level (47% correct), and that health effects would take years to appear (44% correct).

Other data from our survey show that most people underestimate the seriousness of the risk. Only 42% realize that high radon levels are more serious than air and water pollution. If someone in their house became ill from radon, only 15% believe it would be serious or very serious; 20% think it would be minor or somewhat serious, and 65% say they have no idea how serious it would be.

On the other hand, people overestimate the need for immediate action once radon is found. When asked how important it is to act quickly--on a 7-point scale ranging from "not at all important" (1) to "very important" (7)--over half gave responses of 6 or 7. Unfortunately, the belief that little time is available to undertake a protective action tends to increase the likelihood of panic.

Even in our sample--in or near the Reading Prong--people do not see radon as a problem likely to affect them. On a 7-point scale ranging from very unlikely (1) to very likely (7), 71% gave responses of 1, 2, or no opinion when asked whether their own home might be above the suggested action level. There is a clear optimistic bias at work. For example, the number of people who believe their home radon level is below average outnumber those who believe their home level is above average by a ratio of 10 to 1. Although 55 people volunteered that the ventilation level in their home tended to make their risk below average, only 2 thought their ventilation level raised their risk.

Further, we found that our respondents were getting conflicting messages from acquaintances about the radon issue. For example, 38% reported hearing someone say that the radon problem is serious, but 32% reported that they had heard someone say the problem is exaggerated. Whereas 36% had heard it said that radon is in the area, 42% had heard there was no radon in the area.

PANIC, APATHY OR HEALTHY CONCERN

Two indicators of the public's overall pattern of response are the emotions they are experiencing and their actions and plans with respect to obtaining a home radon measurement. Self-reports of emotion showed a moderate degree of concern, along with quite low levels of worry, fear, anger, depression, and helplessness. These data do not indicate panic.

When people were asked whether they have thought about monitoring their homes, we obtained the responses displayed in Table 6.5. Thus, at time of our survey, only 50% of our sample--all in or near the Reading Prong--had ever even thought of monitoring and half of these were undecided. Of the quarter that had made a tentative decision, only 7% had actually taken steps to get a measurement, the same percentage as those who decided not to do anything. The main reasons offered for not monitoring were: don't know which method is best (46%), don't think I have a problem (43%), haven't gotten around to it (36%), don't know how to get tested (31%), and want to see others' levels first (26%).

	Percentage	N
Already monitored/ test in progress	7	41
Plan to monitor	11	68
Haven't decided	25	156
Not needed	7	47
Never thought about it	50	319

Table 6.5: Reported intention to monitor for radon.

Statistical calculations sought to identify the beliefs that best predict the intention to measure one's home radon level. The most influential factors were opinions about: 1) the likelihood of there being a problem in one's own community and 2) the perceived concern or lack of concern among one's neighbors. Fears about radon or beliefs concerning the difficulties involved in radon mitigation did not seem to have a strong impact on monitoring decisions. It does not seem that inaction about monitoring reflects fatalism or feelings of hopelessness.

CONCLUSION

In the region of New Jersey we studied, the public has learned a lot about radon in a short period of time. Nevertheless, additional public education efforts are needed to increase the level of information about the nature of the health risk and the degree of risk. It is important for people to understand not only that radon can cause lung cancer, but also that it does not have other health effects. Without such knowledge people are likely to attribute a wide range of health problems to radon in their homes and may become supersensitive to harmless, irrelevant symptoms. At the same time, information can stress the fact that this is a problem with a solution; if high radon levels are found in a home, they can be reduced. People need also to be reassured that if high radon levels are found, they can take time to work out a solution.

Taken together, the optimistic perceptions of risk, the reports of little fear or anger, the small percentage of those about to monitor, and the types of reasons given for not monitoring suggest that panic in the general population is unlikely. A small segment of the population may become frightened, but because of unconcern or confusion, most people will not even monitor their homes without encouragement. Other recently collected data from a sample of 135 people who do have substantial amounts of radon support this conclusion. Even in this group, 65% said that radon made them feel not at all or only slightly frightened. Like other issues in prevention, where self-protective behavior requires people to expend some effort, inaction by the public appears to be a serious problem.

Our findings suggest a number of ingredients that ought to be included in attempts to encourage home monitoring. People are clearly confused about where radon is found. Communications should clearly mention that radon is not limited to any particular area (e.g., the Reading Prong), list prominently the counties and communities where radon may be a problem, and emphasize that there is no way to be sure about your home without testing it. A number of activities can help people realize that fellow community residents are concerned. Local media coverage, statements by local leaders, meetings held locally, and distribution of informational materials through local channels such as schools, churches, and post offices will all increase home monitoring. Certainly, the level of risk and the number of people potentially affected warrant active public education efforts such as these.

COMMUNICATING RADON RISK: ALERTING THE APATHETIC AND REASSURING THE HYSTERICAL

Peter M. Sandman*

INTRODUCTION

In one way or another, we are all discussing the communications aspect of radon. The piece of it that I want to focus on is the issue of panic on the one hand and apathy on the other. I think one of the metaphors that goes through almost all talking about radon risk management --and we heard it yesterday from Don Deieso, we heard it from Judy Klotz this morning-- is the notion that we are walking a tightrope between the risk of public panic and the risk of public apathy. The question I want to address is, "Is this in fact a tightrope, is there a chasm on both sides, is there a bigger chasm on one side than the other? What in fact, is the likelihood, with respect to radon, of a public that over-responds, that becomes hysterical, that becomes panicky? What is the risk, on the other hand, that the public will pay too little attention to radon, that we need to arouse as much concern as we possibly can?" Much of what Neil Weinstein addressed in his paper speaks to that issue, but I want to narrow that down still further to this question of panic versus apathy.

REASONS FOR UNDERESTIMATING RADON RISK

Let me start by talking about what theory tells us about when people get very frightened and when people tend to dismiss the risk and become too little frightened. I want to list for you some of the reasons or ways in which the risk of radon is likely to be underestimated by the audience.

One reason is that there is no enemy. This is God's own radon; this is not mill tailings. We are talking here, for the most part, about geological radon. We know that alarm and anger are very closely tied in public responses to environmental affronts, and when the public has no one to be angry at, it tends to ignore the problem altogether. One only needs to look at the public response within the radon

*Peter Sandman is Professor of Communications at Rutgers University and a collaborator in radon research sponsored by the New Jersey Department of Environmental Protection.

issue to radon in mill tailings in places like Montclair as opposed to the public response to the same radon, the same kind of risk when it's geological, to see that we have orders of magnitude more fear, more anger, more energy on the part of the pubic when there is someone to blame than when there is not someone to blame.

Second, and closely related, radon is a naturally occurring risk. We know that the public responds with far greater concern to risks generated by human activities than to risks that are naturally generated. You can think about Chernobyl as an example.

Third, the radon risk is widely dispersed in space and time. The radon is found in widely separated places; the cancers come one at a time, years later. To the extent that we find radon clusters, areas of high concentration, that is likely to increase the level of alarm. If radon continues to be widely dispersed, then the level of alarm will stay down.

Fourth, radon is remediable. For most homes, it is remediable by some combination of low-technology, handyman strategies and somewhat higher technology strategies that require bringing in a contractor. As we learned in the first day of the conference, it is not as readily remediable as many of us hoped, and we have hopes that it will become more readily remediable than it is now as the state of the art improves. At the moment, the public perception of this risk, as something that can be solved, ironically enough, greatly diminishes the level of alarm and tends to diminish the likelihood that people will bother to solve it. It is almost as though, if you tell me that I can remedy the problem, then I don't have to worry so much about the problem, and I don't, in fact, bother to remedy it.

Five, the remediation is individual; radon is not a problem that government or the community has to solve or indeed can solve. Individuals have to test their own homes and then have to decide whether and how to remediate. The knowledge that the risk is subject to individual control, again, typically leads people to underestimate its seriousness. Precisely because you can deal with it on your own, whenever you choose, without any government intervention, you don't feel the same strength of concern, the same need to deal with it immediately.

Six, radon is not visible. We can't see it; we can't smell it; we can't feel its effects, and that makes it easier to ignore.

Seven, our homes are comfortable places; they're places where we are accustomed to feeling safe. We have been safe in our homes so far, and nothing feels different. The situation is not so alien as to trigger our alarm system. One of the characteristics that make risks very alarming is that they are alien; they feel unknown. Our homes, indeed, are perceived as very known and very safe territory.

Eight, we have no mental image of the risk. Unlike asbestos, for example, which comes in particles that we can imagine, it is very hard to wrap our minds around exactly what this risk is. If we can't conceptualize it, we find it more difficult to worry about.

Nine, remediation is fairly costly, and potentially difficult, and possibly emotionally upsetting. This provides us with reasons not to want to be alarmed, especially those of us who expect to sell their homes before too many years. Prospective home sellers have less reason to need to know about the risk and more reason not to know about the risk so that they won't face a moral crisis when they sell their homes: "If I don't monitor, and I, therefore, don't find out I have a problem, I don't need to deal with the need to remediate. That is an emotional gain and a financial gain for me, and a major motive not to be alarmed."

Ten, the latency for lung cancer is long; nothing is happening to make us feel unhealthy, and nothing will happen to make us feel unhealthy until far too late to do much about the radon.

RECOGNIZING THE LOW LEVEL OF CONCERN

To be sure, there are some characteristics of the situation that lead to alarm. In particular, of course, cancer in general is a highly dread consequence, and radiation, in general, is more highly feared than most other sources of disease. But Paul Slovic demonstrated that this is not always true for all kinds of cancer or for all kinds of radiation. It is not, for example, true of x-rays. I think as you look through this list of the ten characteristics I've just given you, there is good theoretical reason to expect that for most people the radon problem is going to be one they have trouble paying enough attention to. Those of us working on this issue have trouble paying enough attention to that! Most of the people attending this conference are interested in radon--that is why they are here. Most of the people we talk to about radon are also interested in radon. Many of us are radon professionals in one way or another. Many of the rest of us have become alarmed, have become concerned, and we walk around in a universe of our own creation, of people who are very concerned. If you spend two weeks working on the DEP Hotline, you go home, after two weeks of work, quite sure that everyone in the state of New Jersey is alarmed about radon. We are at natural risk of misperceiving the level of concern that is genuinely out there.

The Swedish went through a comparable period of radon concern to what we in New Jersey and other states are going through now. They went through it a decade ago. Baruch Fischhoff, who works closely with Paul Slovic, commented at a conference I was at two weeks ago that in his judgment, in the Swedish experience, the problem was that people were unable to take the issue as seriously as the authorities wanted. As Fischhoff put it, they wished they could get the public to feel a little bit more panic.

I think that may well be the same experience that we are going to have in New Jersey. To be sure, DEP and DOH and other agencies are encountering non-trivial numbers of very alarmed people: people who are taking the risk very seriously, people who need and deserve reassurance and guidance and counsel as they work their way through this new and very serious problem. I don't want to leave you with the impression that there are no people who are taking the risk very seriously and who are very alarmed. But there are also many, many people (and the evidence suggests more people) who are underestimating the risk.

I make it a habit to go to public hearings about Superfund sites, because they're interesting. I went to the Clinton hearing about radon, and the contrast was remarkable. If people were going to be panicky, this was a good opportunity. Clinton was the first community in New Jersey with very high levels; the State was there in force; the County was there in force; local and state legislators were there; the EPA was there. These are fairly well-off people who are accustomed to taking risk seriously, and demanding attention. In contrast, the mood at Clinton was concerned, serious, sober, but very, very far from panic; they might have been talking about a bond issue. If you've ever gone into a meeting to discuss a Superfund site, you are at no risk of confusing that with a discussion of a bond issue. I think there is evidence in that experience. If the people who turned up on Saturday morning in Clinton--the first community in New Jersey to face a visible serious problem--are bringing their children, as a civics lesson, and sitting around taking it seriously and being very responsible and very sober, then the people who stayed home are not at great risk of panicking. They are at far greater risk of ignoring the issue.

In the data from the survey that Neil Weinstein has described, one question in particular speaks directly to the issue. We asked people how they felt about monitoring their own homes. Of the six hundred responses, 7 percent said they have already monitored or have a test in progress. Another 11 percent say they plan to monitor. That's a total of 18. On the other side, 7 percent say they see no need to monitor. Everybody else is in the middle. "In the middle" breaks down into two categories. 25 percent say they haven't decided. And 50 percent say they have never thought about it. That's important. We are asking 600 people in north Jersey, on or near the Reading Prong, where the publicity has been most serious and most attentive, how they feel about monitoring. We get 18 percent who are doing it or planning to do it, 7 percent who say there is no need to do it, a quarter who say they haven't decided yet, and half who haven't thought about it yet.

This does not suggest an audience at high risk of panicking. It does suggest an audience that is open to communication, that will pay attention to communication and that needs to be alerted to the risk; that will respond to the alerts rationally and with the kind of the sober concern we saw in Clinton. But given a decline, rather than an increase in communication, those people seem to me quite likely to let the issue go. Neil Weinstein and I will be doing future waves of monitoring with these same people -- not monitoring their homes, but monitoring their attitudes, trying to find out what happens to some of these numbers.

CONCLUSION

I think it's worth considering what we want the goal to be. I would like to see virtually every single family homeowner in north Jersey think that he or she wants to monitor. I think that's a reasonable goal. If you live in a house on the ground in north Jersey, I think it would be desirable if you wanted to monitor. We have only 18 percent, at the moment, who want to monitor. We need to reach the other 82 percent with communication. The data suggest that they are open to being reached, that they are interested in the issue, concerned about the issue, not terribly alarmed, not terribly angry, not terribly frightened, and likely to forget the issue if we don't keep the communication up.

CONSIDERING THE PUBLIC'S PERCEPTION OF RISK IN THE DEVELOPMENT OF A RADON COMMUNICATIONS PROGRAM—RECOMMENDATIONS OF A SYMPOSIUM TO THE NEW JERSEY DEPARTMENT OF ENVIRONMENTAL PROTECTION*

Caron Chess**

INTRODUCTION

The contamination of indoor air by radon gas confronts the New Jersey Department of Environmental Protection (DEP) with a unique set of risk management concerns. Although the Department is playing a critical role in investigating the nature and extent of the radon problem, unlike other environmental problems, the DEP cannot realistically undertake most of the radon risk reduction effort. Instead, it is likely that the actions of individual homeowners will be important to protecting public health. This is in contrast to environmental problems such as hazardous waste sites where the state clearly plays a major role in mitigation efforts. Furthermore, the potential numbers of people affected and level of health risk involved with radon contamination may increase the likelihood of fear and confusion on the part of the public.

In light of the above factors, the Office of Science and Research, and experts with whom we have consulted, believe that the Department's efforts to communicate about the radon issue can greatly influence the public's perceptions of the problem and ways in which the DEP is perceived. These perceptions can, in turn, lead to action or inaction on the part of the public that could seriously affect public health in the state.

In order to assist the Department to deal with the communications aspects of this critical issue, a symposium of the country's leading risk communication experts was convened in October, 1985. Although the Department had previously convened scientific advisory panels, consisting primarily of acknowledged scientific experts based in New Jersey, the radon symposium marked the first time the

---

*This paper is based upon Neil Weinstein and Peter Sandman, "Recommendations for a Radon Communication Program," Report to the New Jersey Department of Environmental Protection, November 5, 1985.

**Office of Science and Research, New Jersey Department of Environmental Protection, Trenton, NJ 08625.

Department formally brought together experts from throughout the country. It was also unusual for the Department to consult a panel of experts concerning the communication aspects of a problem rather than technical concerns. The DEP's Office of Science and Research (OSR) took this innovative approach because of the potential magnitude of the radon problem and the difficulties in adequately communicating the issue. In addition, OSR saw such a symposium as a way to explore the vital issue of risk communication while using the radon issue as a case study. Because of their unique combination of practical expertise and theoretical background, Drs. Neil Weinstein and Peter Sandman of Rutgers University were contracted by OSR to convene the symposium and develop several reports.

Prior to discussing the communication aspects of the radon issue, symposium participants received an evening technical briefing from NJDEP and NJ Department of Health personnel as well as a description of Pennsylvania's community relations program from a representative of that state. A full day of discussion followed. Subsequently, participants were contacted by mail to elicit their rankings of findings from the symposium. The following working document is based on a complete report of recommendations from the symposium, developed by Weinstein and Sandman, and available from the DEP's Office of Science and Research.

The goal of the symposium was to provide guidance about many of the issues the state should consider when developing risk communications programs, and the radon situation was used as a case study to focus the discussion. As with any communications program, it is difficult to develop the details of a plan until there is clarity on the basic communications strategies and approaches to be used. This is similar to development of a scientific research project in which a sampling plan cannot be determined until the methodology and approach is clear. Thus, one goal of the symposium was to make recommendations concerning the approach the state might take to radon communications so that the state could later translate the recommendations into a detailed plan addressing specifics such as budget, development of educational materials, etc.

COMMUNICATIONS GOALS

The recommendations of the symposium were based on goals that were so apparent that they did not elicit controversy in discussion among participants. <u>The primary goal was to create accurate understanding and a healthy degree of concern on the part of the public.</u> That is, a radon risk communication program should elicit appropriate responses rather than either panic on the one extreme or apathy on the other. Closely related was the aim of encouraging and facilitating appropriate risk mitigation behavior. While these goals were clearly key, the following symposium recommendations reflect an emphasis on the goals (often underscored by participants) of maintaining the credibility of state agencies and coordinating with NJDEP's monitoring program.

Choosing a Communications Strategy

One of the most problematic questions involved with government developing risk communication programs is: How active a role should government take in communicating information when the information is uncertain or incomplete? It is perhaps understandable for government to be hesitant about taking an active communications role on an issue as potentially volatile and scientifically difficult as radon. There tends to be a healthy concern about unduly alarming the public as well as concerns about actively raising radon as an issue before programs have been fully developed to help the public. In addition, radon is an issue which raises many questions which research has not yet adequately answered. However, after spending considerable time wrestling with this question as it related to radon, the symposium participants expressed an overwhelming preference:

> The state should immediately take the initiative in explaining the problem and the state's plan to the press and those who seek information, including health personnel and elected officials. An active public information campaign would be delayed for a few months (2 to 6 months from the October symposium) to allow time to plan the campaign and to verify from preliminary monitoring that the hazard is genuine.

Participants identified the following factors as key to their determination that the state should conduct an active public information campaign:

1. Fill the information vacuum

While acknowledging the potential reluctance of state officials to initiate a high visibility campaign, risk communuication experts (from an ad agency executive who manages industrial accounts to government officials involved in cancer education) felt that the state would be running greater risk in failing to initiate a high visibility communications program. One of the most salient reasons was that if the state doesn't define the issue, someone else will fill the information vacuum; this information is likely to be less accurate than that the state would provide and is also likely to include a negative portrayal of the state's actions. This cautionary statement may be prophetic in light of many inaccurate press reports and the initiatives of elected officials since the symposium. In addition, participants noted the responsibility of the state to inform citizens of actions which they could take to protect their health.

Despite protracted discussion during the symposium, participants, when indicating by mail the factors that were important in choosing a communications strategy, viewed the possibility of increasing public alarm as insignificant when compared to the risk of the information vacuum.

2. Maintain trust and credibility

According to symposium experts, a low-profile approach is likely to hurt the state's credibility when the public learns of the problem. For example, homeowners who are not on the Reading Prong could become angry and distrust messages coming from the state if they became aware that the state knew they were potentially at risk long before they heard they were at risk. That distrust is more likely to then translate into inappropriate reactions to information disseminated by the state.

3. Provide information

The symposium suggested that it was unlikely that the state could keep information about radon low profile for a protracted period of time; therefore, the state was far better off attempting to control the flow of information by actively acting as the source for it. In addition, the symposium participants felt that, based on responses by state personnel, the state had cogent arguments for its risk assessment and mitigation plans. Although they understood the uncertainty of the information, the experts felt that the information could be presented in a positive context and in such a way that the public could understand it. By using this approach it was thought that the state would increase the accuracy of information, reduce the potential of public reaction characterized by fear and suspicion, and increase the likelihood that people will take recommended actions.

4. Emphasize planning

Participants felt that adequate planning was critical to the success of any communications effort. Thus, despite their general support for a high visibility approach to the radon issue, symposium participants favored an initial 2 to 6 month period of low visibility during which planning might occur.

Recommendations of "special importance" for New Jersey's communication strategy

When participants ranked suggestions raised in the symposium in a subsequent mail survey, these recommendations were ranked as having "special importance:"*

Increase feelings of individual efficacy

Risk perception research strongly suggests that people are less apt to react with fear and confusion if they can easily take some form of positive action, or better yet, exert individual control over a fearful situation. Thus, many of the symposium recommendations relate to increasing feelings of efficacy:

> The program should stress the availability of risk-reduction measures, that action can be taken, and that citizens who want to can take steps on their own to measure and reduce risk.
>
> The program should expect people to seek private monitoring before the State results are in.
>
> The program should prepare citizens to be wise consumers of private/commercial monitoring and mitigation services, perhaps through the development of training programs, instructional literature, consumer warnings, etc.
>
> The program should facilitate training of the public in how to use monitoring devices and interpret findings, perhaps through the use of health officers, the development of instructional literature, etc.

Respond to concerns

Communications research suggests that people are more likely to respond positively to information that meets their needs and concerns. In other words, a radon communications program should not only tell people what the state feels it is important for them to hear but should also address the issues that the public feels are most salient. The following recommendations from the symposium address this issue:

> The program should prepare in particular answers to questions that the symposium experts felt were

*For a complete listing of recommendations, see the original report by Weinstein and Sandman.

most likely to be asked. They include: How do I find out what my risk is? How accurate is monitoring (private and state)? How costly, intrusive, etc. is the monitoring? How costly and how effective are the various mitigation options? What health risk is associated with my radon level?

The program should analyze the questions asked on its information telephone line, and prepare literature, telephone answers, training programs, etc. to respond to the most frequently asked questions.

The program should stress that the radon risk is geographically variable, that your neighbor's high (or low) reading is not a reliable guide to your own risk, even if your neighbor's house is very similar.

Maintain the state's credibility

Communications research suggests that people's response to information is determined, in part, by their degree of trust in the messenger. As a result, many of the symposium recommendations dealt with the state developing and maintaining trust. This is particularly important with an issue such as radon when people might be apt to feel like victims and might want to find a villain to blame. In fact, preliminary results from a pilot survey (conducted by Drs. Weinstein and Sandman under contract with NJDEP) of New Jersey's residents' perceptions of radon indicate feelings that government is one of the parties they hold responsible for the radon problem. The recommendations of the symposium would suggest that the state should work actively to counteract that misperception.

The program should organize the state communication effort so that all spokespeople are giving similar answers; insofar as possible, the state should speak with one voice.

The program should stress that the state is working actively to study which areas are most at risk, what mitigation approaches are most appropriate, etc., that while individuals can act now, the state has not abandoned its responsibility to help.

The program should stress that the radon problem is no one's fault--neither the state's nor the citizen's.

> The program should stress that the radon problem is serious; no effort should be made to downplay the problem or belittle the fears of citizens.
>
> The program should stress that radon is a naturally-occurring problem, that there are no villains with respect to radon.

Define audiences and channels of communication

Communications experts suggest that information programs are most effective when they are developed for defined audiences. While a message should be consistent across audiences, it should also be adapted to the needs of particular audiences in order to maximize effectiveness. Thus, many symposium recommendations dealt with how to define the audience for a radon communications campaign and how to reach those audiences, including the media, which are obviously key.

> The program should try to define the radon topic in the mass media, rather than waiting for others or the media themselves to define it; that is, the program should recognize that the media in particular do not tolerate an information vacuum for long, and that the media program should therefore be pro-active.
>
> The program should include a training program for journalists in the fundamentals of radon and radon monitoring and mitigation; the training should be aimed not only at reporters, but especially at editors.
>
> The program should monitor the information needs of various constituencies and continue to prepare and disseminate appropriate answers, bearing in mind that information needs will change over time.

CONCLUSIONS

The experts spoke in terms of possibilities and probabilities -- not certainties. They indicated that an information vacuum would increase the likelihood of fear and confusion and decrease the likelihood that the public would take health protective responses the state feels are appropriate. Symposium participants felt that a high-profile approach would be more likely to facilitate prudent action.

The extent to which the symposium stressed the need for planning cannot be overemphasized. The participants stressed the necessity of planning the development of informational materials, coordinating activities among agencies, and assessing the success of the communications program in relationship to clearly articulated goals. The type of information conveyed and who it is conveyed to are critical. This is analagous to scientific research in which a sampling plan is vital to ensure that the samples in the aggregate yield meaningful results. Thus, it is possible to develop many brochures and not communicate effectively, while developing fewer materials could yield more meaningful results.

Finally, while the initial symposium report is specific to the radon issue, analysis by Drs. Weinstein and Sandman in a subsequent report suggests that some of the recommendations of the symposium would apply generically to communicating about risk. The symposium raised questions about the extent to which the DEP should incorporate risk communication into its development of policy to deal with other environmental problems, particularly those for which public perception and response are key elements.

MEDIA AND THE PERCEPTION OF RADON RISK*

Michael R. Edelstein and Valari Boyle**

What is the contribution of the media to the public's understanding of the threat from radon gas exposure? To ascertain this, coverage of radon in a northern New Jersey paper was reviewed. A small study was then conducted in which comparisons were made between the responses of students reading a newspaper article about radon which was rated as highly threatening with responses from students reading an article portraying radon in a much less threatening manner.

The second author initially reviewed 41 articles on radon printed between December, 1983 and February, 1986 in The Record, a prominent northern New Jersey newpaper, whose coverage of the radon issue appears to be representative of the general presentation of the issue. She rated statements in the articles as falling into one of the three categories:

a) a threat message (for example, "radon causes lung cancer")
b) neutral statements, and
c) qualifiers (for example, "radon problems can be solved").

These ratings were subsequently compared to ratings by students in an environmental seminar who made similar judgments. Based upon this analysis, several conclusions emerged. The majority of statements in the articles were threat messages. Furthermore, the presentation of the radon issue was in some ways misleading. For example, a possible point of confusion was information about the geologic correlates of radon. Many articles stated or implied that radon exposure in New Jersey was associated with one geological formation, the Reading Prong. Readers would be well justified to assume that they were not at risk if they lived outside of this formation. Furthermore, there was a

---

*A grant by the Ramapo College Foundation to the authors was used to underwrite costs of the research project described here. The assistance of Rebecca Manus in this work is gratefully acknowledged.
**Michael Edestein is an Associate Professor of Psychology at Ramapo College and was a conference coordinator. Valari Boyle is a sophomore at Ramapo College.

noticeable amount of contradiction between articles. For example, one article might qualify the radon threat by noting that it is a solvable problem, whereas an article several days later would indicate that remediation attempts were unsuccessful. The result of reading the collection of articles might well be confusion, reflecting inconsistencies and technical disagreements. Additionally, readers who particularly recalled a given article might have their views of radon dominated by the extent to which the article contained threat messages or qualifiers. If that article minimized the risk, their estimation of risk might be lessened. Conversely, some articles might induce active fear of the threat from radon (see Slovic, Fischhoff and Lichtenstein, this section).

To explore this later point further, a study was devised which sought to compare perceptions of an article which contained mostly threat messages and one that contained many qualifiers. We expected there to be differences in ratings of the radon issue between audiences exposed to the threat article versus those exposed to the qualified article such that the more threatening article would induce greater concern and fear for radon.

The Study

Two articles were selected to illustrate extremes within the range of typical radon media coverage. Both articles were of equivalent length and fit on a single page. Both came from the same newspaper, The Times Herald Record of Middletown, New York.

The high threat article[1] was entitled "Radon may threaten the health of millions," summarized in Table 6.6. Of 14 brief paragraphs, nine contained at least one message which could reasonably be seen as threatening. Only one clear qualifying message is found. Five messages could be interpreted as either qualifying or threatening, depending upon one's perspective. Only one paragraph contained neutral information. Threatening information is placed early in the article.

In contrast, the "low threat" article[2] gave quite a different picture of radon. This article, entitled "Scientists discount radon risk," is illustrated in Table 6.7. The article was printed with two pictures, one of an expert addressing a local meeting; the other of the audience watching. The expert's photo is accompanied by a quotation printed in large type which says that there is little risk involved. Unlike the high threat article, which was a wire service piece, this was written by a local reporter covering a local meeting. As the table shows, of 11 paragraphs, five serve to qualify or lessen the threat of radon, one includes a high threat statement, and two are neutral. Four other paragraphs might be interpreted as either high or low threat. Instructively, the qualifying statements are early

P.1 Federal health officials say millions of Americans are exposed to higher-than-recommended levels of radon, a naturally occurring radioactive gas blamed for as many as 30,000 lung cancer deaths yearly.

P.2 Background to radon issue: engineer who had very high levels of radiation in air in his home.

P.3 A survey in that area discovered 40% of houses over the recommended guidelines and 7% over the stricter occupational levels.

P.4 Quote expert that "sizable percentage" of US homes could be above guidelines.

P.5 Expert quoted "That doesn't mean people should panic" ***but 6 million Americans may have radon levels of concern.

P.6 EPA has set no precise exposure limit but is working on it.***

P.7 Engineer Stanley Watras's family was exposed to 50 times the recommended occupational exposure.

P.8 Radon byproducts inhaled with dust damage lungs. Studies link it to cancer.

P.9 Watras relocated until [remediation could be completed.]

P.10 Reading Prong described.

P.11 Reading Prong stretches through Pennsylvania, New Jersey and southern New York -- lists local towns.***

P.12 Pennsylvania's elevated radon levels may indicate broad national problem.

P.13 Other states with problems listed. Adjacent homes may have different radon levels.***

P.14 EPA has begun testing remediation methods.***

Table 6.6: Statements in 14 paragraphs coded for threat content from the "high threat" article.[1] Threat messages are underlined. Qualified statements appear in brackets. Statements given to varied interpretation are marked with three stars. Neutral statements are unmarked.

P.1 State [health officials] said that local tests indicate that there is [little to fear] from radon in Orange County.

P.2 Expert says that "we don't know everything there is to know about radon..."*** but [it doesn't look like this area has a big problem."]

P.3 Expert spoke at meeting before 90 people.

P.4 Expert did 50 soil samples revealing "about average" radium levels.***

P.5 Further testing said to [not be required.] Samples termed [scientifically dull.]

P.6 Scientists [discounted widespread problems] in area; said to be [unlike Pennsylvania.]

P.7 Radon <u>causes lung cancer</u>. Scientists have said there <u>might be radon in area</u>.

P.8 Description of Reading Prong.

P.9 Description of local areas on Reading Prong.***

P.10 Cites another expert at meeting saying that radon causes cancer in less than 1% of cases while smoking is largest cause of lung cancer.***

P.11 Quotes expert: ["What we are really talking about here (sic) a very small risk."]

<u>Table 6.7:</u> Paraphrases of paragraphs in the "low threat"[2] article. Underlining indicates a threatening statement, qualified statements are bracketed and *** indicates a statement that can be read either way. Neutral statements are unmarked.

in the article while the potentially threatening statements come much later.

A study was conducted in classes at Ramapo College using these articles. Students were given a two-page questionnaire asking them several demographic questions, as well as questions about environmental hazards, focusing particularly on radon gas. This served as a pre-measure of their perceptions. They then were given either the high threat or low threat article to read. Finally, they answered several additional and repeated questions about the radon threat. In all, 44 high threat and 32 low threat questionnaires were collected in four courses.

Results

In order to be certain that differences between the high and low threat groups were due to the articles, a baseline comparison of the groups was done by assessing the demographic data and other pre-measures. Table 6.8 summarizes some results for several baseline indices. As shown, the two groups differ in several ways which may be significant. Specifically, while there were no differences between groups in location of home, in loss of family member through lung cancer, in smoking, or in residential stability, the low threat group is older, contained more women, more homeowners, more married people and involved three times as many children (12 children in five families as opposed to 4 children in 3 families). This comparison suggests that the low threat group might perceive themselves to be more susceptible to the radon threat, a factor not accounted for in the test articles that nevertheless might influence ratings of the radon threat.

	AGE			SEX		OWN HOME		SMOKE		CANCER		MARRIED	
Threat	18-20	21-25	25+	M	F	Y	N	Y	N	Y	N		
Hi	16	56	28	60	40	16	84	20	80	19	81	9	91
Lo	14	32	54	35	65	39	61	25	75	16	84	31	69

Table 6.8: Some demographic comparisons of the two threat groups (reported in percentages).

Do the groups vary in their ratings of environmental hazards? Prior to the reading of the article, respondents were asked two questions about their general perception of environmental hazards. On a seven point scale with low scores indicating a lesser impact, respondents rated themselves as moderately aware of (M=4.77) and frightened by (M=4.13) environmental hazards. In both cases, the high threat group was most sensitive to environmental hazard, as Table 6.9 indicates. They also rated radon as slightly more dangerous. Yet, despite this general sensitivity to environmental hazard within the high threat group, the low threat group indicated greater fright of radon and a slightly higher rating of radon as an environmental problem. The greater sensitivity of the low threat groups to radon may reflect the demographic differences noted above. However, in the absence of statistical tests, unavailable at this time, one must be cautious in assessing the meaning of these differences.

	Hi	Lo	Total
Environmental Hazard Awareness	5.05	4.41	4.77
Fright due to Environmental Hazards	4.45	3.81	4.13
Radon Seen As a Problem	4.67	4.81	4.74
Fright due to Radon	4.07	4.55	4.31
Danger due to Radon	5.26	5.07	5.16

Table 6.9: Mean scores on seven point scales (7=high value) for pre-article perceptions of general environmental and radon hazards.

It is interesting that both for environmental hazards generally and for radon, the abstract ratings of the hazard were greater than the rating of personal fright for both groups. Subjects rated their awareness of environmental hazards as higher than their fright due to environmental hazards, implying that such hazards were not generally seen as personal. Similarly, with radon, while the danger of radon was rated as fairly large and it was seen as a problem of at least moderate size, fright due to radon was quite moderate.

Two additional questions were asked which reflect possible perceptions of radon risk. As a measure of active radon concern, subjects were asked whether their families had tested for radon. Only two respondents in each group had tested for radon in their homes, none finding significant levels. Secondly, respondents were asked whether they lived on the Reading Prong. While about half didn't know, 20% of the high group and 12% of the low group indicated a Reading Prong residence. When their fright of radon was examined in light of their residence on the Reading Prong, there appeared to be some interesting differences within the groups. While the low threat group revealed slightly more fright among those living off the Reading Prong, the high threat groups showed a strong difference between fright among Reading Prong residents (M=4.67) and those living off of the Prong (M=3.35). That is, those living on the Prong tended to be more frightened by the threat of radon than those living off the Prong. Given the low threat group's generally high susceptibility, it may be that they do not perceive location as the critical factor accounting for radon danger. The high threat group, on the other hand, seems to reflect the general belief that the Reading Prong is a direct correlate of danger.

Following their review of either the high or low threat article, the subjects answered an additional set of questions. Several of these focused upon the meaning of the article they had just read. Ratings of the accurateness of the article revealed a strong difference between the conditions, with the low threat article seen as less accurate (M=3.1) than was the high threat article (M=4.73). When asked about the main points in the article they had just read, the most common response for both groups was the association of lung cancer with radon (see Table 6.10). While no difference is seen between the groups in acknowledging the radon cancer link, it is interesting that low threat respondents tended to list little else while high threat people noted quite a few additional points. On one hand, this may simply reflect a greater variety of potential points in the high threat article. On the other hand, the high threat article may have invited greater vigilence on the part of readers, leading them to draw more conclusions. Additionally, the high threat group was more likely to see the article as confirming their own geographic safety and touching on the physical attributes of radon. Interestingly, they also were more likely to interpret the article as reporting that radon was not a problem, possibly indicating a degree of denial on their part. At the same time, 10 high threat readers surprisingly volunteered the view that their article minimized the problem.

Response	Hi	Lo	Total
Lung Cancer	17	14	31
Not in my area	9	1	10
Article minimizes problem	10	1	11
Not really a problem	6	1	7
Radon is colorless and odorless	5	0	5

Table 6.10: List of main points in the article with frequency of mention.

A final question explored what parts of the article were frightening and what parts reassuring. Table 6.11 lists the main points seen by respondents as frightening. Aside from two low threat group people who were concerned by the dismissivenss of their article, Table 6.11 reveals only two sources of fright from the low threat article, lung cancer and geographic location. The high threat article was seen as having only one additional component that was frightening, the discussion of radon levels and statistics. It is interesting that the identification of one's home area as a radon-prone region is perceived as more frightening than the reports of cancer risk. Location may trigger an awareness of personal susceptibility more immediate than the probabilistic random long term risk of cancer. Five respondents indicated that there was nothing at all frightening about their article (3 in the low and 2 in the high threat condition).

Response:	Hi	Lo	Total
Frightening			
Lung Cancer	4	4	8
Radon affects my area	6	4	10
Radon statistics	5	0	5
Reassuring			
Government recognizes the problem	11	0	11
Problem can be fixed	4	0	4
Don't live in affected area	1	3	4
Scientists discount	1	3	4

Table 6.11: Frightening and reassuring points in articles.

Table 6.11 also lists the reassuring points of the articles. The most selected response here was the recognition by government of the problem, selected exclusively in the high threat condition. Also reassuring were the indications that the problem can be mitigated, the realization that one does not live in a vulnerable area and the views of experts who discount the problem. Note that the low threat group found much less reassurance than did the high threat group. Similarly, of 9 respondents who answered that there was nothing reassuring about the article, 7 were in the low threat group. It would again appear that the high threat article invited denial while the low threat article aroused suspicion that something was being hidden.

Several questions were repeated from the set which preceded the article. Respondents were again asked to rate radon as a problem, to indicate their degree of fright and to rate the degree of danger from radon. As Table 6.12 indicates, the articles appear to have had some definite influence upon perceptions of radon. As expected, the rating of the importance of the radon problem dropped for the low threat group and increased for the high threat subjects. Similarly, while the mean rating of the danger from radon stayed the same, the high threat group increased its post article estimation of danger while the low threat group dropped theirs. It is also not surprising that ratings of radon fright decreased sharply for the low threat group which was demographically more susceptible to radon harm to begin with. Again indicating possible denial on their part, the high threat group also indicated less fright after reading their article. Fright is equivalent in both groups post-article. It would again appear that there is a difference between personal threat and the evaluation of the problem as an abstract entity. While reading the articles diminished the personal threat (i.e., fright) for both groups, it produced opposite and expected effects upon the more distant estimation of danger and the importance of the problem.

		Hi	Lo	Total
Radon Problem	Pre	4.67	4.81	4.74
	Post	4.93	4.19	4.56
Radon Fright	Pre	4.07	4.55	4.31
	Post	3.91	3.93	3.92
Radon Dangerous	Pre	5.26	5.07	5.16
	Post	5.52	4.79	5.15

Table 6.12: Mean scores on seven point scales (7=high value) for pre and post article perceptions of general environmental and radon hazards.

Table 6.13 displays several questions which explore respondents' expectations. Neither group saw a radon problem in their homes as particularly likely, a further indication that the problem is not seen as personal. The high threat group indicated a somewhat greater likelihood, as would be expected.

Neither group is strongly dependent upon government to solve the radon issue, with the high threat group the least so. Neither are the groups likely to test for radon, although the low threat group is particularly unlikely to test. This is especially interesting in light of the demographic differences that make this groups more susceptible. Despite their doubts of its accuracy, it appears that the low threat newspaper article reassured this group that testing is unnecessary.

	Hi	Low	Total
Likelihood home affected	3.57	3.04	3.31
Government Dependence	3.68	3.83	3.76
Likelihood will test	3.63	2.66	3.15

Table 6.13: Mean scores on seven point scales (7=high) for post article questions.

Finally, respondents were asked whether they would test if they were buying a home and whether they would purchase a home that had a higher than normal radon rating. These results are given in Table 6.14. In keeping with the prior question, high threat respondents overwhelmingly indicated that they would test a prospective home, appearing to be much more likely to do so than would low threat respondents. Neither group would be likely to buy a home where radon had been identified, however.

		Yes	No	Maybe	Don't Know
Would Test	Hi	24	4	3	0
If Buying	Lo	10	4	5	1
Would Buy	Hi	0	28	1	1
If Radon	Lo	1	17	2	0

Table 6.14: Frequency of response to two questions about home purchase.

Discussion and Conclusion

Overall, it is clear that its presentation in a newspaper article can affect at least short term perceptions of the radon issue. Generally, predictions were confirmed that a "high threat" article will induce an evaluation of the radon threat as more serious than will a "low threat" article. This occurred despite demographic differences between the comparison groups that suggested that the low threat group was more susceptible to the radon threat. The articles seemed to have the greatest impact on abstract rather than personal estimations of the threat. Locational information appears to be more personal and may make the more abstract threat of cancer more or less salient for the reader, suggesting the distorting effect of misinformation regarding the Reading Prong. Some indication of denial in the face of fear-inducing information is also found. It is particularly instructive that even the high threat group is not likely to test for radon, although this group, largely consisting of younger college students, would test if they were buying a home. Government involvement is seen as highly reassuring to the high threat group, a fact that suggests the psychological importance of an effective government response (see Weinstein and Chess, this section).

What does all of this suggest? Clearly, the media may play a major role in determining how the radon issue is viewed and how effectively people respond to the threat. It is almost certain that the media portrayal of radon contributes directly to a possible underestimation of the problem by the public as discussed by Sandman (this section). To understand this outcome, one would need to go beyond this simple study to examine the cumulative impacts of media messages containing contradictory threat messages. At the same time, the fact that any given article is itself likely to contain these contrary messages suggests that the media reflects the ambiguity found in a developing field of inquiry which has additionally become a topic of political sensitivity.

FOOTNOTES

[1]Associated Press, "Radon May Threaten the Health of Millions," The Times Herald Record, Friday November 1, 1985, p. 5.

[2]Ruth Boice, "Scientists Discount Radon Risk," The Times Herald Record, Wednesday July 10, 1985, p. 3.

PSYCHOLOGICAL IMPACTS OF RADON GAS EXPOSURE: PRELIMINARY FINDINGS

Margaret Gibbs, Michael R. Edelstein, and Susan Belford*

Psychologists have discovered that exposure to stress is bad for your mental health. We know this through our study of the stress of disaster. Hurricanes, floods, earthquakes, plane crashes, war, shipwrecks, and nuclear disasters can all lead to enormous human distress and eventual psychopathology in some persons. Recent studies, including some that both the senior authors of this paper have conducted, have found that exposure to toxics, such as in water contamination, can also cause distress and psychopathology.[1-10] One would expect that serious radon problems would also be detrimental to mental health.

One of the problems of this whole body of stress research is that of separating the effects of stress from the pre-existing personalities of the individuals involved. That is, skeptics about the negative effects of stress claim that victims of disaster are sometimes self selected, and that those individuals who show emotional reactions have pre-existing personality deficits. For instance, it is argued that those veterans who come down with Post Traumatic Stress Disorder had pre-existing personality deficits, and that individuals who went to Vietnam and saw combat were less stable than individuals who managed to avoid the experience, thus predisposing them to psychopathology. The literature does not support the skeptics' interpretation, but the argument is nevertheless hard to discount. The basic dilemma of disaster research is that pre-test measures are not available for the victims. That is, we don't know when or where the earthquake, the toxic water crisis, or the plane crash is going to strike.

An exception to this appeared in the case of radon gas exposure. When early findings identified the Reading Prong as a source of radon, a major portion of Orange County, New York suddenly was labeled as a likely site of elevated radon levels. To ascertain the extent of radon exposure both on

*Margaret Gibbs and Susan Belford are from Fairleigh Dickinson University (Gibbs is Professor and Chair of the Psychology Department; Belford is a degree candidate in that department); Michael Edelstein is Associate Professor of Psychology at Ramapo College of New Jersey and a conference coordinator.

and off the Reading Prong, a non-profit, tax-exempt organization, Orange Environment, Inc., sponsored a voluntary testing program. Given that the first two authors are board members of this organization, an excellent opportunity was presented to study stress related to radon exposure. Funding was received for this effort from the Ramapo College Foundation as well as through the Natural Hazards Research Center of Boulder, Colorado.

Two successive groups of residents purchased alpha-track detectors through Orange Environment, placing them in their homes for three-month periods. Some of these families came from the Reading Prong area, while others lived off of the Prong. Following the return of these detectors to the laboratory, another two-month lag time occurred before the residents were informed of their results. The two-step nature of this process invited a longitudinal study comparing participants at the time the detectors were placed in the home with the point at which scores were returned. Results from this part of the study are just being collected.

Two League of Women Voters' chapters were selected as a comparison population for contrast to participants in the Orange Environment program. The League groups were expected to differ little from those testing for radon on important indices related to perception of environment hazards. Virtually none of the League respondents had tested for radon. The comparison of the League members with the Orange Environment sample would thus allow for an assessment of differences between those who respond to the radon threat by undertaking tests of their home and those who do not.

A secondary comparison was intended between the two League of Women Voters chapters, one of which was centered on the Reading Prong, and one which was located ten miles to the west. It was thought that there would be differences betweem these groups stemming from the varied proximity to perceived risk.

Our long-term expectations were that those members of the testing group who found higher levels of radon in their home would show increased levels of stress on the widely-used Horowitz Impact of Event Scale and on questions asking about their happiness, amount of worry, worry about health, sense of control over the environment, perceived competency, trust of others, and social intimacy.

We expected the group testing for radon, prior to knowing their results, to be comparable in most ways to the comparison League groups. We expected that their stress levels would be similar and low (since none yet knew of any radon in their home) on both the Impact of Event Scale and on the above questions asking about current level of adjustment. We did, however, think that individuals testing their homes might be motivated by an increased concern about their own and their children's health, so that we expected differences between the groups on that question.

The questionnaire also explored various aspects of the perception of threat from radon. It was expected that, in contrast to the comparison group, the testing group would rate the risk from radon as higher and that the testers would see themselves as more likely to have high exposure and more susceptibility to harm from radon. It was further thought that proximity to the Reading Prong and type of home (particularly in terms of energy efficiency) would affect estimates of the likelihood of exposure, while such factors as presence of children in the home and length of residence would affect perceived susceptibility from exposure.

The questionnaire contained a further examination of the impacts of radon exposure upon health, the home, the environment, the community, trust in government, and one's personal sense of control. Respondents were also asked about their perceptions of the difficulty of mitigating radon problems, knowledge of radon, who was to blame for the radon threat, government handling of the radon issue and whether their own actions had contributed to their families' risk. Finally, respondents indicated their most and least trusted sources of information about radon. It was generally expected that those choosing to test for radon would have a greater appreciation for its potential effects.

It was also expected that the later phase of research would discover systematic changes on these indices within the testing group according to the level of radon discovered in their homes. Specifically, for "high" scorers, their assessment of the risk of radon would change, the amount of information about radon they obtain would increase, the sources of information would shift, and unless intervention from government occurs, their trust in government and sources of information would decrease.

RESULTS

There is insufficient data at this time to discuss the main research hypotheses. We have just begun to receive copies of our follow-up questionnaire from our first sample of families who had their homes tested. Of this group, 29% had radon readings above 4 pCi/l with the range of readings between 0.14 and 60 pCi/l. Thus, while there is no serious cause for alarm in the area, there are indeed individuals who should take action to remediate the problem in their homes, and who probably will experience stress and a shift in attitudes. These individuals will be followed up and our findings reported at a later date.

The findings that can be reported at this point involve our hypotheses about the type of individual who tests the home for radon. We had two groups of individuals who tested their homes. Each family was given two questionnaires; a total of 236 were handed out, and 76 were returned. The return rate was actually higher than the apparent 32% as some percent of the families had only one adult. Comparison of the responses of the two groups obtained no more differences than would be expected by chance. They are thus combined for the purpose of analysis.

As noted above, members of the two chapters of the League of Women Voters were used for comparison. This group was chosen because we thought it would be comparable in education and environmental awareness to the radon testers. Approximately 120 questionnaires were distributed, and 35 were returned, a return rate of 29%. Comparison of the responses of the two control groups obtained no more differences than would be expected by chance. Thus, expected differences between the League chapters due to proximity to the Reading Prong appear not to have materialized. However, a particularly poor response from the League chapter furthest from the Prong may itself reflect a lower appreciation of the issue by respondents in a locale thought to be "safe."

Demographic differences between the experimental and control groups appeared. We were of course, aware that the League of Women Voters sample was almost entirely female (32 females, 3 males) while the radon testers sample turned out to be composed of 27 females and 48 males. The groups also differed significantly on age (testers 40.2, SD 9.6; League 50.4, SD 13.4; $t = 4.44$, $p < .01$). Testers were more likely to be married (96% vs. 83%, $t = 2.41$, $p < .05$). League members were more likely to have lived in their homes longer (12.9 years, SD 8.7, vs. 7.6 years, SD 7.8; $t = 3.16$, $p < .01$).

On the important dimension of education, however, the groups did not differ, with the mean for both groups falling into the category between "undergraduate degree" and "some graduate training." In addition, as will be presented later, both groups had similar attitudes about environmental hazards.

Table 6.15 summarizes the differences between the testers and comparison groups. As predicted, no differences were obtained on the dimensions of past control, anticipated control, happiness, worry, trust or support from spouse and others. Contrary to prediction, there were also no differences in worry about health and health of children. Also not predicted was the non-significant trend for the radon testing sample to describe themselves as more competent than the comparison group. On the Impact of Events Scale, the groups did not differ significantly.

We tested our hypothesis that radon testers would perceive environmental hazards, especially radon, as more dangerous than would control subjects in the following way. We compared the ratings of the respondent group on ten environmental hazards, such as pesticides, nuclear power, and cigarette smoke. The difference was nonsignificant, although experimental subjects did rate the hazards as more dangerous. When radon alone was examined, the difference between the groups was significant. An analysis of covariance was performed to test whether, when the somewhat higher overall rating of environmental hazards by the

	Radon testers		Comparison		
	M	SD	M	SD	t
Past control over life*	2.6	1.0	2.8	1.1	.94
Anticipated control	12.6	1.1	2.8	1.2	.72
Happiness	2.3	1.3	2.5	1.0	.65
Worry	4.1	1.7	4.6	1.8	1.38
Worry about health	3.4	1.7	3.9	1.7	1.42
Trust of others	3.3	1.6	3.3	2.4	.15
Competence	2.0	0.8	2.4	1.2	1.73
Social support	1.8	0.6	1.9	0.6	.45
Support from spouse	1.1	0.3	1.0	0.2	1.21

*Except for the last two items, higher numbers indicate less of variable.

Table 6.15: Comparison of radon testers and nontesters on several indicators of psychological adjustment.

testers was separated out, the radon hazard alone was still perceived as more dangerous ($F = 7.3$, $p < .01$, $n = 89$) by the testing group, while comparing the group means on the covariate of overall perception of environmental hazard still showed a nonsignificant result ($t = .28$, $p = .78$).

SUMMARY AND IMPLICATIONS

In future research, it will be useful to include additional comparison groups that provide greater similarity in gender and age. It would also be of interest to examine less educated and affluent groups. However, the comparison presented here confirmed the expectation that the testing group would be essentially similar to the League comparison groups in certain key indices. The most important of these was the similar perception of general environmental hazards. This basic tendency to respond to environmental hazard may prove to be an important disposition useful in predicting concern about environmental threats. The actual decision that radon is a particularly important threat which justifies voluntary testing of the home is reflected in ratings of fright from the possibility of radon gas exposure. Here, the testing groups ratings were significantly higher than those of the comparison group. Thus, while both groups are sensitive to environmental hazards in general, they differ in their assessment of radon. The only other factor distinguishing the groups which might account for the fact that one group tested and the other did not was indicated by a greater rating of competence among the testing group. This may reflect an increased likelihood that this group of people are more likely to take actions to identify risks.

Overall, the fact that the comparison groups are emotionally similar suggests that in the second phase of the study, the identification of stress reactions among the testing group can clearly be related to the impact of concern over radon levels. Thus, the most important aspect of this study is the provision of baseline data for the second phase of analysis.

At this point in our analysis, the data suggest that the decision to have one's house tested for radon is not a function of being a worrier, or an untrusting human being, but rather a simple function of one's perception of radon as more dangerous than other people see it.

FOOTNOTES

[1]Edelstein, Michael R., The social and psychological impacts of groundwater contamination in the Legler section of Jackson, New Jersey," report prepared for the law firm Kreindler & Kreindler, 1982.

[2]Edelstein, Michael R., "Stigmatizing aspects of toxic pollution," report prepared for the law firm Martin and Snyder, Fall, 1984.

[3]Edelstein, Michael R., "Contaminated Children: Toxic Exposure in Jackson, New Jersey." Childhood Quarterly. January, 1983.

[4]Edelstein, Michael R., The Social Impacts of Residential Exposure to Toxic Waste." Social Impact Assessment, 79/80, April-May, 1983.

[5]Edelstein, Michael R., "Social Impacts and Social Change: Some Initial Thoughts on the Emergence of a Toxic Victims Movement," Impact Assessment Bulletin, 3, 1984-85, pp. 7-17.

[6]Edelstein, Michael R., Toxic Exposure and the Inversion of Home. Journal of Architectural Planning and Research, August, 1986.

[7]Edelstein, Michael R., "Psychosocial Impacts of Toxic Exposure: An Overview," pp. 761-776, Hank Becker and Alan Porter, Eds., Impact Assessment Today, Vol. II, Utrecht: Jan van Arkel, 1986.

[8]Edelstein, Michael R., Inverting the American Dream: The Social and Psychological Dynamics of Residential Toxic Exposure. Boulder, CO: Westview Press. Forthcoming, 1987.

[9]Edelstein, Michael R. and Abraham Wandersman, "Community Dynamics in Coping with Toxic Exposure," in Irwin Altman and Abraham Wandersman, Neighborhood and Community Environments, Volume 9 in the series Human Behavior and the Environment, Plenum Press, Forthcoming, 1987.

[10]Gibbs, M.S., "Pschopathological Consequences of Exposure to Toxins in the Water Supply." In Lebovits, A.H., A. Baum, and J.E. Singer, Eds., Advances in Environmental Psychology, Vol. 6: Exposure to Hazardous Substances: Psychological Parameters. Erlbaum, Pub., Hillsdale, NJ, 1986, pp. 47-70.

[11]Horowitz, M.J., N.J. Wilner, and W. Alvarez, "Impact of Event Scale: A Measure of Subjective Stress." Psychosomatic Medicine, 42, 1979, pp. 209-218.

IMPACTS OF LIVING IN HIGH LEVEL RADON HOMES

Kay Jones and Kathy Varady*

Since early in 1985, our lives have been fundamentally changed by the discovery of high levels of radon gas in our homes. We are residents of a small township called Colebrookville, which is near Boyertown, Pennsylvania and neighbors of the Watras family, whose home was the discovery point for the current national concern for radon gas. In this paper, we attempt to summarize the devastating effects of living in a home containing elevated levels of radon. This is a different yet vital perspective to bring to the study of radon. While scientists, health physicists, radiation experts and government agencies discuss the disastrous possibilities attributed to high levels of radon, many families are living with grave uncertainties. The economic impacts, social ramifications, and mental anxieties that we and our neighbors have experienced suggest the need for an aggressive approach by government to addressing the radon issue.

LIVING WITH THE KNOWLEDGE OF RADON EXPOSURE

The levels of radon in 65 homes in our area have been described as leading to at least a 50 percent chance of lung cancer for our families. How has this affected us? We have been robbed of any sense of security which may have been present prior to our knowledge of radon. The only constant remaining in our lives is fear. This fear is always present, and it always will be with us in the future. Little did we know that the silent killer which permeated our lives would drastically change not only our lifestyles, personalities and priorities, but would strike at the very core of our existence as well.

As parents and homeowners, we were struck a devastating blow. We worry about what will happen to our children. In the meantime, our feeling for the homes that are so central to our family life has been changed. We might leave, but financial commitments plus stigma bind us to our homes. In our minds, we realize that we cannot assume total

---

*Kay Jones and Kathy Varady presented this paper during the session on perceived risk and psychosocial impacts of radon. They are President and Vice President, respectively, of PAR (People Against Radon), a national grass roots organization formed to address the radon issue.

responsibility for this situation, but in our hearts, the grief, pain and anguish will be ever present.

It is particularly hard to watch the effects of this on our families. Children absorb much of the stress brought on by the drastic changes in daily routines occasioned by attempts to mitigate or avoid radon levels. Through our own personal experience, however, we have found that the bulk of the mental anxieties have fallen upon our husbands. For them, the guilt for placing their families in an environment now determined to be cancer-causing has been almost unbearable. After all, their life's dreams and ambitions were suddenly being shattered.

It is not always possible to deal with this situation in an individual way. We have had to commit ourselves to extensive research and self education. Our feelings of shyness and timidness were soon overpowered by feelings of anger and sheer disbelief that a problem of this magnitude was not being addressed in the same manner as issues of lesser significance. One way that we have tried to cope is by banding together to form an organization to represent affected families. This organization is called People Against Radon (formerly, Pennsylvanians Against Radon). PAR seeks to provide a united effort in attracting the help of state and federal agencies. Why is this so urgent?

THE FRUSTRATION OF GOVERNMENT INACTION

It has been 16 months since our homes were discovered to be unsafe. Yet, while the levels of radon gas discovered in some 60 percent of the homes in our community reached levels unheard of at that time, there has been a disappointingly minimal amount of assistance for those affected.

By EPA's own admission, radon is the second leading cause of lung cancer after cigarette smoking and may cause up to 20,000 radon-related lung cancer deaths per year. The EPA has projected that up to 200,000 may die over a five-year period due to radon contamination. At the same time, this federal agency plans to undertake a five year national assessment to identify areas of elevated radon. While we realize that this study is essential, there is an immediate crisis at hand that no agency, be it federal or state, is addressing -- THE IMMEDIATE REDUCTION OF RISKS ALREADY BEING INCURRED BY FAMILIES LIVING WITH EXCESSIVE AND UNHEARD OF LEVELS OF RADIATION!

This lack of government action is galling to us for many reasons. First, we now have many methods which will at least reduce the exposure levels for these families. The cost of implementing these methods should not be the determining factor in assisting these innocent victims.

Second, the federal government has spent millions of dollars cleaning up homes which have much lower cancer risk than do high radon homes. For example, in the mill tailings projects in the west, it is not unheard of to spend in excess of $100,000 per home. With one home in Colorado,

$20,000 is being spent to reduce exposure from 0.187 WL (approximately 37.5 pCi/l) to 0.17 WL (approximately 34 pCi/l), the state standard. Compare this to the levels found in high radon homes, where peole are being exposed in many cases to levels exceeding 5 WL (1000 pCi/l).

Third, those experiencing high radon exposure levels have been denied access to Superfund (CERCLA) assistance because it is an EPA policy to exclude naturally occurring contaminants. Yet, the law states:

> If it is determined that the release or threat of release constitutes a major public health or environmental emergency and no other agency or person has the authority or capability to respond in a timely manner, a CERCLA response may be initiated.[1]

Fourth, we have also been denied access to assistance under the Federal Emergency Management Agency (FEMA), which commonly assists victims of natural disasters. This is a natural disaster unlike any natural disaster known to mankind. It is one which destroys the true value of a home and subjects its inhabitants to life-threatening elements. Yet, because the devastation and loss of life is caused by an invisible killer, it is not classified as a traditional disaster and, therefore, is seen as unworthy of immediate assistance.

Fifth, there is a grave contradiction in agency response to our plight. Representatives of both the federal Centers for Disease Control (CDC) and the Pennsylvania Department of Health (PDOH) admit that the radon decay product levels in a large number of tested homes constitute a major public health problem that requires immediate attention. How can the CDC declare an emergency at 0.1 WL (approximately 20 pCi/l)(as they did in the Montclair area of New Jersey) yet completely ignore a situation where levels are 10 to 100 times greater? For a major health agency to take a back seat merely because funding is not immediately available is beyond our realm of comprehension. Somehow, the wheels of justice always turn back to how much it will cost, not how many lives can be saved.

Sixth, the EPA is responsible under the Atomic Energy Act for protecting public health from environmental sources of risk due to radiation and has the authority to advise the president on radiation matters which affect health and welfare. The majority of health scientists and radiation experts are in full agreement regarding the public health impact of radon gas on the citizens of the United States. Every day spent in a high level home is critical. Despite this, no one has responded in what we feel is a "timely manner." The EPA is collecting statistics before action is taken! What does the EPA consider in this case to be a "timely manner"?

Seventh, at the state level, the Three Mile Island (TMI) experience was very frustrating for the radiation department within Pennsylvania's Department of Natural Resources (DNR). Our Governor showed a deep concern and commitment to victims of TMI. Such concerns seem ironic to us now. We are told that the radon concentrations found in hundreds of our homes have the radiation equivalency of having a TMI accident occur in the neighborhood once a week. How is it possible that similar attention is not being focused on radon? Surely Governor Thornburg must be aware of the hazards and impending health complications these homeowners are living with on a daily basis. Action at TMI was immediate and epidemiological studies are plentiful. Yet all that our state has forthcoming 16 months after the shock of radon discovery is a proposed loan program that has already been seven months in coming.

It is hard for us to understand and accept the fact that in the United States of America, with all of our numerous agencies and vast resources, the innocent victims have been abandoned. Some argue that a precedent not be set for an invasion of the private domain -- the home. We feel that it is a greater violation of our constitutional rights to ignore the health, happiness, and welfare of innocent victims who are pleading for a technical and financial solution. Realizing the magnitude of this problem, it is inconceivable to expect that either the state or federal governments can deal with it solely through their own resources. A cost sharing program by federal and state agencies should be initiated to at least address urgent needs of homeowners facing a life-threatening environment.

CONCLUSIONS

For victims of high radon homes, this is a major blow to our families. We need desperately to find some way to come to grips with the threat which invades our homes. We need government assistance to accomplish this. The confused and inadequate response to date merely adds to the frustration and fear. Not only have we been invaded by an unseen enemy, but the government that we look to for help is looking the other way.

It is our feeling that an "emergency" situation currently exists in many homes. Although the urgency to reduce radon levels is not at a critical stage for the majority of homeowners, a definite emergency does exist for those whose homes far exceed the national guidelines.

In many homes across the United States, time bombs are ticking. When will they be defused? Ask yourselves this question -- HOW MANY OF YOU WOULD LIVE IN A HOME FOR ONE DAY, ONE HOUR, AT LEVELS NOW BEING IDENTIFIED IN MANY HOMES?

FOOTNOTES

[1]Comprehensive Environmental Response, Compensation Liability Act, 1980.

HIGHLIGHTS FROM THE DISCUSSION: PERCEPTION OF RISK AND PSYCHOSOCIAL IMPACTS OF RADON EXPOSURE

COMMENTS ON THE SESSION

M. Edelstein*: I think one of the things that has struck everybody in this room after listening to Kay and Kathy, as opposed to the rest of us, is that there is difference in the framework of thinking, a difference in what is reality for two sets of people up here. Caron referred to this earlier when she talked about "the real people" versus the rest of us. One of the distinguishing factors here, which may not be the only one, is that the rest of us are part of the group that is somewhat comparable to the public at large. We know that there may be a risk. Some of us have tested, some of us haven't. We may have varying concerns, but to my knowledge no one else up here, besides Kay and Kathy, is living and have their families living at an extraordinary high level of risk because of radon. So there really is a major difference between people who have come to define themselves as victims of radon and radon exposure and people who are flirting with the possiblity that there might be some problem out there that is undefined. What I want to try and do is to build on that for a moment and see if I can touch on a number of issues that have come up in the session.

We started with Paul Slovic giving a fundamental and extremely good overview of how people perceive risk. But in light of what I just said, I think there is a real difference that can be built into this analysis, the difference in risk perception between people who are at the point of potentially accepting the risk and those who have realized it. The first group have the possibility of accepting that they may be at risk. They may not have tested, or if they have tested, they may not have fully come to understand the implications of whatever tests have been done. There is a difference between those people and people for whom that reality has set in; who are no longer flirting with the possiblity of a potential risk or for whom the risk, to make a distinction that is important in the literature, is not just a risk out there, but is now a personal risk for them. Once it's a personal risk, once it's an identified risk, once it's a real risk that you can't deny anymore, or not very easily, you then move into a second phase. You are no longer dealing with acceptance of

*Michael Edelstein, Associate Professor of Psychology, made these comments in the role of discussant in this session.

the risk, as you do initially when learning about the potential threat, you are now dealing with coping with the reality of the risk. And I would submit that there is a difference in perceiving risk between people who are in the public who may or may not be affected and people who have been defined as already being at risk. I have done eight years of work with toxic victims from chemical toxins. I see the same process there as I see here. We are outsiders, most of us, to the situation that Kay and Kathy are part of, because we don't see the situation the same way. We are not victimized the same way. It is hard for us to really sympathize or to really understand the situation that they are in. There are two fundamental differences in reality. They're coping with the situation that we are flirting with. That is the first point I wanted to make, and I think it has to do with Paul Slovic's comment about to what extent are people frightened by information. I think there is a difference between information when it may not apply to you, and information when it helps to define a situation that you already know you are in. So the uncertainties that Kay and Kathy just spoke of make a difference to them, they increase their stress level. The uncertainties to someone who is flirting with the possibility of risk somehow make vague the possibility of that risk being real, and may help explain some of the things that Neil Weinstein and Peter Sandman were talking about.

It also explains why, as Paul noted, experts tend to be very much caught up in the quantitative analyses of risk, whereas citizens use qualitative analyses of risk. It has been my observation that when we are affected by risk, we trip into an all-or-none type of reality. We are either at risk or we are not. Experts can calibrate that and measure it and quantify it to their heart's content. At what point, if you are an expert and you happen to be living in a house with 1,000 pCi/l, does your reality change? It becomes a qualitative kind of issue rather than a quantitative issue; it can't be subjected to a quantitative analysis. My point is that experts not only have a quantitative way of thinking, but as experts, we are often outside of the situation. We are invited, therefore, to think differently than the people who are in the situation. In approaching people who are in a situation where they are under extreme risk, and concerned and stressed by it, we should be aware of that difference. It puts into perspective the 4 pCi/l guidance level. What is the reality of that level? What is an acceptable risk? Acceptable to whom? There is a difference in reality between it being acceptable to someone sitting in the Office of Mangement and Budget, certainly, to take the most extreme case, and someone sitting in Kay and Kathy's situation.

I'd also like to touch on Neil's comment that his data doesn't reveal much misinformation but rather a lack of information. It's pretty hard to talk about what is misinformation and what is not. For example, Neil spoke about his fear that people who were here yesterday might get the wrong idea that in fact mitigation is hard to do. Well I've gone to enough radon conferences, and spoken to enough radon experts and listened pretty carefully to know that, in fact, I would view it exactly the opposite way that he does. My concern is with what I call the "mitigatory gap," or the lag time that it may very well take to mitigate all of the radon problems. Thousands and thousands of people are testing. But we are not ready to help them lower their home levels when they get their results back. We have to expect a lag time until we are going to be able to really help those people. Yesterday the comment was made that mitigation isn't at the point where you can do it yourself. Well, I have news for you, look at who is out there to do it for you! How many mitigators are there? The comment was made to me by Harvey Sachs that at a conference a week or two ago, when informed about the EPA testing program for mitigators, that Art Scott said he'd like to take the course. Well, when Art Scott says he's got to take the course on how to do mitigation, who are we to collude in having people believe that mitigation is ready to come on line and help them? Now people may be reassured at this point in time, because they believe that this is a solvable problem, but how are they going to feel a year from now, or six months from now? Kay and Kathy have been at it for 16 months. During that lag period, and depending on how they come to define an acceptable level of risk, even if their level is low, we are sitting on a major potential for panic.

One other comment on misinformation. We found evidence of misinformation in the study I've been doing with Margaret Gibbs, although she didn't get to speak about that portion of the research. The average person who knows about radon believes that radon is only on the Reading Prong. We worked very hard to set up a session yesterday to debunk that myth completely. The session certainly did that. I hope we get consistent press coverage on that, because if you do an analysis of newspaper reporting on this issue, you will discover that it's almost impossible to have an article written that doesn't at some point say something about the Reading Prong. Now, we've created a belief system that if you are on the Reading Prong you are at risk, and if you're not on the Reading Prong, you are safe. How we change that is not such an easy thing to deal with because, as Paul noted, once beliefs like that become ingrained, they're not so easy to change right away -- even for the newspapers. I was quoted last week in a newspaper article as saying, "We have to get rid of the myth of the Reading Prong." The article goes on to define the prong as the radon-emitting band of rock crossing Pennsylvania, New York, and New Jersey, which is the cause of the whole problem. These myths die hard.

Let me make one last point. Peter said that he is surprised by his comparison of superfund site meetings to radon meetings. I think that you will find less of a difference in the long run. The reason is this. There is an evolution of how a community or family or even a person deals with coping with the reality of some type of toxic exposure. Initially you have to learn a lot. You don't really know what to think. You are not real emotional. These people may not have even had tests done on their houses, and they are finding out about the problem. They are acting very serious, but they are not real bent out of shape, probably. Compare that to the situation that Kay and Kathy are talking about, which is much more analogous to the situation that someone in a superfund site experiences. They had the problem identified a long time ago, in some cases, 4, 5, 6, 7, 8 years ago. They have been sitting with it a long time, trying to get action. They have realized that government isn't there to help them, or if government is there to help them, that maybe it will be the year 2040 before it acts, and then it's going to put a clay cap on the site. So, that is somewhat analogous to what you are facing. There is a developmental trend. I see the possibility of a lot of community organizing, a lot of real developed concern on this issue after it matures and people have had a chance to deal with coping with the situation, realizing that they are not going to get help. The expectations that Neil showed are real important, because they help us outline what the public response is going to be in the near future when a lot of the expectations that you identified cannot, I submit, be met realistically by where we are right now.

RISK COMPARISON

Question: In communicating risk of radon exposure to lay persons, what are the advantages and disadvantages of comparing risk of health affects from radon to risk of health affects from smoking?

P. Sandman: In general you don't want to do risk comparisons if you can avoid it because risks vary on so many dimensions that almost any comparison usually doesn't work. The folks who are usually comparing risks, and the folks who are usually comparing their risks to smoking, are the people who are trying to minimize the risk. I think we've all heard it said about a whole lot of risks, "My God, why are you worrying about this, it's enormously less important than smoking." It may be that for radon that the comparison with smoking will cut the other way, because we can say, "My God, you'd better worry about this, it's almost as important as smoking." This may be the exception that proves the rule, and comparison here may actually prove to be useful.

ACCEPTABLE LEVELS OF RISK

Question for Kay Jones and Kathy Varady: What have you done to date to reduce the levels of radon in your home?

K. Jones: Kathy and I were both chosen to be in the initial 18 home demonstration project that was initiated by the EPA to come up with low-cost methods. Arthur Scott is in charge of that project. Arthur was in my home six times with different techniques. At one point with the subfloor slab system, my levels had actually gone up to 15 WL (3,000 pCi/l). It did a reverse on the house to what it was supposed to do. We are now at relatively safe levels. The levels in my home are anywhere from a 0.02 (4 pCi/l)to a 0.1 WL (20 pCi/l). I do have peaks when I use water, because I also have extremely high levels of radon in the water. We are at the point now in our home that the water is a major contributor to the levels in the home.

K. Varady: I'm also relatively safe right now. They did a complete wall ventilation system where they vented all the walls in my basement, plus I have a crawl space. They had tacked me on as house #19 after they decided they better do something with me because I was still out there fighting. They were back 5 times. My levels dropped to 0.02 WL (4 pCi/l) in summer testing. Mr. Scott closed me up in the fall and retested me, and I was back over 1 WL (200 pCi/l). So the system which worked great in the summer did nothing for me in the winter, which would verify the fact that summer readings are not always year round readings. You should be tested in each season to get the right test results. Right now, we are down. The only problem with this is when we talked with Tom Garusky of the DER, I said, "You know, Tom, it's very easy for somebody to sit there and say to you, "Why didn't you get out of your home?" But if you've got children and a family, it is not easy to go somewhere else and continue to pay your mortgage. The bank says "If you default on one mortgage payment, that's it, we take the house." Right now these people have no option. They are told to get out, but there is no agency, no federal agency or state agency that is going to be there for them once they leave. I think this is important. I think more people would leave if they had that option, even housing or something.

RADON AND LOVE CANAL

Question for M. Edelstein: Do you think that this may be the next Love Canal?

M. Edelstein: I think there are some major differences between what we are talking about and Love Canal. My own expectation is that radon is likely to be a long-standing issue of extreme public concern. The radon issue could cease to be a major stressor if it proves to be an easily solvable problem that most people can afford to solve and have access to the solutions. However, I would have some question about whether this will occur. Assuming that radon persists as a crisis, then it must be seen as differing from toxic exposure incidents like Love Canal in several important ways. First, radon is naturally occurring and, thus, blameless. Secondly, radon's boundaries are ill-defined. Radon is pretty pervasive. It's all over the place. At the same time, it has some perverse characteristics. It will crop up in one house, but not in the house next door. One of the things that characterizes Love Canal and other toxic waste sites is that there is a bounding of the area that is contaminated delineating the contaminated area and areas outside the boundaries that are supposed to be clean. At Love Canal they put up a fence which made that delineation. When I worked in Jackson Township, New Jersey, the neighborhood of Legler came to be defined around the boundaries of contamination. If you were outside of Legler you were clean, even if you shared the same water source. So, with toxic incidents there seems to be a real development of boundaries separating those defined as being contaminated from those not so defined. Radon is complicated because those boundaries are harder to define. There are exceptions. In a case like Boyertown, you have a situation that is maybe more analogous to a Love Canal situation. The same thing may be true in Clinton, where you have a community that seems to be particularly impacted, and impacted at a high level. In such cases, it is possible to have stigma, a devaluation of an area because it is defiled by radon. However, over the long run, that is not going to be sustained if in fact we continue to find that radon is very pervasive. It will be everywhere, not just Clinton and Boyertown. Radon will mirror Love Canal in terms of the degree of long-term concern that will be evidenced, but I think that it is a very different issue. Incidentally, you also have lots of 4 pCi/l people as well as some people with 1,000 pCi/l. They are in two very different situations, and I think they may respond in the long term in very different ways as this issue evolves.

EASE OF MITIGATION

N. Weinstein: In his discussion, Mike talked a little bit about the mitigation issue and obviously drew a different conclusion from his experience than I do from mine about the possibilities of mitigation. What that points out is that we both think that it is an important issue to be very careful about how you present the difficulties or possibilities of mitigation. My own particular point is that we need to be careful not to go from a belief that it is often difficult and not simple and cheap to conveying an impression that it is impossible and hopeless. That's very important to be sensitive to that possibility.

M. Edelstein: In line with that though, it is important to begin conveying some reality information that it may not be such an easy task so that people are more prepared for that possibility. If you look to Kay and Kathy, each talked about Art Scott visiting 5 or 6 times to their home. Kay mentioned that at one point her level went up to 15 WL (3,000 pCi/l). What was it before that?

K. Jones: Before that it was anywhere from 2 WL (400 pCi/l) to 8 WL (1600 pCi/l).

M. Edelstein: This is clearly something where some experimentation is going to occur, maybe even on each site. People need to know that. Otherwise they are going to expect a contractor, if they can even find one, to have easy success. Maybe in the long run that will occur. Hopefully that is the case. Maybe at the next radon conference, if we have one in a year or two, we will be talking about how we have solved these problems. We have "tech fixes". But we need to alert people, because your data scares me, frankly. People have a belief that this is a different type of problem than what I hear from a conference like this.

P. Sandman: I just want to second that. I think it is not only important that people know some of the problems they are going to encounter, but it is important that they know that when they are trying to decide whether to monitor. Otherwise, we really set people up. I expect to be involved in the next several years in public education - trying to urge people to monitor - and that's terribly important. But lord knows, the last thing we want to do is urge millions of homeowners to monitor and do so with the promise, "Well, once you've monitored, if you've got a high level, all you are going to have to do is a few simple things that we can explain to you in this handout, and you will be fine." If we do that we are setting up lots of people who are going to be justifiably angry a couple of years down the road, because they will have been mislead. We have at least three different communciation problems. One is to get people to monitor with information that levels about the extent of the risk, levels about the difficulty of remediation, and yet also encourages them not to bury their head in the sand. That is problem one. Problem two is what do you do with people who have monitored and come up with these horrifyingly high levels, and that's a social problem of really needing to address that emergency, and not give them the runaround that they're getting. The third problem, and the one we have said least about, and I think the one we know least about, is what to say to people who do monitor, and find they haven't got the kind of levels found in Boyertown. They've got 4 pCi/l or 6 pCi/l or 8 pCi/l, or even 2 pCi/l and they say, O.K., I haven't got an outrageous risk, but I've got a real risk. What are we going to say to them? Those are three different communication campaigns, and involve public policy issues that I think need to be addressed independently and not confused with each other.

Section 7

Socioeconomic Impacts of the Radon Issue

INTRODUCTION AND SUMMARY OF SECTION 7: SOCIOECONOMIC IMPACTS OF RADON

Michael R. Edelstein*

The radon threat has been continually redefined over the past two years to make it increasingly all inclusive. It no longer is seen as affecting only a few states or geological formations. It is no longer a problem of houses that are too energy conserving. Instead, it is seen as an issue affecting millions of buildings in every state in the U.S.

It is not surprising that so pervasive a phenomena as radon exposure would generate a wide variety of impacts. The most widely recognized of these involve the health threat and government concern with potential panic. But a third impact area deserves attention because of its complex implications and because it too has played an important role in the evolution of government policy. This involves the various socioeconomic impacts of the radon issue.

The socioeconomic impacts of radon are many. They include the development of a new industry focused on mitigation and testing, budget expenditures by government, and the fiscal impacts upon the homeowner. Building construction guidelines and codes may increase the costs of new buildings. New building materials may be developed and old ones put to new use, generating profit. Uninformed beliefs about energy efficiency and radon may hamper energy enterprises already affected by changed government policy and an oil glut. The real estate industry must contend with a complex set of issues which also affect mortgage banking, insurance and legal aspects of realty. Stigma associated with radon may cause some to lose land value or sales. Prohibitions against construction in certain areas may have similar effects. Legal questions regarding liability are suggested by radon, despite its recognition as a natural occurrence. Homeowners may lose the use of spaces, such as basements, that formerly were important living areas. Municipalities may have property tax impacts. And, given the combination of public concern and unfamiliarity with radon in light of the paucity of mitigation contractors, the potential for business fraud also exists. Real estate fraud, stemming from the ease of falsifying radon tests and

---

*Michael Edelstein is Associate Professor of Psychology, Ramapo College of New Jersey, and a conference co-organizer.

the impact on real estate values, is another concern. Such issues promise to generate new areas of case law. Finally, the health-related costs associated with a failure to address the radon problem have not been calculated. The list of socioeconomic issues, hardly exhausted here, is indeed a long one.

The Ramapo conference included a session on the socioeconomic impacts of radon chaired by two aides to Senator Frank Lautenberg, Jeffrey Morales and James McQueeny. Contributions were selected to highlight some of the more intriguing questions about radon's socioeconomic impacts, as reflected in the papers included here.

In the "Legal Aspects of Radon," Philadelphia attorneys David Toomey and David Sykes, of the firm Duane, Morris and Heckscher, provide an extremely comprehensive review of this topic. Because radon gas is naturally occurring, its treatment in the legal system based upon assignment of responsibility is unclear. Furthermore, the radon case record is sparse. The record includes cases which support government efforts to regulate radon-producing wastes, limitations on confidential handling of radon results by government agencies and the limited conditions under which government can be held responsible for not warning those potentially at risk. The difficulty of proving that lung cancer is caused by radon exposure is the major issue in several lawsuits. Only in the case of uranium miners is there sufficient documentation to get the courts to side with the plaintiff. Finally, there is a paucity of case law regarding home exposures, with all the cited cases involving situations where someone was responsible for the exposure. Among the promising areas for future litigation, the authors identify architect and contractor liability and cases involving real estate transfer. Some of the real estate cases will involve situations where neither party knew of the radon problem, and other cases may involve situations where the seller was aware of the contamination. In the former condition, the courts may force a reformulation in the contract of sale to cover costs of remediation or may even void the contract. When the seller hides his knowledge of radon, however, the seller may be guilty of fraud and liable for punitive damages as well. Radon in new homes may be found to violate "warranties of habitability," where they exist. Overall, Toomey and Sykes strongly recommend testing of newly purchased homes. While the seller may lose value if radon is found, they have more to lose in a subsequent law suit if they do not make a full disclosure. Full testing and disclosure is advocated as a means for protecting both the buyer and the seller. Finally, the authors do not see the radon issue as diminishing home sales or home lending. Radon codes will assist all concerned in assuring a uniform level of response.

The impacts of radon on the construction trades is addressed by builder and mitigator Terry Brennan. Brennan discusses the perspectives of both the small and large-scale builder concerned with short-term construction questions. For the small contractor, the radon issue opens up a new type of work, although the extent of the market for mitigation services depends upon the as yet unknown willingness of people to act on an unseeable threat. All builders face increased costs with radon-proof construction, although the impact may be more noticeable with the bigger developers. The resistance of builders to address the radon issue is often overcome by concern over liability for a documented radon problem. Other radon impacts on construction may stem from limitations on building on certain soils and other regulations. Liability concerns may help to convince a diverse body of building inspectors to take radon codes seriously. Liability issues are particularly relevant to the builder because many high radon houses are characterized by poor masonry work. To meet their responsiblity, builders will have to design and construct houses which limit entry of soil gases and which are easy to remediate if radon is detected. A national training program for builders has begun to inform contractors about radon and to train mitigators. Building associations are also serving to disseminate radon information to the building industry.

The two top officers of a new trade association, the American Association of Radon Scientists and Technologists (AARST), provide a thoughtful overview of the developing radon industry. Donald Schutz discusses the impetus for organizing the industry in the first section of "The Role of Scientific Professionalism in Meeting a Public Need." The state of New Jersey, in realizing that it could not address the radon issues alone, looked to radon-related businesses for assistance. A partnership with government is emerging which can address the scope of the radon problem. The private sector has responded by conducting more radon tests than have all government sources combined. AARST has already reached companies across the United States. Committees of the organization have addressed questions of professional standards and public information. A major concern has been to weed out fradulent services by limiting membership to firms which share high professional standards. A benefit of such activities may well be the lessened need for government to regulate an industry that has shown maturity in assuring its own quality.

In his contribution to the above paper, engineer Harvey Greenberg addresses the issues faced by a radon contractor in dealing with clients. Greenberg profiles the testing business in some detail. Most tests requested are for private homes and most of these are done as part of home transfers. The clients are typically middle or upper middle class. Areas of New Jersey and nearby New York that have been heavily tested reveal between 15-30 percent of tests in excess of 4 pCi/l. For the testing firm, contact with the client is often awkward. As they advertise their services, firms must be careful not to act as "opportunistic parasites." After carrying out their services, the tester may be blamed as a "messenger of gloom" in the case of high results or for wasting the family's money if results reveal no radon. Testers are also caught in a bind by the possibility of fraud by prospective home sellers.

How can the question of fraud be addressed when virtually all testing methods can be easily biased by direct interference, usually without the knowledge of the testing firm? In "Fraud-resistant Radon Measurements for Real Estate Transactions," Harvey Sachs draws upon his long experience in the radon field to provide two concrete suggestions. One involves the use of an escrow account to cover possible mitigation costs and the shifting of the testing burden to the buyer. Unlike the seller, the buyer has little to gain from distorting radon test results. Efficient use of the escrow account to lower radon levels can benefit both parties if funds left over are split between them. The second approach involves a "technical fix" using a passive perflourocarbon tracer to assess ventilation rates at the time of testing. Excess ventilation would cause the tester to suspect the validity of the radon test. These approaches address directly what appears to be one of the major roadblocks to the active use of radon testing in the field. Particularly the first can be directly implemented without delay.

As these papers on the socioeconomic impacts of radon suggest, such impacts are extensive and complex. In the discussion following the conference session, some of the additional impacts emerged. Harvey Greenberg noted that the role of the realtor is seen differently in different states. Specifically, he noted that the role of the New Jersey realtor in preparing contracts of sale has led to a much greater awareness of the radon issue than is evident in New York State, where realtors are principally salespeople. Various panelists discounted the model of the "perk test" for preconstruction radon monitoring because of technical difficulties in accurately depicting the potential for radon infiltration to a building on the basis of soil tests. Testing is more effective as part of the process for issuing certificates of occupancy and for selling the building. It is likely to be less expensive to employ preventative construction techniques than to measure radon levels on the site. It was further recognized that a house with a low radon level might over time increase radon entry. Testing over time will be necessary, complicating questions of liability where original tests revealed little radon. Finally, there was a discussion of confidentiality as an element of the client-tester relationship. Given the questionable ability of government to protect client identity, efforts to centralize data collection in government hands run counter to the premise of the professional approach to testing.

THE LEGAL ASPECTS OF RADON

David C. Toomey, Esq. and David T. Sykes, Esq.*

The title, "Legal Aspects of Radon," may be a bit misleading, because it suggests that there indeed have been legal developments of significance. In fact, perhaps the most significant aspect of the legal issues surrounding radon is that very little can be reported.

To date, most action in the legal arena has involved public policy questions and the effort to adopt regulations of value. Because a number of prominent and informed speakers from the legislative and executive branches of government will be on the program, each of whom is scheduled to address the activities and plans of their aspect of government, this paper will address questions of private rights and liabilities involving radon.

Even in the public arena, however, government bodies have accomplished very little to date in addressing the radon problem. The reason for ineffective government action reflects the reason for inactivity in the private sector as well -- it is very difficult to regulate activities which are not well understood and for which no person or other entity created by humans can be held responsible. The legal system works comfortably only when responsibility can be assessed on some basis, whether it be for improper conduct or due to the imposition of social rsponsibilities on otherwise blameless entities.

In the total jurisprudential history of the United States, there have been very few published cases which have even mentioned the word "radon" in a substantive context. Only a handful of those were decided before 1980. Thus, when one attempts to speak about the legal consequences which flow from radon, one speaks more in the area of prediction and prognostication than in the area of actual results and accomplishments.

Nonetheless, we think it may be useful to assemble and review the cases that have been decided, and we do so in the first portion of this paper. The decided cases illustrate both how little has happened, and, to some extent, the

---

*David Toomey and David Sykes are partners in the law firm of Duane, Morris and Heckscher based in Philadelphia, Allentown, and Paoli, Pennsylvania, and Wilmington, Delaware. Their firm is affiliated with Jamieson, Moore, Peskin and Spicer of Princeton, New Jersey.

problems which the legal system will have in dealing with radon in the future.

The bulk of our comments, however, will relate to areas in which we predict the law will develop in an effort to cope with radon-related problems.

REVIEW OF DECIDED CASES

The cases which have involved radon break down into four groups, including decisions involving government regulation, claims against the United States, and workplace exposure cases. A final group amounts to little more than a miscellany of cases which have merely referred to radon in passing. These four groups are reviewed, in turn.

Decisions Involving Government Regulation

Of the cases that have referred to radon in a substantive context, several have involved technical questions of administrative law. Largely, these cases have involved attacks on the adequacy of government regulation where radon has been involved. For example, in Baltimore Gas & Electric Co., et al. v. National Resources Defense Council, Inc., 462 U.S. 87 (1983), the Court held that the Nuclear Regulatory Commission's decision that nuclear power plant licensing boards should assume that permanent storage of some nuclear wastes would have no significant environmental impact was within the bounds of agency discretion and consistent with the requirements of the Administrative Procedure Act. 462 U.S. at 90. Two of the cases were unsuccessful efforts to set aside EPA regulations concerning the handling of uranium mill tailings.[1] One case involved the failure of the Mine, Safety, and Health Administration to adopt regulations protecting miners from radon exposure[2] and two were state decisions involving the siting of radon-emitting process wastes.[3]

Perhaps the most significant aspect of these cases is that the regulatory agencies have won every one of them. While the particular regulations and facts in each case differed somewhat, the thread running through them is the difficulty in understanding and establishing effective regulation, and a willingness on the part of the Courts to give administrative agencies wide latitude in grappling with a problem which is difficult to define and control.

One case of considerable significance to testing professionals is Robles v. U.S. Environmental Protection Agency, 484 F.2d 843 (4th Cir. 1973), which involved a request under the Freedom of Information Act for the release of the results from the testing of homes for radon pollution arising from uranium mill tailings. The EPA took the position that these records were obtained on a confidential basis and could not be revealed. Id. at 844. The Court rejected that position on the ground that the public interest in important matters of public health far outweighed any claim to privacy that might be made by the owners of irradiated houses. Id. at 848. While this case applies to an agency of the United States, it suggests that any private testing company which supplies its data to EPA may find the privacy of that data to have been removed once it is in the agency's hands.

Claims Against the United States

Generally, the United States and its agencies are immune from suit for damages unless that immunity has been waived by statute. See, Kawananakoa v. Polyblank, 205 U.S. 349 (1907). To a limited extent, the United States has waived its immunity in the Federal Tort Claims Act. 28 U.S.C.A. Sec.2671 et seq. In essence, the Federal Tort Claims Act provides that the United States will be liable as if it were a private party for non-discretionary negligent acts. 28 U.S.C. Sec. 2680. For example, the United States can be held liable under the Federal Tort Claims Act for an employee's negligent operation of an automobile; it cannot be held liable, however, for the failure to pass regulations, or the allegedly improper content of those regulations.[4]

There are a small number of cases which have been brought by persons who developed cancer or other injuries allegedly caused by radon exposure, most of whom were exposed on the job. Virtually all of these cases have failed. For example, in Begay v. United States, 768 F.2d 1059 (9th Cir. 1985), uranium miners suffering from lung cancer and other diseases related to radiation exposure brought an action against the United States alleging that the government negligently failed to warn them of the dangers involved in uranium mining. The Court held that the decision of the Public Health Service not to warn the miners of the radiation dangers to which they were exposed was a discretionary matter, and, therefore, the United States was not amenable to suit under the Federal Tort Claims Act. 768 F.2d at 1066. The Court reasoned that by maintaining immunity for discretionary acts, "Congress wanted to prevent the courts from deciding in tort actions the policy and regulatory types of decisions that have been delegated to the agencies." 768 F.2d at 1064. See, Johnson v. United States, 597 F. Supp. 374 (D. Kan. 1984)(action against government as supplier of radon-emitting material dismissed based on immunity for discretionary activity).

There is one important exception to the government's otherwise successful defense of these cases. That exception may suggest not only a basis for the future liability of the government, but may also indicate the basis upon which private parties might be held liable. The case is Allen v. United States, 588 F. Supp. 247 (D. Utah 1984) which is the District Court decision holding the United States liable to certain of the persons who were exposed to radiation in the form of fallout from above-ground nuclear testing. The government's liability was predicated on the fact that it knew of the danger from exposure to the radiation, and failed to warn those who were in the area who would be exposed. The case suggests that when the government or a private party has a reasonable basis for knowing that certain persons will be exposed to radiation from its conduct and fails to warn them, liability will result.

The Allen case has another aspect of future significance, however; that is the failure of a significant group of the plaintiffs to prevail due to their inability to establish that their particular physical problems were a result of the radiation. This aspect of Allen has occurred in some of the workplace accident cases described in the next section of this paper. It illustrates that, even when the conduct of the government (or, for that matter, a private party) has involved the knowing exposure of persons to radiation, persons exposed will still have a difficult time prevailing due to the difficulty in establishing a causal relationship between the exposure and the allegedly-resulting physical problem.

Work Place Exposure Cases

Several of the published cases concerning radon involved claims by employees for cancer or other injuries due to exposure to radon at their places of employment.[5] These cases have all arisen in the context of state workers compensation laws, since most states provide that any recovery for a job-related injury must be sought exclusively within the workers compensation system.[6]

In some cases, employees (uranium miners) have successfully obtained workers' compensation awards based on exposure to radon in the workplace. Death of Garner v. Uanadium Corp. at America, 194 Colo. 358, 572 P.2d 1205 (1978); Climax Uranium Company v. Death of Smith, 33 Colo. App. 337, 522 P.2d 134 (1974). In other cases, awards have been foreclosed by procedural defects in the employee's claim. See, McCormick v. United Nuclear Corp., 89 N.M. 740, 557 P.2d 589 (1976).

Several workplace-related radon exposure cases have a common theme--the inability of the employees to show that their physical problems were caused by radon exposure. For example, in Harrison v. Industrial Commission of Utah, 578 P.2d 510 (Utah 1978), the Court upheld the denial of workers' compensation benefits to survivors of a uranium miner who died of lung cancer. The decedent miner had smoked one-and-one-half packs of cigarettes a day for many years. The Court affirmed the denial of benefits based on the inability of the claimants to establish that the decedent's condition was caused by occupational exposure to radon gas. See, Olson v. Federal American Partners, 567 P.2d 710 (Wyo. 1977)(same result where uranium miner was a heavy smoker); Garner v. Hecla Mining Co., 19 Utah 367, 431

P.2d 794 (1967)(same result where uranium miner smoked for 13 years).

The failure of the plaintiffs to win these cases reflects a significant fact: the sciences of epidemiology and toxicology are frequently too inexact to satisfy legal standards of proof that require a plaintiff, to succeed, to establish as a matter of fact that his illness was caused by his exposure. While these disciplines do give us valuable information upon which public policy decisions can be reliably based, the very nature of the disciplines frequently means that scientists are unable to connect a particular cause to a particular problem in a particular person.

Thus, while a legal vehicle for establishing recovery for these people is available in the theoretical sense that concepts of recovery are well established, one can reasonably predict that injured persons will continue to have difficulty in establishing a causal connection between their exposure and the injury. The fact that only a few persons in Allen v. United States have obtained a judgment--and that the only persons exposed at work who have prevailed in workers' compensation cases were uranium miners who experienced intense and constant exposure to radioactivity--indicates that, for the foreseeable future, personal injury cases arising from radon exposure may be difficult cases to win.

Other Cases

There are several reported cases which substantively mention radon in a variety of factual and legal contexts. Three of these cases deal with the obligation of an insurer to indemnify its insured for losses incurred due to the latter's liability for damages based on radiation exposure. Two early cases held that the issuer was obligated to cover such losses. American Alliance Insurance Co. v. Keleket X-Ray Corp., 248 F.2d 920 (6th Cir. 1957)(business loss policy may cover contamination of plant by radium dust); Canadian Radium & Uranium Corp. v. Indemnity Insurance Company of North America, 411 Ill. 325, 104 N.E.2d 250 (1952) (comprehensive liability policy covered claim based on long term radium exposure). The other insurance-related case deals with radiation exposure in a medical malpractice context. Broadbent v. U.S. Fidelity and Guaranty Co., 25 Utah 2d 430, 483 P.2d 894 (1971).

Wade Lewis Sr. & Elkhorn Mining Co. v. Reader's Digest, 162 Mont. 401, 512 P.2d 702 (1973) was an attempt by a uranium mill which advertised exposure to radon gas as a health remedy to recover a libel judgment. The case of Navajo Tribe of Indians v. United States, 9 Cl. Ct. 227 (1985) was an action based on claims of mismanagement of uranium and other mineral resources on Indian land by the U.S. government.

Of particular interest are three cases addressing the problem of radon and/or radiation exposure in or around the home. In Wayne v. Tennessee Valley Authority, 730 F.2d 392 (5th Cir. 1984), cert. denied 105 S.Ct. 908 (1985), a homeowner brought a products liability action against TVA, based on the fact that TVA had supplied radioactive phosphate slag which was incorporated into concrete blocks used in the basement of the plaintiff's home. The case was dismissed based on the untimeliness of plaintiff's claim under the applicable statute of limitations. Another homeowner who found his home contaminated by buried uranium mill tailings sought compensatory and punitive damages based on claims of forcible exclusion (constructive eviction) and increased risk of cancer. Brafford v. Susquehanna Corp., 580 F. Supp. 14 (D. Colo. 1984). The Court denied the defendant's motion for summary judgment, leaving open the possiblity of recovery under the plaintiff's legal theories. Finally, in Freed v. Ozard-Mahoning Co., 208 F. Supp. 93 (d. Colo. 1962), the Court found a mining company liable for downstream radioactive contamination of a creek which ran through the plaintiff's property, based on principles of common law water rights.

FUTURE DEVELOPMENTS

As we indicated previously, the principal problem that the law has in dealing with radon is that, in most cases, no person or organization is responsible for radon pollution. The typical radon case involves deteriorating radium, uranium, or phosphates emitting radon gas naturally into the environment. Unfortunately, for the innocent homeowner exposed to radon by this process, there may be no remedy absent a public solution. There are some exceptions.

Personal Injury Cases

Although, as we have shown, personal injury plaintiffs or workers' compensation claimants have had some lack of success to date in winning cases involving radon exposure, Allen v. United States, the Silkwood case, and the uranium miner/workers compensation cases demonstrate that these cases can occasionally be successful. Considering the general litigation-conscious nature of the American public, it can be reasonably anticipated that resourceful lawyers will develop theories of recovery, but probably very few of the cases to be decided in the near future will be personal injury cases not involving worker exposure.

Worker Exposure Cases

For the same reason, worker-exposure cases may be difficult cases for an injured employee to win. Nonetheless, the relatively confined nature of the work environment (as contrasted, for example, with formaldehyde exposure in the home), and the relatively relaxed standards of proof in worker compensation and occupational disease cases compared to the standards of proof in court cases, suggest that the worker-exposure cases are more likely to develop on a fast track than cases involving persons who are not exposed on the job.

Architect and Contractor Liability

Except for the Tennessee Valley Authority case mentioned above, no decided cases to date have involved architect or contractor liability. However, architects and contractors can undoubtedly expect litigation against them in the future.

The liability of architects and contractors is based upon the standard of conduct generally exercised by similar professionals in the particular area in which they practice. See, Bloomsburg Mills, Inc. v. Sordoni Construction Company, 401 Pa. 358, 164 A.2d 201 (1960). Thus, an architect who designs a house to be built in a known radon area without adequate basement or crawlspace ventilation could well expect to find himself in court. Similarly, a contractor in such an area who cuts corners in his basement and concrete work may find himself in court when cracks develop. We think that this is a fertile area for liability in the future.

Cases Arising Out of Sale of Radon-Affected Real Estate

Although none of the decided cases have involved efforts by buyers of houses to set aside the sale when radon was discovered, it is almost certain that such cases will develop. The cases effectively fall into two categories: the case where neither party knows of radon pollution prior to the settlement and the case in which the seller knows, but the buyer doesn't.

Where neither party knows that heightened levels of radon can be found in the house, the parties may have been operating under a "mutual mistake of fact" that will permit the buyer to rescind the transaction, return the title to the seller, and retrieve his money.

> Where a mistake of both parties at the time a contract was made as to a basic assumption on which the contract was made has a material effect on the agreed exchange of performances, the contract is voidable by the adversely affected party unless he bears the risk of the mistake [usually by agreement].

Restatement (Second) of Contracts, Sec. 152(a)(b)(1981).

Thus, in order to avoid his contractual obligations under an agreement of sale, the buyer must establish that neither he nor the seller knew of the high radon levels at the time the agreement of sale was signed, that the absence of such condition was a basic assumption of the agreement and that such condition has a material effect on the agreed-upon exchange of performance. However, as the Restatement recognizes, if alternative relief is available, such as reformation of the contract to allow for a reduction in the purchase price to reflect necessary remedial measures, a court may employ such alternatives rather than allow the buyer to avoid the contract altogether.

A different situation arises where a seller knows of the radon contamination, but the buyer does not. In such cases, the buyer may be able to set aside the transaction on the basis of fraud, recovering resulting monetary damages, and perhaps punitive damages as well. A claim of fraud could be based upon an affirmative representation of the absence of radon upon which the buyer relies _or_ upon active concealment of contamination by the seller. _See_, _Mancini v. Morrow_, 312 Pa. Super. 192, 458 A.2d 580 (1983); _Eckrich v. DiNardo_, 283 Pa. Super. 84, 423 A.2d 727 (1980). In addition, in some jurisdictions, a seller may be under an obligation to disclose to the buyer serious latent defects known to the seller or suffer liability for failure to do so. _See_, _Quashnock v. Frost_, 299 Pa. Super. 9, 445 A.2d 121 (1982). However, despite the potential liability of the seller for fraudulent affirmations, concealment or non-disclosure, a buyer is unwise to rely solely upon his own assumption that no radon exists, or that the seller will fulfill his duty to inform the buyer if it does.

Furthermore, in some jurisdictions there may be some assistance to the buyer of _new_ housing through a judge-created doctrine known as the "warranty of habitability." Under this doctrine, a person or entity which constructs and sells a new house impliedly warrants that the home he has built and is selling is fit for its intended purpose of habitation. _See_, _Elderkin v. Gaster_, 447 Pa. 118, 288 A.2d 771 (1972). In the _Elderkin_ case, the Supreme Court of Pennsylvania stated that the doctrine of implied warranty of habitability may be applied based on the unsuitable nature of the site selected for the house. _Id._, 228 A.2d at 777.

While no cases involving a warranty of habitability have addressed radon, we see no reason why the doctrine would not apply to a transaction in which highly-elevated levels of radon are found after the buyer has settled.

We believe that any buyer contemplating the purchase of a house in an area known to be affected by radon should include in the agreement of sale the right to test for radon, and the right to withdraw from the agreement if radon is determined to exist at certain specified levels. Better yet, depending upon the bargaining power of the parties, the buyer might insist on the seller providing him with a certification from a testing agency concerning the levels of radon, coupled, of course, with a right to rescind if the levels exceed a certain amount.

We would urge sellers not to resist such an approach, although we know that real estate brokers and sellers have tended to oppose it. The defense of lawsuits is expensive and bothersome under any circumstance. A seller who conveys a radon contaminated house will very likely find himself looking down the barrel of a lawsuit no matter how innocent he may have been. Even though the discovery of radon may reduce the amount of consideration he will get for his house, there is no indication to date of any slowdown in sales of houses in the "Reading Prong" (an area in Pennsylvania, New Jersey and New York in which radon is prevalent), and his disclosure of the facts may very well protect him from a lawsuit which would absorb any increment in value that he may have obtained by failing to disclose the existence of radon in the house.

Financing Implications of Real Estate Transactions

In analyzing the threat of radon to the financing of commercial enterprises and residential purchases, one must keep in mind that financial institutions are driven by the balancing of risk to reward; risks of unknown proportions are unlikely to be taken in the face of uncertain rewards. Banks, savings and loan and other thrift institutions, and sources of financing of every type, need markets for their product. In areas affected by radon pollution, banks which refuse to lend on collateral affected by radon will diminish their financial marketplace and thus will diminish their prospects for reward. All these considerations suggest that institutional requirements for radon testing should not be viewed as alarming and as potentially destructive of consumer financial sources. Radon testing, however imprecise, could be used by lenders to assure that borrowers take corrective or preventive measures to deal with the presence of radon in their homes or businesses.

Building codes and local ordinances should be modified to require the kind of construction and ventilation which will reduce the impact of radon if the gas is present in the construction area. Such modifications would be of great benefit to the consumer seeking home financing and, indeed, to the home builder who, with compliance, will acquire a measure of protection from the inevitable litigation over indoor radon.

CONCLUSION

The existence of substantial quantities of radon in heavily populated areas has led to extraordinary (and justified) concern, but the legal system has not yet responded in a fashion satisfactory to those burdened with the problem. The few legal precedents which exist suggest that there will be little recourse except for extreme examples of purposeful behavior. The development of statutory law (primarily in the area of financial support for those affected by radon -- see, e.g., 35 Pa. Stat. Sec. 7110.101 et seq. (1984)) and judicial decisions will be of great interest to those who are concerned about the lack of an effective legal remedy for a very serious problem.

FOOTNOTES

[1]American Mining Congress, et al. v. U.S. Environmental Protection Agency, 772 F.2d 640 (10th Cir. 1985); American Mining Congress, et al. v. U.S. Environmental Protection Agency, 772 F.2d 617 (10th Cir. 1985).

[2]Oil, Chemical & Atomic Workers International Union v. Zegeer, 768 F.2d 1480 (D.C. Cir. 1985).

[3]Teledyne Wah Chang Albany v. Energy Facility Siting Council, 298 Ore. 240, 692 P.2d 86 (1984)(en banc); Teledyne Wah Chang Albany v. Norma Paulus, 295 Ore. 762, 670 P.2d 1021 (1983).

[4]Of course, the United States can be sued in equity to be required to pass regulations, and the content of those regulations can be challenged through the administrative appeal process. However, the United States cannot be liable for monetary damages due to its inappropriate regulation of an industry. See, Dalehite v. United States, 346 U.S. 15 (1953).

[5]The famous case of Silkwood v. Kerr McGee, 769 F.2d 1451 (10th Cir. 1985), cert. denied 54 U.S.L.W. 3729 (May 5, 1986), is not included in this discussion since it involved plutonium contamination rather than radon exposure. The Silkwood case, most recently, has been remanded to the district court for a new trial on the issue of punitive damages. 769 F.2d at 1462. Plaintiff's personal injury claims were barred by the exclusivity of the Oklahoma workers' compensation law. However, punitive damages were held to be available to the plaintiff under Oklahoma law.

[6]It is noted that the exclusivity of state workers compensation laws may be overcome if the harm to the employee was caused by the intentional conduct of the employer. However, "even if the alleged conduct of the employer goes beyond aggravated negligence, and includes such elements as knowingly permitting a hazardous work condition to exist, knowingly ordering a claimant to perform an extremely dangerous job, willfully failing to furnish a safe workplace, or even willfully and unlawfully violating a safety statute, this still falls short of the kind of actual intention to injure that robs the injury of its accidental character." 2 Larson, Workmen's Compensation, Sec. 68.13 (1983).

THE IMPACT OF RADON GAS CONCERN UPON THE CONSTRUCTION TRADES AND CODES

Terry Brennan*

Before discussing the impact of radon on the construction trades, it may be useful to generate some empathy for builders. Houses are large-ticket, one-of-a-kind items. There are a million decisions to be made, and a million things to go wrong. As anyone who has ever had a house built realizes, when you are building something that costs more than $100,000, there are many, many things that can go wrong -- and will! You have to expect that.

The first time a given builder ever hears about radon gas is probably a Sunday morning. They have already spent three hours talking to their subcontractors. They have also just spoken with three customers calling to complain about problems (i.e., "the mason showed up today and couldn't stand straight," or "I thought we decided to put the shower on the other side of the bathroom," or "the kitchen cabinets stick out too far and block off part of the doorway"). Those are the kinds of problems that a builder faces every day. If he's lucky, he hasn't heard from his liability insurance company recently. And then someone calls and asks, "Geeze Terry, have you ever heard of radon gas?" I'm likely to be more concerned about the mason than about some increased risk of lung cancer 17 years down the road for the homeowner. I know that the homeowner is likely to call me tomorrow about the foundation and that if the mason goofs up, we will have problems all the way through the rest of construction. I am not happy to learn that there may be an adverse health impact on the residents; no one wants to build a house that is going to kill someone. But, as a builder, I am more concerned that the house not fall down on them than that there not be a health risk from an indoor air contaminant.

There are really two types of residential builders in the United States. The previous concerns mostly belong to the small builder who erects less than 20 houses per year and accounts for at most 40 percent of the total construction. The remaining building is done by approximately 10 corporations using factory panelized houses shipped to tract sites. Clearly, the two types of builders will be impacted differently by the radon gas issue.

---

*Terry Brennan is a builder and radon mitigator, and principal of Camroden Associates.

The big builder faces even more aggravating problems involving large amounts of money. Rather than practical construction problems, he faces political and legal problems on a daily basis. The small builder having trouble with a 2-by-4 can get mad at it, saw it, hit it or throw it out the window. But when things go wrong at the zoning board or the road in your hundred-unit development wasn't put in properly, these are much bigger problems than stairs falling down in someone's home. As a result, the big builder is going to be even less thrilled with the radon problem. For both types of builders, radon is likely to be treated as just one more problem in a problem-filled environment.

This doesn't mean that many builders will not be concerned. Small builders are likely to be more worried about homes already built where high radon levels are found. Even in the absence of lung cancer, the builder may have liability because they built a house that had a high radon concentration.

On the positive side, radon mitigation itself promises to open up a new area in home-improvement contracting. There are a number of people in New Jersey and Pennsylvania right now who are making a living doing mitigation, and they are busy. I think that a lot of it has to do with how we as a society perceive radon. We may decide, despite the EPA, that the problem is not there; it would be easy to do because radon is so hard to detect. I see a lot of people who have tested and found large exposures who are not doing anything about it. They say things like, "I already have 15 years exposure, so what can I do about that?" It's such an unglamorous thing for a homeowner to test and then do something about a high reading. It's not like putting in a new bathroom where they can show everybody their new tub. A lot of people may decide, for example, that 16 pCi/l is okay. If they do, 70 percent of the buildings over 4 pCi/l would be eliminated from mitigation. So, while radon is generating a lot of work right now, it's hard to predict how extensive the demand will be. Mitigation is likely to be spurred by realtors, mortgagers and home buyers who will force the issue of testing and fixing the home. But, as a basic form of "home improvement," while a new market has been created, it remains to be seen how large a market there is.

New construction suggests different issues. Such institutions as the National Association of Home Builders (NAHB), which has set aside a research budget to work on radon, offer the potential for discovering and disseminating solutions that will assist the large developers. At the same time, new construction techniques that add a few hundred dollars to the cost of a house will have less impact on the small buidler than on the large contractors. A developer building 400 units and spending $200 on each building is laying out $80,000 for radon proofing. This is in a trade where a savings of 25 cents per door knob on that many units would be considered to be a major savings. One can expect resistance to such expenditures.

There is also a natural resistance by builders to the notion that something that they do, and have done for a long time, can be causing a major problem. But where the problem surfaces in a big way, they do take notice. For example, in the Syracuse area, publicity about radon problems led the local home builder's assocation to circulate an article that I co-authored which details radon-proof new construction steps. It only takes one member of such an association to get interest going in an area. This group is even moving to develop its own construction detailing. For big builders, action was also inspired by actually finding radon problems. One of the largest builders in the country recently encountered average radon levels of 20 pCi/l in a major development that was half completed. They were forced to undertake remedial work on a large scale and to alter their plans for the remaining buildings. In a situation involving so much potential liability, these builders wake up. It took an incident on this scale to motivate the NAHB to action.

There will be other impacts on construction as well. For example, there will be dozens of areas in the United States where development will have to be limited because of soil radon concentrations. Much as one avoids building in swamps, there may well be avoidance of high radon areas. Following the Swedish model, soil types might be divided into a number of risk levels according to such factors as radium content in the soil and permeability. Of course, in the Swedish regulations, many of the criteria end with question marks because they just weren't sure of the relationships of different factors. So you look at the criteria and it says "high permeability?"

In line with the above, it would not be surprising to find that building codes will shortly dictate new construction techniques which minimize radon entrance to the structure. This has already come under consideration at the local level. For example, in Clinton, New Jersey, there are some radon regulations which detail new construction. It is likely that local areas which are very concerned about radon, such as Clinton, will take the first steps toward code development. Later, the consensus codes, such as the Uniform Construction Code, will address radon. Finally, these national code groups will disseminate their work back to the community level, where other municipalities will begin to pay attention. It's going to be a slow and uneven process. There are tens of thousands of code jurisdictions in the United States. It's a very diffuse body of enforcement. Each little community has its own zoning and enforcement officers. It's difficult to reach all of them and, thus, to get a uniform response. It's going to be piecemeal. And, while B.O.C.A can write the best radon resistant details that they can think of, it's John the building inspector out there in the field who has to sign the certificate of occupancy. If he just winks at the contractor, there's going to be a problem. Perhaps the legal liability developments will dictate how much actually gets winked at. If local building inspectors are liable for faulty construction that lets radon into the building, then they may be less likely to overlook violations.

By the way, the quality of building has a lot to do with the radon problem to begin with. If there is a careless installation of masonry, the chances of a problem are greater. Many of the high radon houses that I've seen have poor masonry. For example, does the mason fill the joints on a block wall or do they parge or damp proof the sides of the wall? Or do they say, "It's out-of-sight, nobody will know!" Obviously, the more radon entry points that they provide, the more of a problem you will have. So, while I have no statistical proof of this, we see poor masonry over and over again in houses with radon problems.

Radon-proof construction techniques are currently being identified. There are two generic operating principles for keeping radon out of the house:

(1) The foundations can be made so as to keep the radon out, by careful sealing or by diverting the soil gas away from the building through passive ventilation.

(2) The building can be designed so that it is easy to fix if it does have a problem.

Ideally both approaches are used. Such techniques are further elaborated elsewhere.[1,2]

Of course, the development of the means to prevent radon entry to buildings raises the question of how we reach builders with this information. Because most builders constructing 20 or more houses a year belong to the NAHB, reaching the large builders is fairly easy. The NAHB will reach the mainstream builders who are concerned with high volume construction. But interestingly, the fastest innovation can be expected through smaller builders who belong to various progressive trade organizations. For example, in organizations like the New England Solar Energy Association Quality Building Council, or the Midwest Energy Efficient Builders Association, radon has been discussed for some six years now. These builders were aware of what had happened in Scandinavia because many import quality prebuilt homes from there. More importantly, because of their concern with energy issues, they had taken the time to learn how buildings work. As a result, they are capable of understanding the radon problem. I gave my first talk on indoor air pollution at a meeting of one of these groups about five years ago.

Beyond such associations, a means of disseminating radon-limiting methods to the small building industry will be challenging because small builders learn their trades largely by apprenticeship. One attempt to further the radon mitigation and prevention field is a radon mitigation course which I am staffing with a group of other skilled mitigators. The course is being funded by EPA through a grant to the New York State Energy Office. Some 600 people have so far attended 20 sessions held in several states. Of these, a third were government officials responsible for radon issues, a third were measurement and diagnosis professionals, and a third were contractors expecting to actively go out and do radon mitigation work. There were a few realtors, bankers and insurers in attendance as well. This was a good start in the effort to disseminate information necessary to promote the prevention of radon exposure. It will be interesting to see the extent to which such efforts help to involve builders directly in the creation of a radon-free generation of buildings.

FOOTNOTES

[1]Turner, William and Terry Brennan, "Radon's Threat Can Be Subdued," Solar Age, May 1985, pp. 19-22.

[2]Brennan, Terry, and William Turner, "Defeating Radon," Solar Age, March 1986, pp. 33-39.

THE ROLE OF SCIENTIFIC PROFESSIONALISM IN MEETING A PUBLIC NEED

Donald F. Schutz and Harvey Greenberg*

MEETING THE NEED FOR A RADON-RELATED PROFESSIONAL ORGANIZATION**

So that everyone in the audience may breathe a sigh of relief, let me give you the good news that this building has been tested and no unacceptable levels of radon have been found.

I should add, further, that you can count on this evaluation because the testing was done by a member of the American Association of Radon Scientists and Technologists, the organization of which I am president and an active member.

You should also know that I and other members of our organization, as business and scientific professionals, consider ourselves called by the disciplines we represent and by the economic objectives of our companies to bring expertise and ethical standards to this new-born industry.

Were it not for the dramatic discovery of unprecedented levels of radon in a Pennsylvania residence that has made radon a household word practically everywhere, we would probably not be here now talking in such professional and business terms.

Were it not for the energetic concern of outstanding public officials like U.S. Senator Frank Lautenberg and New Jersey State Senator John Dorsey, both of whom recognized the problem as one of potentially major proportions, the situation may never have arisen in which business and government would be called upon to attack the problem together.

Through their efforts we are now moving ahead on a national and state level to provide funding for an organized

---

*Donald Schutz is president of Teledyne Isotopes, Westwood, NJ and of the American Association of Radon Scientists and Technologists, Inc.(AARST). Harvey Greenberg is Engineering Manager of Radon Engineering, a Division of PSI Engineering, Mahwah, NJ. He is Vice-President of AARST.

**The first part of this paper was prepared and delivered by Donald Schutz.

assessment of the dimensions of the problem and for the development of methods of remedial action.

Until that incident in Pennsylvania became the focal point of attention for radon everywhere in this country, few people outside of those professionally involved had even heard of radon. Few were aware of the possible health problem within their own homes. Yet, radon and its potentially harmful threat to the public health had been identified some years before in studies of the incidence of lung cancer in uranium miners.

Understandably, the main reason it escaped public attention was that it was thought to be a threat to only a very small part of our total population. It was considered an industrial health problem. Few people could relate to a uranium miner, and, after all, the uranium miners were all safely out west.

But the experience in Pennsylvania just 18 months ago changed that picture entirely. Suddenly, entire families saw themselves as potential victims of a life-threatening disease. It was not in my neighbor's backyard; it was right in my own.

In New Jersey, the State Bureau of Radiation Protection, under Dr. Gerald Nicholls, met the emerging crisis with intelligence and action within available resources. In quick order, the state set up a hot-line to respond to citizen questions about the problems of radon, followed by meetings, literature explaining radon, and well-publicized statements urging the residents not to panic. And very few did, to the credit of Dr. Nicholls and his associates, in spite of some highly inflammatory publicity including, but not limited to, bold headlines such as "RADON: Lethal Gas," etc.

The red flag was being waved. And what was recognized as a public health threat was also recognized as a business opportunity. For some, it was a natural extension for well-established businesses in radiological measurements and health physics services. For others, it was an opportunity to exercise marketing and distribution skills. Unfortunately, in some cases, these particular business skills were not backed up by the professional, scientific and technical expertise required by a highly complex and difficult problem.

Suddenly, the state was confronted with reports from citizens that their homes had been found to contain radon with truly excessive levels.

Others told stories about estimates for correcting the problem at costs in the thousands of dollars.

Realizing that the state could not possibly handle, much less control, the exploding nature of the radon problem in many different areas, Dr. Nicholls and his staff started meeting with representatives from firms in the radon testing, measurement and remediation businesses to exchange information and to bring order to the problem.

Out of this grew the suggestion that the wisest move of all would be for the industry to organize to form an organization that would respond in a professional and exemplary business fashion to the needs of the public urgently looking for help.

Likewise, the inability of a single entity to respond adequately, such as a government agency or a university, which traditionally have been the organizations concerned with the measurement, study and correction of radon problems, focused increased attention on the need for business to become involved.

Leaders from reputable business firms and knowledgeable members of the scientific community responded by joining forces to organize the American Association of Radon Scientists and Technolgists.

Its objectives were evident almost immediately to the organizing group when it sat down to decide on how to proceed. Uppermost in the minds of all was the recognition that no business can succeed over the long term unless it provides a true public service and accepts all of the responsibilities that accompany that goal.

The main responsibilities were really those of any business: to offer a quality product at a fair price and at the time and place required by the customer. That sounds fine, but, when examined in detail, complicated and difficult business decisions remain: What is the product? Who needs it? For what purpose? What is a reasonable price?

Private enterprise, though, can, and has, moved rapidly, sometimes supporting government initiatives, and sometimes independently, to provide a needed diversity of products and services. There have probably already been more than 40,000 houses monitored for radon by private sources, although it is difficult to tell how many of these involved some sort of government sponsorship.

By comparison, the U.S. Environmental Protection Agency is planning a nationwide research program that will involve 110 houses. The State of New York is just beginning to test 2,000 houses; the State of New Jersey 6,000; and the State of Pennsylvania 20,000. By the time the results of these studies are available next year, private industry will have surveyed many tens of thousands of additional houses.

The aims and purposes of the government-sponsored programs are and should be different from those of the private sector. There is no way, however, that we can look to any source other than business to meet the need which may involve up to 1.4 million residences in northern New Jersey alone and many more millions of homes nationwide.

This reinforced our belief that business had an essential and inevitable role to play. New companies were being formed to provide legitimate services based on well-established expertise, and the need for an organization to bring together responsible companies and members of the scientific community was increasingly evident.

In emerging markets like this, however, new opportunities often attract an undesirable fringe element, and there were reports of unscrupulous business practices. While most of the reported abuses tended to be illusive and highly exaggerated when subjected to close scrutiny, we were intent on overriding any indiscretions by formulating standards of the highest scientific, technical and ethical order.

The Association also has other objectives. An important one is to cooperate with the State Bureau of Radiation Protection to help formulate testing, measurement and remediation standards so that the public will gain a better understanding of the radon problem as well as to insure confidence in the industry.

From the beginning, it was our intention to make this a national organization. While we are based in New Jersey, and New Jersey was the first state to develop significant private industry involvement, we are reaching out to attract persons of national and international stature to work with us to attain our goals.

Our membership now includes 20 companies and 22 consultants and associates. Our geographic coverage includes members from nine states spanning the continent from California to New England. Our membership is open to those companies and individuals with expertise in radon problem-solving and also to individuals and organizations in fields such as public health and real estate which have an interest in the results of our work even though they are not involved in doing such work themselves. We are also extending invitations to students who will be confronting the problems of radon in the future.

We believe that there is a great need to bring the scientists who have studied the radon problem in the laboratory together with federal and state government officials, land developers, and contractors so that a homeowner's real needs are met.

The Association wants to help educate the public about radon hazards, to improve the investigation, measurement and remediation of radon, and to promote high business standards.

Already we have set the wheels in motion with the establishment of committees to achieve these goals. We are in the process of forming a Technical Standards Committee with a primary focus on developing training courses for our members as well as for others who want to improve their technical abilities.

Our Professional Standards Committee is meeting to develop the way the membership responds to the needs of the marketplace. Here we are guided by anti-trust legislation and fair trade laws.

Our Public Information Committee already has a number of projects in the works to inform responsible public officials and the public at-large about radon. One of our major projects is to develop standards for advertising and promotion so that the seal of the Association will be recognized and respected.

We intend to win the public's confidence in this rapidly growing industry. We have assumed this responsibility.

Some of our members, already, are governed by strict codes adopted for their particular field of endeavor. Among our members are certified health physicists, certified industrial hygienists, professional engineers and many with doctorates in relevant fields such as geology and radiation science.

We recognize, however, that academic and professional stature are only part of the picture. This expertise must be used by technologists and craftsmen in various fields. And the scope of our organization is planned to be inclusive of all of the disciplines and technologies needed to understand, quantify and solve the radon problem.

PROBLEMS OF PROVIDING RADON-RELATED SERVICES TO THE PUBLIC*

My name is Harvey Greenberg, and I am the First Vice-President of the American Asociation of Radon Scientists and Technologists. I am a licensed professional engineer in both New Jersey and New York and am employed right here in Mahwah as the engineering manager of Radon Engineering. Don Schutz, the prior speaker, was correctly introduced as the President of the Association, but his modesty prevented him from mentioning that he holds a doctoral degree in geology and is president of one of the nation's most prestigious radiation measurement laboratories, Teledyne Isotopes.

We both are typical of the many local and national members of the American Association of Radon Scientists and Technologists in qualification and purpose.

I'm here to particularly address the role of the firms and/or individuals offering measurement or remediation services and to share some of the observations of these firms as they relate to this afternoon's topic, that is, "the socio-economic impact of the radon issue."

As many members of the Association represent commercial "for profit" enterprises, they normally request a fee from their customer, and in exchange, they provide a service of comparable value.

Testing/measurement services must be based on the use of accurate, and when possible, economical equipment operated by experienced, qualified personnel. Appropriate testing protocols should be strictly adhered to and the reliability/quality of the data must be assured. Participation in the U.S. EPA's Radon/Radon Decay Product Measurement Proficiency Program is highly recommended. Professional and comprehensive written reports should be provided to each client after testing.

---

*The second portion of this paper was prepared and delivered by Harvey Greenberg.

Organizations providing abatement/mitigation consulting services and/or equipment must likewise provide quality. Any advice offered a client should be appropriate for the particular circumstances, effective, and worthy of the fees charged. Ventilation, particulate removal and similar equipment must be safe, functional and offer a reasonable service life.

Services rendered do not end with the completion of the work offered the client.

We must serve the community and the science itself. Continuing education of the community, being careful not to alarm and only to enlighten, must be a priority. Providing detail (without specific client information) to government agencies as requested and to the scientific community should be considered equally important.

Advertising our services is a particularly sensitive area. Commercial organizations can only survive by informing the public that they exist and are capable of providing a particular service. It is important, however, that we do not act like or even appear to be opportunistic parasites. We must be careful not to make false or misleading claims, including offering services that cannot be provided at the current time. Alarmist statements or any type of "scare tactics" shall not be utilized. Our mission should be to alleviate people's fears, not exaggerate them or feed upon them.

At this time, I'd like to change course slightly and share with you just a small number of the many observations of Radon Engineering and several other member firms of AARST. The observations chosen particularly reflect the direct social and economic impacts of the radon issue in our local area.

DEMOGRAPHIC PATTERNS

If we begin by examining demographics, the structures tested for radon/radon progeny are typically:

> Private homes - representing 80 to 90 percent of all field tests
> Institutional Buildings (schools, hospitals, etc) - 5 to 10 percent
> Commercial Structures (office buildings, factories, etc.) - 5 to 10 percent.

Testing for radon in water is still uncommon in this area. It is only done when homes with high radon gas levels have private wells.

Soil testing is even less common. Although a very desirable test for builders, no particular device or field methodology has proven to be totally reliable. Samples sent to a laboratory for analysis, where very accurate screening can be done, cannot be considered as representative of an entire lot.

Who contracts for testing? Approximately 60 percent of the field tests provided by private firms are purchased by home or building buyers, the majority having built in a contingency to their purchase contracts requiring "acceptable" levels of radon. These customers are as equally concerned with the health risks of radon as with the effect on real property value.

Approximately 20 percent are purchased by concerned homeowners to establish their health risks (due to radon exposure) at home.

The remaining balance of the tests are purchased by school and other institution administrators, commercial building operators and tenants. Additionally, homeowners selling their properties often want to pre-establish (and, if required, remediate) any radon levels that might affect the sale of their property. Likewise, many clients who had previous tests which indicated high levels request a retest for confirmation or to prove out remediation measures.

The average value of homes tested is well above $200,000. Clients are typically middle to upper middle class, well educated, with environmental and/or health interests. Many are physicians, lawyers, teachers, other professionals, or business men and women. All are concerned with protecting property values and minimizing health risks. We have observed that engineers and other scientists typically prefer to purchase passive detectors so as to perform their own testing. Upon the discovery of high levels of radon, they do not hesitate to purchase a full field test to confirm their own findings.

In our local experience, we find that 15 to 20 percent of the homes tested in Bergen, Rockland, and Orange counties have radon concentrations above the 4 pCi/l or 0.02 WL "remedial-action" level. In Passaic and Morris counties that percentage increases to 20 to 25 percent, and in Hunterdon County to over 30 percent of all homes tested. The highest readings recorded have been in Hunterdon County with southern Morris and western Passaic counties not far behind.

The data above, however, must be recognized for its prejudiced basis. Obviously, now that Clinton has been identified as a prime area of concern, many more tests are being requested by residents of that area, and the results tend, not surprisingly, to also be high. This obviously biases the statistics far to the high side. We previously saw the same thing happen in the Mendham-Chester area of Morris County, and before that, in the West Milford-Sparta area.

Client and homeowner reaction to reported results is very interesting. Homeowners tend to be angry when high levels are discovered. Homebuilders are particularly hostile. As the messenger of "gloom," the testing firm is normally the recipient of that anger. Customers who receive reports of very low levels normally are pleased, but a small group of those are angry that they purchased a service that "in hindsight" was not really necessary for their particular circumstance.

Many homes are tested for clients who are purchasing homes, and who, therefore, are not resident in the structure being tested. This presents us with the possibility that our equipment could be tampered with or results prejudiced by one method or another. Real time data acquisition equipment and partial witness tests provide the testing firm with a means of recording initial accurate data, and determining to what extent windows or doors might have been open or other prejudicial actions committed after our representative has left the site. We are not, however, policemen. How do we handle a situation of tampering? Do we accuse the home resident, notify the client, retest - and at whose expense?

It is certainly worth noting the number of tests performed daily in the state of New Jersey. It is estimated that private firms perform 100 to 150 field tests each day. Add to this another 50 to 100 tests conducted by individual homeowners using passive detectors supplied by Terradex (through their local distributor), Teledyne Isotopes, and others. Now add again to the above figures an almost equal number of tests performed using the University of Pittsburgh's charcoal canisters (we estimate that the University of Pittsburgh is the single largest vendor of radon monitors in the state of New Jersey).

At this rate, the quarter of a million homes built on or near the Reading Prong can be tested in just two years. All private homes in northern New Jersey can be tested in just 6 to 10 years (depending on the number of retests required after remedial action or upon the re-sale of a home). As you are aware, the New Jersey DEP offers free confirmatory testing to residents who purchased private testing and received reports of "remedial action levels." The state is already having problems keeping up with the demand. Now consider the impact of over 100,000 tests each year, with up to 10,000 homeowners possibly requesting DEP retests.

In closing, let me reassure you that the member firms and individuals of the American Association of Radon Scientists and Technologists are fully aware that we are still climbing a very steep learning curve. Please note, however, that we are acquiring as much relevant data and "hands on" experience as all government agencies combined. The Association intends to behave much like, in computer jargon, the CPU, or Central Processing Unit. Our inputs are our members' prior education, day-to-day experiences and quantitative field data, and our output is reliable, much needed public information and continued professionalism.

FRAUD-RESISTANT RADON MEASUREMENTS FOR REAL ESTATE TRANSACTIONS

Harvey M. Sachs*

INTRODUCTION

Background

In regions where too many houses have too much radon, a common response by governments, bankers, and other regulators is to require "radon certification" as a condition of house sales. Such programs are analogous to existing requirements for termite certificates and well-water testing. Another parallel is the formaldehyde testing that was required for houses with urea-formaldehyde foam wall insulation.

Unfortunately, the programs that have been proposed or used in the USA can give very misleading results: They use inadequate technology and have insufficient safeguards against fraud.

Present Approaches

In some proposals, one or more grab samples would be taken by a technician for laboratory or on-site analysis. The cost of field technician time mandates that these samples all be taken during the same visit. Since radon values in houses can vary by factors of 3-10 over short time intervals, these grab samples are unlikely to give a good estimate of average concentration in a given house. Radon risks are proportional to total exposure, not peak exposure, so certification programs should be concerned with average concentrations. Grab samples insure that there will be unacceptable numbers of false "passes" and false "failures," both of which can lead to real problems. As important, because the occupant knows when the visit is scheduled, it is too easy to thoroughly ventilate the house shortly before the "test," so that not enough time is available for the radon value to "grow in."

As an alternative, an early certification program in the U.S. northwest required 24 hour samples. Unfortunately,

*Harvey M. Sachs is a radon consultant and former faculty member at Princeton University's Center for Energy and the Environment.

there is anecdotal evidence that there were frequent systematic attempts to produce low readings by dilution with low-radon outside air. An additional opportunity for fraud is to move the instrument outdoors for most of the test period, unless the instrument can be locked in place.

Goals

Our goal is to propose alternatives to these methods, ones that approach the following objectives:

*Costs should be as low as possible.
*The system should be as fraud-resistant as feasible.
*The system should have built-in incentives for compliance, in contrast to the present methods in which the seller, who is responsible for the testing, has strong incentives to mislead.

We have explored two ways to satisfy these criteria:

1. An escrow-backed system, in which the buyer does the testing during an appropriate season. In this proposal, the cost of mitigation, if required, is largely or completely covered by an escrow account established in the sales contract.

2. A technical fix, based on development of a package that simultaneously measures both the integrated radon concentration and the integrated ventilation rate of the house.

AN ESCROW-BACKED SYSTEM

The best way to measure radon for purposes of estimating risks in houses is with long-term integrating monitors. This is required because of the large temporal variability of radon concentrations, from day to day and from season to season. Thus, one wants a multi-day to multi-month measurement taken during that season when concentrations are likely to be greatest. The instrument should be one that can be fixed in place, is inexpensive, and is reliable. Either activated charcoal or Track Etch type monitors may be appropriate, if the packages are designed with care. Both are passive, so no electricity source is required, and both are analyzed in the laboratory by technicians with no interest in the outcome.

As noted above, the seller has great incentives to reduce readings. However, if the buyer knows that reasonable mitigation costs could be recovered from an escrow account, then he or she has no reason to fear testing or to try to bias its results.

Thus, this proposal is simple, and requires no new technical development. In radon-prone areas, an escrow account is established by contractual agreement at the time of closing. This account might be $2,000 - $4,000, depending on the complexity of the house and local mitigation experience. During the next heating season, the buyer is required to conduct an approved long-term radon test. If the results indicate levels in excess of an agreed threshold, mitigation is undertaken and paid by the escrow account. Monies not spent are divided between buyer and seller by an agreed formula, giving the buyer an incentive to seek the least-cost mitigation strategy. Where radon concentrations are acceptable without mitigation, the escrow account and its interest are returned to the seller.

Pros and Cons

The great advantage of this proposal is that it requires no new development or validation of technology: everything is commercially available. In addition, it maximizes use of available market methods (contractual arrangements, escrow accounts). Its major disadvantage is that most realtors dislike anything that prevents a clean and final sale, or that requires additional explanations or lengthens the contract. I can only note that most alternative "certifications" are nearly worthless, and could lead to critical legal exposure for the realtor.

A 'TECHNICAL FIX'

The Method

If the average ventilation rate of the house can be measured during the same interval as the average radon concentration, then successful fraud is much more difficult. This can be shown quantitatively, but it is also intuitively true: If the ventilation rate is unusually high (an unusual amount of outside air is being brought in), it will dilute the radon and lower its concentration.

The passive perfluorocarbon tracer (PFT) technique for ventilation rate measurements[1] is ideally suited to this application. The instruments are very small, entirely passive, and can be deployed by relatively untrained persons. There are enough different PFT sources that it is unlikely that fraud could be perpetuated by seeding the house with excess PFT. It has been estimated by Dietz (personal communication) that the commercial cost of routine in-house measurements would be in the range of $100; some automation of the laboratory analysis is possible.

In this technique, a technician would install the PFT sources and estimate the house volume. The PFT detector would be sealed in the radon detector package. Thus, if the detector spent time outdoors, or if the house were artificially ventilated for the duration of the measurement, the PFT devices would flag the anomalously high apparent ventilation rate of the house.[1]

Pros and Cons

The major advantage of this technique is that it allows pre-sale radon certification with reasonable assurances of integrity. There are still limits, of course, in that measurements made in non-standard seasons would require interpretation that could cause disputes. The major disadvantage is that the PFT technology, although routinely used in experimental studies of houses, is not yet a mature commercial technology. Commercial applications would probably require additional calibration and validation before using it for work with important legal and health ramifications.

DISCUSSION

This paper offers some bad news and some good news. The bad news is that I have been unable to justify any certification program based on the use of grab samples. In addition, any programs for long-term (average) measurements by the seller must have strong built-in features to discourage fraud.

The good news is that it is possible to design programs which change testing so the incentives are aligned with the responsibility (the escrow system) or which offer good resistance to fraud (simultaneous ventilation and radon measurements). These proposals meet the minimum standards outined above. Less carefully implemented programs will not serve public health, may mislead (and defraud) the public, and may subject their sponsors to unwanted legal attacks.

House transactions are the largest investments made by almost all people. It is our obligation in the technical and policy communities to provide the best possible assurance that our test requirements protect both their financial and their health interests.

ACKNOWLEDGEMENTS

Discussions with Russell Dietz, Gautam Dutt, Thomas Hernandez, and Joel Nobel were instrumental in formulating various ideas proposed here.

FOOTNOTE

[1]Dietz, R.N., and E.A. Cote, Air Infiltration Measurements in a Home Using a Convenient Perfluorocarbon Tracer Technique, _Environment International_, _8_, 1982, pp. 419-433.

HIGHLIGHTS FROM THE DISCUSSION: SOCIOECONOMIC IMPACTS OF THE RADON ISSUE

RADON AND REAL ESTATE TRANSACTIONS

Question: Harvey, how is the real estate industry responding to the radon situation? Are they doing a good job or are they doing a bad job?

H. Greenberg: I can relate just the New Jersey and lower New York real estate response, and it is probably consistent with everything else that goes from A to Z. We see, especially in New Jersey, where real estate sales people have a little bit more responsibility in their charter, that they're helping write a contract for the home buyer. We see some realtors who haven't heard the word radon yet. We see others who are aware, but are taking a wait-and-see attitude and would just rather not participate in anything about the subject right now. We also see some who are very aware and responsible as far as informing their clients. I believe many of the multiple listing agencies which write contracts that are used by the realtors have incorporated a radon disclosure statement. It is basically just a warning for the buyer to be aware that radon gas and progeny have been found in some dwellings in some areas of the state. The comparison of this response to that in lower New York State is between night and day. I don't mean to insult the Rockland or Orange County realtors, but in New York State, sales people deal strictly with purchase binders. They have been trained only to show real estate and to create a purchase offer from a buyer and an acceptance from a seller. While some act far beyond that scope, New York realtors basically turn issues like radon over to the attorneys and duck away from it. We are having a great difficulty in educating the market in Rockland and Orange Counties right now that the radon problem exists.

J. McQueeny: When Senator Lautenberg organized a task force of people to go to Sweden in January, the real estate people jumped at the chance. They were helpful when we were over there trying to find out what programs worked, what did not work, what could be applied back in New Jersey and what had to be dismissed. They have been very helpful in trying to deal with radon in New Jersey on that score.

Question: Does "Let the buyer beware" apply in real estate transactions in this case?

D. Toomey: The answer is no. If you know information and you don't disclose it, and the person who buys the property asks you or you have reason to know that he is relying upon your truthfulness, he can set the deal aside. He can sue you for consequential damages. "Let the buyer beware" doesn't exist in most places anymore.

Question: Three people are interested in a home that is untested in an area of homes with elevated levels. The realtor advises the purchaser that two other parties are interested and any contingency clause for radon could essentially jeopardize the chances to buy the home. Is there any legal recourse for the perspective buyer when he loses the chance of buying a home because they want the test?

D. Toomey: The honest answer is I don't know. Certainly, if it was a true statement, the answer is clearly no. There is no legal recourse; the realtor is just telling what the facts are. But the assumption in the question is that the statement is untrue, and that the realtor is trying to hammer the buyer into signing his agreement of sale. I have to tell you that in that circumstance I am not sure of the answer.

D. Sykes: Being a litigator, I will go out on a limb and say I would rather have the seller's case than the guy who didn't get the house.

RESPONSIBILITY FOR SECONDARY RADON PROBLEMS

J. McQueeny: Senator Lautenberg and people from EPA have held some meetings on radon around New Jersey. One of the most asked questions is the hypothetical one given by individuals, "When high levels of radon are found in a second-story condo, is there a legal responsibility for the first-floor owner to take remedial action to correct the problem, particularly if this may be the only way to correct the second floor condo?"

D. Toomey: That has been written by a bar examiner. I guess the answer would really depend on another question, "Can we determine why the radon is in the second floor and not in the first floor?" If there is some defect in the first floor, cracked walls or some other thing of that nature, I think it wouldn't take a very smart lawyer very long to construct a way to get the guy on the first floor to fix it. If you couldn't pin down how the radon reached the second floor, then the fellow on the second floor is probably out of luck.

Question: What about the condominium association and their responsibility, resulting from common ownership property?

D. Toomey: It depends on whether the problem exists in the area of the common ownership property. This varies according to the agreement from place to place. The association might have a responsibility or it might be the individual owner. Again, it rests on whether you can establish what is causing it.

H. Sachs: As a dumb building scientist, I might not be able to establish that without going into the first floor.

D. Toomey: That would very well be a matter of local and state law. It would be a difficult problem, I agree.

RADON TESTING ORDINANCES

Question: Do you foresee that radon testing could be mandated by ordinances similar to the perk test situation?

D. Sykes: I think there will be ordinance improvement certainly in Berks and Lehigh counties in Pennsylvania. When it is going to happen and what form it is going to take, I really couldn't predict at this time. But I think that that could be one of the areas that would be dealt with.

H. Sachs: There is a distinction I would like to draw between the perk test and the radon test. A perk test is required as a pre-construction test to see if the soil can withstand a drainage system or if it will back up. A number of us made the point yesterday that we are by no means convinced we know how to build a radon test that would evaluate a site cost-effectively. We don't think that we know how to predict that a site will be "dangerous" or not. Certainly not without doing an engineering investigation that would cost much more than making the basement radon resistant. To most of us, the question is not the perk test analogy of testing pre-construction. Rather, testing is best done at the time of transfer and when granting the certificate of occupancy. We would ask our friends, the builders, to assure that the first occupants really can get good values when they run that first winter's test. That certainly could be a Certificate of Occupancy contingency, sort of on a delayed basis.

D. Sykes: I don't quarrel with that at all, but I will observe that I don't think that even at this late stage in the development of termite testing, we have a very scientific termite test, if you have ever seen the litigation over that particular issue in local courts.

H. Sachs: But we don't determine that test before construction.

D. Sykes: I am talking about ordinances that would apply across the board.

ENFORCING PROFESSIONAL STANDARDS

Question to D. Schutz: A Bergen County company recently made a mass-advertising mailing replete with red headlines warning residents that they lived in a government-designated danger zone. This company also boasted of belonging to your trade organization (American Association of Radon Scientists and Technologists, AARST). It seems like policing is not quite up to par yet.

D. Schutz: The policing usually comes after the crime, and the state took the action of not listing that company. That mailing was done contrary to the bylaws of AARST which required consultation for advertising and promotion plans with the Association. As far as I have been able to determine, the telephone of that company has been disconnected. I have not been able to confirm exactly whether they are still in business or not. That wasn't an Association police action, but as far as I know, that is no longer a problem.

Question: What do you do to a member to police a them?

D.Schutz: Jawboning, primarily. For the simple reason that we have not been organized long enough to have a well worked out set of procedures for disciplinary action. We have been advised by council that we have to be very careful about that because of the anti-trust considerations of restrictions on who can be a member and the possible damage to a member if they are disciplined unfairly. It is a tricky proposition to run a disciplinary operation in a professional association, especially one that is only three months old.

USING PRIVATE TEST DATA

Question: What will be done to create public data sets which present the summary of the radon testing being done privately?

D. Schutz: The State of New Jersey has organized a repository for that data. The legislation which is now under consideration would assure confidentiality for such data. A number of our members have been very circumspect about providing data which they generate on a very confidential basis. Even though there are assurances by the state that it will remain confidential, some of our members have been dubious and have not complied with the volunteer program of providing that data. A lot of them have. Some have not cooperated just because it is a lot of work to do; nobody is paying us to do it, and we have other things to do. We should be careful to note that such data is not research quality data because it may be gathered under a variety of different sampling conditions for a variety of purposes and not be of the quality of data that you would want to analyze for epidemiological purposes. It still may be useful to our understanding.

DO HOUSES CHANGE OVER TIME?

Question to T. Brennan: Is it possible for a house to be within the limits of radon levels at one time only to develop a high level later on?

T. Brennan: I guess the answer is yes, but that's the sort of thing that happens in a cyclic kind of way. In the New York State research we have seen differences as large as a factor of 5 between summer and wintertime measurements. Lawrence Berkeley Labs is reporting something on the order of 20, which means you could take a summer reading of 4 and have it be 80 in winter. That's a pretty big spread. That means you could take a three-month measurement in the summer and a three-month average in the winter and have it be very different. So there are daily cycles, as Harvey showed, and seasonal cycles in radon concentrations that have to do with what the weather is doing and what the occupants are doing.

D. Schutz: On the question of variability, some of Naomi Harley's data where she did hourly measurements for 4 years showed annual averages which differed from one year to another by a factor of 2. I think that you just have to accept the fact that radon is not going to be entirely predictable.

H. Sachs: What that says is that we have an education problem. We are getting into voluntary situations such as many real estate transactions are today. It is not just a question of establishing a hard line that says if it's under X it's OK, and if it's over X it's bad. It is much more subtle than that, and it requires a great deal of public education for the public to understand that. Before I turn it back to Terry, I want to go back to what I heard as the meaning of that question. Let's assume a hypothetical case of a house for which we have a perfect one-year continuous reading at, say, 2 pCi/l. Is that house, 5 years later, going to be at 20 pCi/l? Is that house, which we can characterize very well today as not having a problem, ever going to develop a problem? The bad news is that there is no data on that question. The good news is that when you think about the basic physics of what is happening under and around the house, the odds against that look like they are pretty good. I think that most of those from Lawrence Berkeley Labs, Princeton, Teledyne Isotopes and from all other organizations that have been doing radon measurements for a long time would be very surprised to find a mechanism which changed good values in lots of houses into bad values over a 5-year or 10-year span. That is different from the question of whether one year that was dry happened to be different from the next year that was moister.

T. Brennan: Actually what I was going to say was that houses do change. They start out as good as they are going to get, and they go downhill from there. That's how everything works in our universe. You may have a frost heave; you may have a crack in the basement wall. I have been in plenty of houses where the wall was pushed in three inches and there is a one-inch crack. It wasn't built that way. It happened sometime after it was built. Houses do change in their characteristics over the years. There is some chance that that could happen. I think that I agree with Harvey that this is not going to be the normal situation.

H. Sachs: I think Terry has made a very important point. If you've got a house that has been built poorly or even half-way well, and it starts developing a lot of cracks, and the only thing that has been keeping radon out of your basement was the integrity of that system, yes, it could change.

Audience: Is it possible that some mitigation technique, for example a neighbor who ventilated the soil, might change the pressure characteristics of the radon? And, if somebody builds something next to it, say another structure, cannot that affect the radon distribution in the soil and might you get changed radon readings?

T. Brennan: Yes, and yes. It could happen, and how often that would be the case is unknown.

PRIVACY OF RADON RESULTS

Q: How do you know how radon affects anybody else in the neighborhood, aside from your own house? Does the public have access to any of the reported test results of other individual properties?

D. Schutz: The public does not have access to any data on individual properties from any source. A great deal of care is taken to make sure that it remains that way.

T. Brennan: The only way they can get that information is from the homeowner themselves. If you live in a small town, that means everyone knows probably.

D. Toomey: Unless the government does the testing, in which case you may be able to get it under the Freedom Of Information Act. One of the cases I referred to forced the government to reveal its data. It involved Grand Junction, Colorado.

D. Schutz: That is one of the reasons that some of our companies have been reluctant to provide the data to government agencies, even with assurances of confidentiality. It is almost a doctor/patient relationship.

WARRANTIES OF LIVABILITY

Question: Is the Warranty of Livability a legal concept that exists now, or is it one that is developing?

D. Toomey: It is called a Warranty of Habitability in Pennsylvania. It does exist in Pennsylvania. It applies, for example, if you buy a house. Suppose that during the first rain storm the roof leaks like a sieve, and it turns out that the roof is improperly built and always will leak. The answer is that you have a claim against the seller for at least the cost of replacing the roof. I have to say to you candidly that I don't know whether such a concept exists in New Jersey, but it grew out of judge-made law, not statutes. It was a response to people who had claims that they bought a house on total good faith only to find that it had been defectively built in a way that a reasonable inspection would not have revealed. The answer is yes, it is an extant legal concept.

SOIL TESTING

Question: Please repeat your comments on soil testing reliability.

H. Greenberg: I think my specific comment was that it is not sufficiently predictive at this time. Right now, it is probably the most demanded test that we have. Builders, homeowners or lot owners are about to make major investments in improving a lot, and they would much rather not even start from square one and bring a bulldozer in if they are going to have a problem. There are three types of tests that we are aware of. The first involves using an alpha track film, burying it down fifteen inches in a cup. That particular device looks at approximately 6 square inches of soil area when it is down below. That is a very small sample dose. It was brought out yesterday, and it was a very good point, that if you decide to take a number of those tests around the lot or even just around the perimeter of where your footings are going to be mounted, it is going to cost you somewhere between $500 and $1,000. That money is probably much better spent on a mitigation system below any poured concrete that you may decide to put down. Another way of testing is actually taking a soil sample and sending it to a lab. Again, this is a very quantitative test. You will get the result; you will know how much radium and possibly radon is in that soil. But again, it is not representative for the type of question that was put forth. The question wasn't, "Is that soil high?" but "Is the lot high?" You cannot tell on a large lot from a single test. The third method is literally taking a gamma scan, looking for high gamma radiation on the lot as a whole. But you do not have a direct mechanism for comparing gamma to radon. You can have an idea, a qualitative judgment that the entire lot has a very high background as compared to lots in some other area that is clean. You may want to make the judgment that that is not where you are going to build your first house, if you are a builder.

H. Sachs: There is another very important point to be made. We tried also to do a fourth technique which is actually to measure the flux, the amount of radon coming out of the soil so you get it in picoCuries per square meter per hour. It's also a limited technique. The point I want to make is that even if you know the concentration of radon in the soil gas, with any of the techniques you have referred to, you don't care. What you care about is how available that radon is; how readily a given pressure gradient in the house you built can bring it into the basement or in through the slab. I do not believe that there will be, within the next five years, a cost-effective (meaning less expensive than the cost of making a radon-proof or radon-resistant basement), widely deployed, accepted flux measuring technique which incorporates your reservations about wanting multiple tests, so you really characterize the site, and has some way of understanding the effect of the bulldozer when it cuts into the soil. I just despair of ever having useful information of that kind.

D. Schutz: The big problem is that the bottom of the basement is not usually the soil level with which you are confronted before construction starts. It is really a measurement that has to be made when you have established where the bottom of the basement is going to be. There is a fifth method which is technically somewhat complex, but essentially is based on using radon daughter products, the long-lived products which can accumulate in the soil if an extraordinary flux of radon has existed there for some period of time -- 40 or 50 years. I won't go into that right now, mainly because there is not sufficient experience to establish whether the measurement of Pb210 (lead), which I am referring to, is actually predictive of the radon which will be in the building which is ultimately placed on the property that you make the measurement on.

Section 8

The Role of Government in Responding to Radon Gas Exposure

Introduction

SUMMARY AND INTRODUCTION TO SECTION 8: THE ROLE OF GOVERNMENT IN RESPONDING TO RADON GAS EXPOSURE

Michael R. Edelstein

In light of the general issue of who is responsible for addressing the radon problem, the role of government becomes extremely salient. The final session of the conference was devoted to exploring this question. In planning the session, it was recognized that the term government stretches across many boundaries. There are local, state, federal and international dimensions in discussing the governmental role. One can further distinguish between regulatory and political dimensions of the radon issue. Furthermore, in discussing government, one raises the question of what the politics of radon are.

An examination of governmental response invites some basic questions to be asked. For example, in addressing other hazards, there is often a "passing of the buck" within and across different levels of government. Is this evident with regard to radon? Furthermore, as a "natural hazard" which has previously been addressed in human-caused form, radon tests the boundaries between the private domain and public responsibility. Where do we draw the line? In addition, when people expect government to assist them during a crisis, their trust is quickly lost if officials respond in ways that appear to be unhelpful, untruthful, incomprehensible or not in keeping with the facts. Has government acted so as to invite continuing trust for its radon programs? Particularly given the importance of public perceptions of the problem, as discussed in Section 6, government's radon profile has major significance in the recognition of the hazard and the willingness of people to take protective actions. What image has the government projected? One final aspect is particularly intriguing. Given the various clues of the potential radon problem that emerged in the past, why is it that the threat was not earlier framed as a major public health problem? Was there a coverup, as some suggest, or merely an understandable lag in the recogniton of the problem's scope.

This section is divided into five parts which introduce the topic and then detail the federal response and those of three states, Pennsylvania, New York and New Jersey.

(1) In the first part of this section, Richard Guimond of the Environmental Protection Agency (EPA) presents an analysis of the evolution of the radon issue. Using a comprehensive model of environmental policy development,

Guimond suggests that radon has moved quickly through a stage of agreement that there is a problem and into a period of exploration of policy alternatives. Possible solutions have yet to be evaluated and implemented.

Guimond provides a history of radon health concern associated with mining. As health evidence accumulated against radon, radon policy evolved to a point where maximum occupational standards were set. While residential radon exposures often exceed those found in mines, there was a much slower recognition of the problem in buildings. The first hints of building exposures came from studies of homes built atop industrial residues in the United States, such as uranium mill tailings in the west, phosphate slags in the south, and radium contaminated soils in the northeast. In looking for control houses to which these contaminated sites might be compared, it was discovered in the 1970's and early 1980's that even "normal" soils generate radon, although residential levels over these soils were still believed to be lower than levels in buildings atop comparatively more radioactive industrial residues. This misperception was not corrected until late in 1984 with the discovery of "hot houses" near Boyertown, Pennsylvania.

Finally, Guimond reviews radon as a global issue. Some other countries have moved much earlier than did the United States to assess and address their radon problems, most notably in Scandanavia. Interestingly, standards set by national and international bodies demonstrate a great deal of variability in risk assessments and thresholds for action.

Variability within the United States was also found in a survey of state officials responsible for radon described by Edelstein. The findings suggest that, in the absence of a major documented in-state radon problem, state response to the radon issue has been minimal. Perhaps the same perceptual barriers that affect the homeowners' appreciation of radon operate, as well, for politicians and officials. The assumption that a narrow federal radon role with major responsibility shifted to the states will be adequate to proactively address the issue is called into question by these findings.

(2) The second part of the section focuses particularly on the federal radon response. In his conference keynote address, Senator Frank Lautenberg of New Jersey praises the New Jersey radon program as a possible national model. With some 45 states aware of radon problems within their borders, Lautenberg sees an active role for the federal government in addressing radon. This effort is constrained by the federal budget crisis, however, so a corollary state involvement is needed. While the federal government can assist in developing means of quality assurance, mitigation techniques, and, in the case of hot spots, actual direct assistance, states will need to involve their local governments and the business community in fulfilling their end of the responsibility.

In fact, the federal program is predicated on a limited role built upon strong partnerships and a basic responsibility for costs to be borne by the homeowner. Christopher Daggett, EPA Region 2 Administrator, suggests in "The Radon Problem: A Federal Perspective" that the federal radon program should not be regulatory. This is because there are no culprits causing the threat, the danger is decentralized, and the problem invites shared approaches between the central, state and local governments. Daggett details the EPA national action program for radon as consisting of assessment, mitigation, prevention and institution building for public communication.

Further detailing of the EPA program can be found in the paper by Daniel Egan of the agency. Egan argues that radon cannot be regulated because it occurs in the private home and it is also subject to a large element of uncertainty. The EPA programs emphasize cooperation with state and local officials and with the private sector. It is with these groups, Egan argues, that responsibility for addressing the issue will ultimatelty lie.

While the EPA is responding to the need for a public radon program, another federal responder is the Department of Energy (DOE). Wayne Lowder of the Office of Health and Environmental Research, DOE, discusses his agency's efforts to describe the varied environmental conditions which affect radon exposure and risk. His agency seeks to identify scientific models for understanding radon migration and entry to buildings, indoor atmospheres, radon epidemiology and, most urgently, cancer induction due to radon exposure.

While Senator Lautenberg decries the budget policies of the current administration and their effects on the radon issue, Egan and Lowder argue that the federal program is adequately funded to meet existing needs.

(3) A third part of this section reviews the radon issue in the Commonwealth of Pennsylvania. State Senator Michael O'Pake provides a legislative perspective in response to the discovery of high levels in his district. He criticizes the lack of a promulgated federal standard or guideline for radon, despite promises from officials that such rules were forthcoming long ago. O'Pake also reviews Pennsylvania legislation, passed and pending, to address radon.

A representative of the Pennsylvania Department of Environmental Resources, Jason Gaertner, provides a detailed review of the DER radon program. He reviews criticisms of the program by the citizens' group Pennsylvanians Against Radon (PAR) and by realtors. The DER field offices are deeply involved in providing a variety of services to targeted areas. These include testing, school diagnosis, information efforts, a low interest loan program for mitigation and lung cancer mortality research.

Acting as discussants for the Pennsylvania program, Kay Jones and Kathy Varady of PAR discuss the "I'm not affected" perceptions of people who stigmatize radon victims. Given the stress of exposure, victims require support from the general population. They further note inconsistencies in the federal treatment of "natural" and "made" radon problems. In open discussion, radon expert Harvey Sachs presented a well-documented attack on the Pennsylvania program, citing a long delay in the response as a likely source of excess cancers that might have been avoided.

(4) The New York State program is described in three papers. William Condon of the Department of Health reviews New York's evolving radon program. Joseph Rizzuto of the New York State Energy Research and Development Authority discusses a number of long-term studies of radon distribution and mitigation techniques being conducted by his agency. A paper contributed by John Reese of the State Energy Office (SEO) subsequent to the conference provides an update on state legislation and a review of the radon mitigation workshop offered for EPA by the SEO. This training effort is a major attempt to address the lack of diagnosticians and mitigators.

Another added paper by Michael Edelstein and Liana Hoodes expands on conference discussion to describe the efforts of a non-profit, tax-exempt organization, Orange Environment, Inc., to develop a radon program in lower New York State. The program was initially a response to inaction and inaccuracies on the part of the New York Department of Health, an attempt to collect accurate incidence data for Orange county, and an effort to help concerned citizens to test and otherwise address the issue. The importance of a grass roots organization which can implement programs not readily forthcoming by government and which can provide a method of collecting data necessary for advocating a firm government response is suggested.

(5) While Pennsylvania's program is well underway and New York's is just emerging, particular attention is given to the well-funded and highly developed New Jersey program. State Senator John Dorsey, a key architect of the program, describes it in detail, noting that the State's radon effort is better funded than the entire federal response. Of particular note in the Senator's presentation is his description of close working relationships between himself and key state officials administering the radon program. This heavy legislative involvement in New Jersey provides a contrasting situation to the more slowly evolving New York situation. It is understandable that Pennsylvania's effort would be led by legislators representing the "hot" areas, such as Senator O'Pake. New Jersey brought its program on line before hot areas were found as a well-planned proactive effort. In New York State, comparatively distant from the center of radon concern, more difficulty has been found in creating a cross-partisan support for action.

Senator Dorsey's pride in his state's program was echoed by other presenters. Donald Deieso, an Assistant Commissioner of the Department of Environmental Protection (DEP), profiles the New Jersey program in his paper, based upon the opening address to the conference. He particularly highlights some of the unique policy characteristics of radon and the implications for the program design. In their paper, Gerald Nicholls and Mary Cahill of the DEP radon program, provide some history to New Jersey's efforts and detail program goals, activities and findings. But the real indications of the success of these efforts is found in the discussion provided by Clinton Mayor Nulman and local health officer Dan Jordan. Nulman talks about the pressures of finding that his town had a cluster of extremely high radon homes. With the state already geared up to address radon, Clinton residents were the recipients of fast and effective state aid in the areas of testing and information and EPA assistance for mitigation. It is clear that the effort won the trust and gratitude of residents who avoided panic despite the level of hazard found. Jordan describes his local efforts to provide for radon testing and the state effort to use officials at his level to communicate about the radon issue. Summing up these efforts, Judith Klotz, of the State Department of Health, describes her agency's collaboration with federal, state and local efforts to address radon exposures. The DOH is conducting health research while using its local health officer network to provide information. The result of the overall state effort, in Klotz's view, is a model of success that may well be copied by other states.

Acting as a representative of the New Jersey branch of the American Lung Association, Linda Stansfield raised some perceptive issues relating to the scope of the radon effort. In particular, she calls attention to the danger in isolating radon from a group of indoor air pollutants which might best be addressed collectively. She further stresses the radon/smoking connection as collectively representing a source of unparalleled risk. Children in schools or concerned homeowners might well be told that one concrete step in reducing radon threat without having to mitigate would involve cessation and avoidance of cigarette smoke.

Discussion and audience questioning for each part of the section further provides for an examination of issues related to the role of government.

In summary, the Federal government has developed a comprehensive radon program the backbone of which involves state, private sector and homeowner involvement. The ultimate success of the program rests less upon the federal effort than upon perception of the issue by these other groups. What is particularly important about the New Jersey model for state response, in this context, is the provision of the teamwork necessary to make the EPA program design work. The team involves officials and legislators who worked to fund the program as well as officials at various levels who must cooperate to implement it. As the Clinton case study illustrates, disaster can be averted when such a program is in place and functioning. However, the brief survey of states suggests that, at least to date, the New

Jersey response is not representative of government's handling of the radon problem. It remains to be seen whether the ingredients for such a program will develop elsewhere in the timely manner demanded by the level of risk associated with radon gas exposure.

THE HISTORY OF THE RADON ISSUE: A NATIONAL AND INTERNATIONAL PERSPECTIVE

Richard Guimond*

There is a lot of new information being developed on radon. However, to fully understand the significance of the radon issue. I think it helps to have some historical perspective on the topic. Consequently, this paper is designed to review some of the key historical and international developments concerning radon. I would also like to discuss some of the philosophical concepts that I think we have all got to come to grips with ultimately and see if we can't get some unification around some of these issues.

RADON AND THE ENVIRONMENTAL POLICY CYCLE

The first figure (Figure 8.1) was presented at a recent meeting on radon in the Netherlands by an official from the Dutch government.[1] The point he made was that there are cyclical events in the development of public policies on environmental issues. He pointed out that during the early recognition of the problem, we first must develop some kind of an early understanding of the problem. At this time, there is usually a significant amount of disagreement on whether there is a real problem and what should be done about it. The disagreement is illustrated by the wide band on Figure 8.1. As time moves on and more studies are completed, you get to a point where people begin to generally acknowledge that there is a real health problem. It is not speculation; it is not hearsay; there is a problem. As the issue develops further, you reach a point where people start to ask, "What are we going to do about it?" "What kind of policies might we create in resolving the issue?" As formulation of policy begins, different policy alternatives are examined, debated, and talked about between the private sector and government. Next is the solution phase, where people are coalescing around a particular solution or set of solutions. From there we proceed to the implementation phase of dealing with the problem. That is the sort of cycle followed by public policy that the Dutch official foresaw. He used examples of

*Director of EPA's Radon Action Program, Office of Radiation Programs, U. S. Environmental Protection Agency, Wash., D.C.

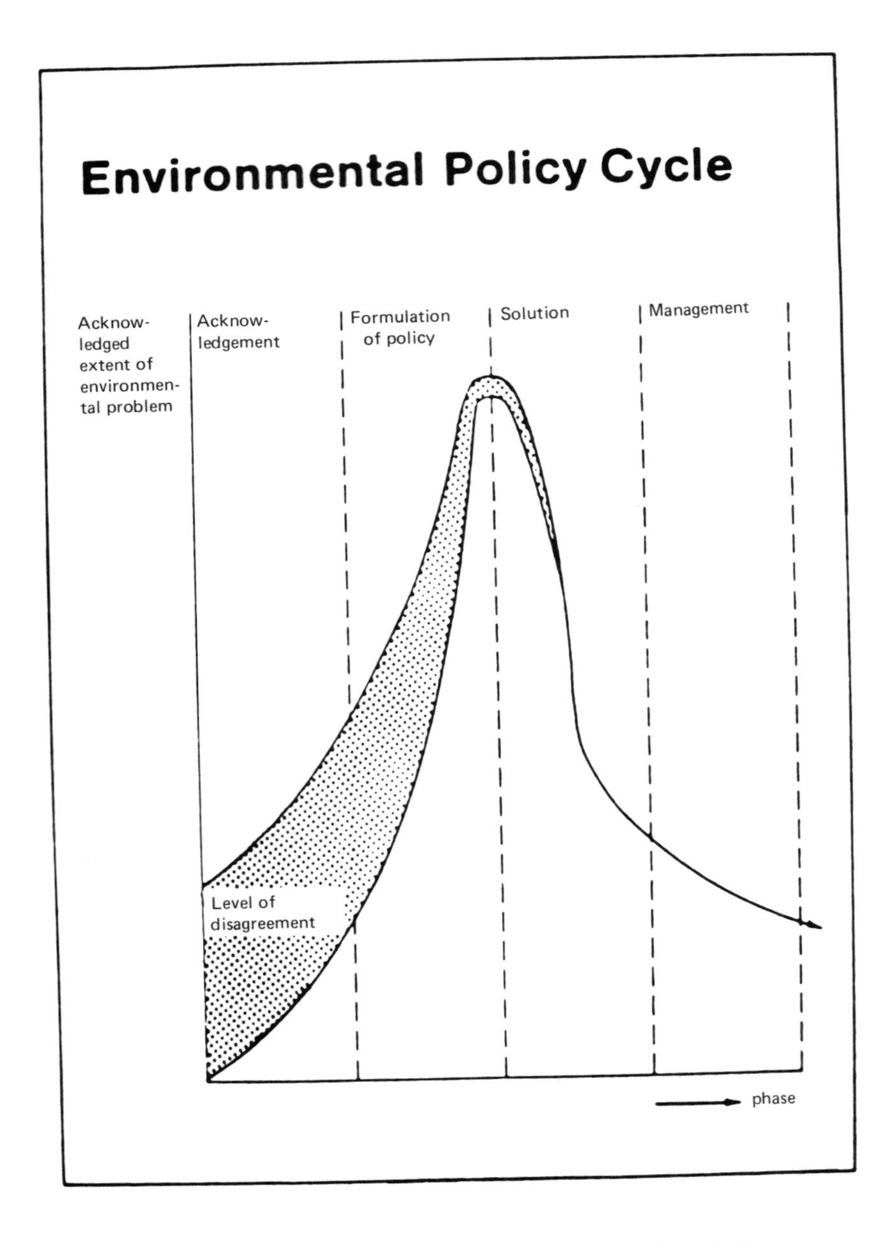

FIGURE 8.1 -- ENVIRONMENTAL POLICY CYCLE

some potential radiation problems in Europe that might fit in various parts of the cycle.

I am not sure that this assessment is directly applicable to all situations in this country, but frankly, I think that we have observed this cycle with a lot of the problems that we have had. With respect to radon, the model has much to suggest. We are pretty much past the "just understanding the extent of it" phase, although there are some aspects we still need to study. People are acknowledging that radon is a problem. With respect to radon, we are just getting into the "formation of policy" phase. Certainly we are trying to look at policy alternatives and solutions that ultimately take us into a management phase. I think as I talk about the history of the radon issue in the U.S., you will see this environmental policy cycle come into play with the various elements of the radon problem, whether it be with respect to uranium miners, or contamination problems, and now normal housing situations. So keep the cycle in mind. We all want to get indoor radon very quickly into routine management so that it is a common problem with an easy solution. Unfortunately, going through the environmental policy cycle takes many years, and it may take decades. Quite frankly, with respect to the indoor radon problem, I think we have been moving quite rapidly through this cycle.

BACKGROUND ON RADON AS A HEALTH ISSUE

How did we begin to understand that radon presents a health hazard? The first evidence of this hazard is data on pulmonary disease in uranium miners in Europe. I think it was first documented around the 1600's. Clearly, we've had hints on radon problems for centuries. This is not a new phenomenon. Later, during the seventeenth and eighteenth centuries, they observed that much of the pulmonary disease was lung cancer, and they were starting to attribute that to the high radiation levels in the mines. They weren't quite sure what aspects of the radiation caused the problems, but they certainly attributed it to something associated with the radiation levels. As we moved from the European experience to the experience in this country, we find that the mining of uranium ores began around the turn of the century. However, it was very limited at that point in time. Uranium was originally mined largely for use as a coloring agent, such as the orange glazing used in "Fiestaware" earlier in the century. In the late 1940's, with weapons production, the amount of uranium mined increased dramatically, and then we started to get more documentation of radon and exposures to radon and what the mining environment looked like. During the early 1950's, people first started talking about establishing permissable levels for exposure to radon. Various permissible levels were suggested, but frankly, nobody took them very seriously. There was such a push at that time to get the uranium out of the mines. Many of these mines were "mom and pop" operations, very small mines operated by two to ten people. These were very low budget operations, using unsophisticated techniques, and so there was not much done to control radon exposures. Control efforts were really quite minimal until sometime during the sixties when the public health service first documented increased lung cancer risk in uranium miners. Throughout the mid-60s they documented an epidemic of lung cancer in miners. At that point, people first started to take radon exposure seriously.

If we take a look at what has happened with respect to guidelines for exposure of radon for uranium miners, there has been a fair amount of evolution. The first really true guidelines were those developed by the previous organization called the Federal Radiation Council, back in 1967. They recommended a limit of 12 WLM per year. If I make a rough conversion to what that might be in a home environment, it would be roughly about 60 pCi/l. That level was quickly perceived as too great based on the evidence they were getting from the epidemiological studies. So very soon the public health professionals wanted to be able to lower the exposure level as quickly as practicality would allow them. Now at this time there was a sizable amount of resistance from the uranium mining industry, which argued that it was just not practicable to achieve levels substantially lower in mines because of excessive ventilation costs. However, during this time we at EPA did some extensive work and were able to show, in effect, that lower levels could be accomplished. Consequently, in 1971 EPA recommended a modification of the 12 WLM standard. The recommendation was adopted by the president and the value was reduced to 4 WLM per year. This translates to about 20 pCi/l in a home. That is where the standard for underground uranium miners currently stands, although it is now under review by the Mine Safety and Health Administration with the assistance of the EPA. Miners' unions have petitioned for a reduced standard, but it is too early in the active process of rule making to tell if this can be achieved. The point is that, as we have gotten more evidence from following miners over their lives, we are able to provide statistically significant documentation that there are lung cancer risks at lower and lower exposures to radon.

Back during the 1960's, the epidemiological evidence was only good enough to say that there was a statistically significant risk if you had a cumulative exposure of greater than about 1,000 WLM. Today, I think that you would get most scientists to agree that the statistics for smoking miners are good enough to show that you've got an increased risk in the 50 to 100 WLM range. It is more difficult to develop enough statistics for non-smoking miners because there are fewer of them, and the absolute risks are somewhat less. The evidence indicates that there are statistically significant risks at a few hundred cumulative WLM. So, needless to say, in a home environment it doesn't take all that much time to accumulate 50 to 100 cumulative WLM. In fact, in many homes in Pennsylvania and New Jersey, you can accumulate this much radon exposure in only a couple of years. This is why we have so much concern over exposure to people in homes throughout the nation. Let us turn from the mining environment to how we begin to focus upon radon in homes.

THE HUMAN-CAUSED RADON PROBLEM

The concern with residential radon began in response to the mining situation. People had begun to take tailings away from mines and mills, particularly in a community called Grand Junction, Colorado. They took the tailings and used them for fill around their homes in concrete aggregate, in their sand boxes, and for a variety of building purposes. The tailings were a nice, sandy construction material that was inexpensive to obtain. I first joined the U.S. Public Health Service in 1970, and one of my first assignments was to go to Grand Junction, Colorado and go from house to house to measure radon and gamma levels around these homes and speak with the people in those areas. That is where our experience with radon in homes first started. But of course that was a man-made problem. Towards the late 1970's, we saw that there were a lot of other communities in the west that had these same kind of problems. There were between 20 and 30 communities where tailings had been used. The radon problem was by now well-established.

We can relate this back to the environmental policy cycle. We were into the mid-to-late 1960's when we started to know that there was a problem; people started to focus on it and to find out what could be done about it. It took until 1978 before Congress passed a law to control the problem. The Uranium Mill Tailings Control Act required EPA to establish standards for the disposal of mill tailings and for the cleanup of contaminated homes. The Department of Energy was required to provide the funding for the cleanup of these homes. During the 10-year period between the late 1960's and 1978, there was extensive debate over who was responsible for addressing and paying for the problem. Because most of these mill tailings were used for weapons in the government's nuclear program, in 1978 Congress concluded that the government would accept responsibility. The Department of Energy was assigned the task and the funds to clean up contaminated homes. The EPA was assigned the task of providing public health standards for the DOE.

There were other radon incidents as well. In the late 1960s, the Department of Energy found high ambient levels of radon in Florida. In the 1970's, EPA also started to do radon studies in Florida. We got there sort of by accident. We were developing effluent guidelines for a variety of industries to comply with The Clean Water Act. One of those industries was the phosphate mining and milling industry.

They had high levels of radium in the discharge water which they were generally just discharging into the environment. We originally went down there only to look at those discharge waters. But once there, we recognized that there might be additional implications of the sizable quantities of wastes, byproducts, and various other things that were going directly into the environment. So we did a pretty comprehensive study; we looked at all the products and reclaimed land. Our original view was that, in all likelihood, homes built on phosphate lands were not going to be as much of a problem as were the homes built on uranium mine tailings. Quite frankly, we were surprised when we found a lot of homes built on reclaimed land with quite high radon levels. We realized then that high radon levels indoors were possible even when the soil had only slightly elevated radium concentrations.

Soon we found other radon contamination problems to investigate in places such as Idaho. As we started to see radon problems associated with various kinds of industries, one of our staff got the idea that maybe we ought to go back and check past records. Companies that have come and gone unrecognized may have left contaminated materials behind. When we went back and started to do this historical check on records, we found many old processing plants throughout the country. For example, we found a New Jersey site where there had been an old radium processing facility. This has turned out to be the Montclair site which is now a Superfund Site. Other Superfund sites were found in the Denver area, and the DOE has been cleaning up sites across the country where uranium and radium ores were stored or processed.

THE NATURALLY-OCCURRING RADON PROBLEM

Now, how did we start to learn about high radon levels in homes on normal land? As we did a lot of the Florida studies, we picked control homes to compare against the homes on reclaimed land. The data from these houses showed that many control houses had radon levels that were just as high as houses that were built on the reclaimed land. We went back and checked, and in a lot of those areas there were unmined phosphate minerals in the ground. Those areas had higher than normal uranium or radium fairly close to the surface. Similarly, in the Butte, Montana area, radon in houses originally attributed to phosphate residues turned out to be coming out of fractured rock beneath the buildings. The DOE's Lawrence Berkeley Laboratories, throughout the late 1970's and early 1980's, was starting to do a variety of scientific studies from which they concluded that normal houses in a variety of places may have high levels of radon. During this period, there were other Department of Energy projects and a number of European studies in Sweden and elsewhere that were starting to hammer home the same conclusions. Consequently, in the early 1980's, scientists started to believe that normal land may present increased risks of radon expsoure. Now, even at this point, I don't think we really understood the significance. For example, we believed that we were talking about substantially lower levels in the natural areas than in the contamination sites. Houses built with mill tailings exhibited levels up to one-half of a Working Level or rarely one Working Level (100 to 200 pCi/l). That was the worst that we had seen in all the houses throughout the west. I think that everybody felt that we were not going to see anything anywhere near that high on homes built on normal soils, because by the early 1980's, the highest levels observed on these homes were about a tenth of that, between 10 and 30 pCi/l. It really wasn't until the Boyertown experience in the Reading Prong that we started to see several thousand picoCuries per liter in "normal land houses." Then we realized that we had underestimated the problem tremendously.

RADON AS AN INTERNATIONAL ISSUE

Having briefly examined the historical record in the United States, it may be interesting to look at how other countries have addressed the radon issue. Many of the European countries have already conducted radon surveys as partially illustrated on Table 8.1. A recent talk with scientists at the CEC revealed that average levels in Europe are on the order of 1 pCi/l or less. Frankly, many of the countries don't have that much in the way of a radon problem. Sweden, Norway, and Finland may have the most extensive problems. Even in Sweden they have not observed the phenomenally high levels that we have observed in this country. So in general, the Europeans don't seem to have as extensive a public health problem as we have. Nonetheless, they are still giving considerable thought to how to address their affected areas.

With respect to international guidelines for radon, at present there are no truly rigorous standards. A few countries have some recommended levels which serve as the basis for implementation. In some cases, they have reached the top of the environmental policy cycle; they are debating alternatives and potential actions. With that in mind, Table 8.2 lists a few countries and some organizations that have programs and guidelines. There is a relevant group called The International Commission on Radiation Protection (ICRP) which is an international group of scientists from many countries that meets and develops radiation protection recommendations. They came out with a report a few years ago that tried to focus on naturally-occurring radiation problems. The ICRP has suggested 10 pCi/l as a guideline for remedial action in existing houses that are relatively easy to mitigate. In more difficult situations, larger levels might be appropriate. For new housing, they suggested a maximum level of around 5 pCi/l, although they would encourage design actions which try to do better where

EUROPEAN RADON SURVEYS:

- Countries include: United Kingdom, Ireland, Germany, France, Italy, Denmark, Belgium, Sweden

- Average indoor radon concentration in Europe is about 20 Bq/m^3 (0.6 pCi/l)

Table 8.1. Radon Surveys in Other Countries.

INTERNATIONAL COMMISSION ON RADIATION PROTECTION (ICRP)

Remedial Action in Existing Houses	
Relatively Easy	10 pCi/l
Severe and Disrupting	larger value may be okay
New Houses	
Upper bound	5 pCi/l

Canada	
Remedial action in existing houses	20 pCi/l
Design limit for new houses	none
Sweden	
Remedial action in existing houses	20 pCi/l
Remedial action in houses be improved	10 pCi/l
Limit for new houses	4 pCi/l
Design objective for new houses	~ 1 pCi/l
United Kingdom	
Proposals by the National Radiation Protection Board:	
Remedial action in existing houses	20 pCi/l
Design limit for new houses	4 pCi/l
Belgium	
Discussion Group Suggestions:	
Unsuitable for habitation	40 pCi/l
Remedial action in existing houses	8 pCi/l
Design limit in new buildings	4 pCi/l

Table 8.2. Review of Radon Guidelines: ICRP and in Other Countries.

possible. In Sweden, they use about 20 pCi/l as a guideline for remedial action in existing houses, about 10 pCi/l remedial action in houses being improved, and for new housing, a limit of about 4 pCi/l and a design objective of about 1 pCi/l.

In Belgium, they are further along with policy development. The Belgian government asked a panel of scientists to develop recommendations. The result was a suggestion that at 40 pCi/l the house is unsuitable for habitation, at about 8 pCi/l remedial action was recommended, and new buildings are to be designed under 4 pCi/l. In Canada, they are currently using 20 pCi/l for remedial action in existing homes, and they have done nothing about a design limit for new homes at this point in time.

CONCLUSION: A PHILOSOPHY FOR RADON GUIDELINES

As can be seen from this discussion of various countries, there is really no uniformity in response to radon. Clearly, from our standpoint, neither the EPA nor anybody else is able to create a "safe/unsafe demarcation point" with respect to radon. Given that we are dealing with people's individual homes, where they are responsible for making the decisions of suitablity for living and how much they want to spend to reduce risks, much must be left to individual choice. People have a lot of individual freedom in deciding what to do about risks in their lives. They choose whether to smoke or not and whether to drive a race car or not. Nonetheless, from extensive discussions with people throughout the country who participate in EPA focus groups, it is clear to us that people need to have guidelines and recommendations that are as clear as possible. These provide benchmarks about when there is a hazard, what kind of action they should take and how rapidly they should take that action. This poses a dilemma because when you do that, you can create some artificial benchmarks. We want to find ways of insuring that people don't put too much rigidity in those benchmarks. We have talked often about levels at EPA of 0.02 WL or 4 pCi/l. Obviously, 0.019 WL is not much safer than 0.021 WL. So to think that you are safe at 0.019 and you are unsafe at 0.021 is a fallacy. Therefore, in communicating with people, we have to be sure we are not leading them astray.

Finally, it should be noted that in some cases low levels may be very, very difficult to reach, particularly in some of the existing homes that have quite high radon levels (over a few hundred pCi/l). Where a house starts out at 10 pCi/l, it may be a piece of cake to bring it under 4 pCi/l or even under 2 pCi/l. But if a house starts out at 2,000 pCi/l, it may be quite a chore to get it to be under 4 pCi/l. These are some of the practical considerations that have to be addressed in the philosophical debates that will occur as we consider where we are all going to go from here.

FOOTNOTES

[1]Bosnjakovic, B.F.M., and B. Vos, "Policy Developments in the Netherlands Concerning Enhanced Exposure to Natural Radiation", The Science of the Total Environment 45, (1985) Proceedings from the Seminar on Exposure to Enhanced Natural Radiation and its Regulatory Implications. Elsevier Science Publishers, Amsterdam.

THE STATE RADON SURVEY*

Michael R. Edelstein**

As part of the work in an undergraduate Environmental Seminar, a mail survey was sent to representatives of all 50 states as listed in the "Directory of Personnel Responsible, Radiological Health Programs," January, 1986. The objective of the survey was to identify the extent to which radon was recognized as a threat and to ascertain what radon responses were planned or implemented.

Sixteen responses to the mail survey were returned reflecting 14 states, the District of Columbia and one EPA region. Despite the lack of data on the entire country, the findings are very instructive. They provide a spot-check on the operating assumption that states will step in to the radon issue in the absence of a major federal initiative. It should be kept in mind that the surveys were returned in spring, 1986 and were filled out by an official from the principal agency dealing with radon. In one case, where two surveys were returned from the same state (one from an EPA regional officer and the other from a state official), surprisingly different ratings are found between the two respondents.

The extent to which radon was rated as a current public concern varied widely, with Pennsylvania indicating the greatest concern, followed closely by Maine, Nebraska and then North Dakota and Idaho. In contrast, radon was seen as the least issue in Nevada, Texas, West Virginia and the District of Columbia. That different respondents from a similar area might differ in their assessments was indicated by EPA Region VII in Kansas' response of 6 on the seven point scale, indicating that radon is a major issue, as contrasted to a response of 2 by the representative from the state of Kansas.

Likewise, the extent to which radon was seen as a government priority also varied widely. Pennsylvania, Maine and North Dakota officials idicated the maximum priority, followed by Idaho and Nebraska. The lowest priority was

*A grant from the Ramapo College Foundation was used to underwrite the costs of this study. My appreciation of Michael Lesh's work on this survey is gratefully acknowledged.

**Associate Professor of Psychology at Ramapo College of New Jersey and conference co-coordinator.

indicated for Ohio, Iowa, Kansas (EPA rating), Nevada, the District of Columbia and West Virginia, followed closely by Texas, Delaware, and Washington State. In six cases, public concern, was seen as matching the government priority, in five cases government was seen as lagging behind public concern and in the remaining five cases, government was seen as prioritizing radon higher than the public.

Dramatic differences also occurred in estimates of the number of homes tested in the state. Pennsylvania, Maine and Washington State reported thousands of state tests (in Maine, these were paid for by individuals) and, in Pennsylvania's case, additionally large numbers of private tests. Most other states reported limited testing efforts ranging from several hundred (Idaho, Wisconsin, Ohio, Nebraska and Kansas), to less than one hundred (North Dakota, Kansas by state estimate, and Delaware), to virtually no known tests (Texas, the District of Columbia, Iowa, Nevada and West Virginia). Knowledge of private tests was usually sketchy, and it was generally estimated that there were fewer private than government tests.

Results of testing programs also varied. Indication of few tests over the 4 pCi/l air guideline were found in Texas, Kansas, Delaware, Iowa and Washington State, with results unknown in Nebraska, Nevada and Washington, D.C. In contrast, Pennsylvania indicated the most severe ratings, followed not too closely by Maine, Ohio, Idaho, and Kansas (by EPA estimate). Maine was the only state reporting large percentages of significant water contamination. Six states reported the location of radon hot spots. Test results did not appear to correlate with government priority for many of the states. Data on air counts was spotty, with Pennsylvania and Maine the only states indicating any degree of problem. Most indicated use of the 4 pCi/l guideline, although several indicated no guideline. Ohio indicated past use of an 8 pCi/l level now revised to fit with the EPA guideline. Maine reported a 25,000 pCi/l guideline for water.

Respondents were asked to indicate their primary concerns regarding radon. The dominant concerns were to address the public health issues (by 13 respondents) and to search for "hot spots" (by 7 respondents). Generally, the radon problem was seen as belonging to the homeowner rather than to government or the private sector. The focus of state response was to investigate the problem and then assist the homeowner in addressing it (12 respondents). A few respondents thought that the problem was entirely the homeowners', that radon should not be separated from the general concern with indoor air pollution, or that radon was simply not a problem in their states. The Nevada official indicated no government responsibility to address radon because radon gas is naturally occurring. Others felt that the radon threat was placed in perspective by other more severe threats (smoking and highway deaths were mentioned) and by the perceived ease with which the radon problem could be solved.

The objectives of the state radon policies varied. Of the states indicating a policy, Nebraska, Texas, Delaware and Washington focused upon education, advice, consultation and distribution of information; Ohio, Washington, D.C. and Wisconsin stressed identification of the scope of the problem; North Dakota identified follow-up screening and the Kansas EPA office was concerned with identifying cancer clusters. Only Pennsylvania and Maine indicated the development of a comprehensive set of policies.

Respondents were asked whether their state engaged in a full and open sharing of information or whether they were more concerned with keeping a lid on panic. Most respondents indicated an open approach, although three admitted a policy of panic control. One respondent commented that the best way to control panic was openness.

Regardless of concerns, targeted funding for radon was not allocated in most of the states for fiscal years 1985, 1986, or 1987. Allocations of a few thousand dollars were made in Wisconsin, Idaho and Washington State. Nebraska and Maine each made $20,000 available for radon use in 1986. Only Pennsylvania, of the responding states, had made a substantial funding effort, with expenditures of $1.5 million in 1985 and $3.5 million in 1986. In nearly all instances, responsibility for radon issues was placed with the state department of health (in Pennsylvania, the Department of Environmental Resources has the lead and in Washington, D.C., the Department of Consumer Affairs). Given the funding levels, it is not surprising to discover that radon programs are not heavily staffed. For 1986, while Pennsylvania fielded 25 radon staff, the next highest state was Washington with 2. All other reporting states had one staff member, several part-time staffers or one part-time radon employee.

Within the last year, 5 states have created radon hotlines to answer public inquiries (Pennsylvania -- 800-23RADON, Washington -- 206-586-3303, Nebraska --402-471-2168, Wisconsin -- 608-273-5181 and Washington, D.C. --202-717-7195. Only Pennsylvania indicated programs for widespread testing and low interest mitigation loans, with most other states fielding only limited testing efforts combined with the use of fact sheets and referral to private companies. Universities have become involved in testing projects in six states and community groups were reported to be active in three. One of these groups, People Against Radon, has become a national radon organization. State radon legislation has passed Pennsylvania, but failed in Washington. No other states reported legislative initiatives.

The development of the radon private sector seems to correlate to the above considerations. Other than Pennsylvania, where 20 firms were doing business in 1986 (up from two in 1984), only Delaware and Kansas reported private activity (one firm each). Of the five states that had given thought to consumer protection, all were relying upon the development of EPA standards.

Ratings of the federal response to radon were at the midpoint of the seven point scale. Responses indicated that the federal government has done well in the areas of information, training, coordination, discussion with state officials and the loaning of equipment. It was indicated that the EPA response was very good given the level of federal resources allocated to radon. On the other hand, the federal government was criticized for the slow pace with which documents and guidelines have been issued, for "unrealistic" risk estimates and for having delaying its activities until 1985. One official found the EPA response to be totally inadequate, commenting, "What have they done?"

The state respondents tended to see the radon problem as at least moderately difficult to solve, with one optimist being the regional EPA official. The Maine respondent also saw it as easy to solve, probably reflecting that state's focus on radon in water. The most difficult aspect of the issue reported was the difficulty of panic control in discussing the risks, followed by such problems as inadequate resources and funding, the lack of firm standards, the absence of proven remediation techniques, the health risk, and the lack of adequate screening of the problem.

Overall, the survey indicated a great disparity in state response to radon. The problem is differentially understood. State programs and allocations differ. Federal assistance in screening, setting standards, developing remediation techniques, monitoring the private sector and bringing down the cost of testing was seen as needed by respondents.

What conclusions can be drawn from this survey? Clearly, perceptions of radon vary greatly across the states, depending upon local experience. The federal government has not succeeded in creating a national consensus on radon which can guide the responses of individual states. The notion that in the absence of a strong federal response, states will adequately generate their own programs is not supported. A substantial documented threat within the state appears to be a necessary precondition in motivating state action. It is interesting that in the wake of the discovery of radon at the Watras house in eastern Pennsylvania, there has been a strong response from that state that has spread to the neighboring states of New Jersey and New York, as reviewed elsewhere in the section on government response to radon.

The Federal Radon Response

CONFERENCE KEYNOTE ADDRESS—RADON: THE FEDERAL RESPONSE

Senator Frank R. Lautenberg*

Good afternoon. It's a pleasure to be with you today. This is the first radon conference held primarily for the public. I commend the faculty at Ramapo College for organizing it.

I know this conference has been underway for a few days. Many of you have been exposed to a lot of facts and practical suggestions. So today my purpose is not to give a "how to" speech. My purpose is to raise some of the very tough issues we face due to the presence of radon in New Jersey.

It was just one year ago that I first raised the issue of radon with the head of the U.S. Environmental Protection Agency. He did not really know just what it was, or why I was asking him about it. But believe me, he knows what it is now! And I've almost got him believing the "R" in Frank R. Lautenberg stands for "radon."

THE RADON MITIGATION EFFORT

In the past year, a federal radon research and mitigation program that I sponsored has been established within EPA. And I want to announce today that EPA is within days of bringing this program to New Jersey.

EPA has selected ten homes in Clinton for full federal attention in its radon mitigation demonstration project. The radon problems in these homes will be assessed and then corrected, bringing the radon to safe levels.

In addition, EPA will provide full technical consultation to another twenty homes, and advise the homeowner on remedial action.

This program has great significance. Not only will these homes be made safe, but effective, affordable mitigation techniques will be developed and improved. These techniques can be applied to homes throughout New Jersey and the nation.

EPA is coming to New Jersey because late in March we identified the first radon "hot spot" in our state. A couple in Clinton had their home tested. When the tests

---

*Senator Frank Lautenberg presented his keynote address during the session on the Role of Government. He is Junior Senator from New Jersey.

were done, they discovered radon levels in their home which are the highest recorded in New Jersey.

The results of their test led to further testing in the community. Now, the picture is complete. In Clinton there is a cluster of homes that--as a group--have the highest concentrations of radon detected anywhere in the county.

I single out Clinton to make a point. What has happened in Clinton is important. Clinton residents have not panicked. They have not despaired. They have pulled together and drawn on state and federal resources to meet their problem head on.

Mayor Nulman of Clinton is here with us today. He is providing strong and responsible leadership for his community. And Clinton can serve as a model for our state and nation.

Mayor Nulman moved quickly. He made reliable information available to Clinton's citizens. He made sure that testing equipment was at the ready. He reached out to state and federal agencies for help. Above all, he did not panic, but moved ahead constructively.

The result? A significant devotion of resources channeled to Clinton. And discernible progress in a short period of time.

Because we are a progressive state, we often identify our problems earlier than others. And that kind of knowledge serves us well.

New Jersey was one of the first "frostbelt states" to understand the changes underway in the American economy. And, with foresight, and the proper investments in education, and transportation, and our infrastructure, New Jersey now has one of the fastest growing state economies in the nation.

So it can be with radon. We are ahead in the race to identify our problems. And we will be the first with solutions.

KEY QUESTIONS - THE SWEDISH EXAMPLE

The questions that face us are difficult and complex:

How extensive is the radon problem? And what causes it?

What is the health threat it poses?

How do we reach the hundreds of thousands of homeowners who may be affected?

Do we need limits on radon incorporated into building codes?

How do we assure that radon detection does not hurt local economies?

Now, it may surprise you - it surprises me - but, I found some answers to these questions in Sweden. The Swedes have known about their radon problem for twenty years. And Sweden has made significant strides. The Swedes have an important story to tell, a story that New Jersey can take advantage of and benefit from.

In January, I travelled to Sweden. I can personnaly attest that this was not one of those junkets you read about! It was only light in Sweden for about six hours a day, and the temperature never rose above 15 degrees!

I went to Sweden with other federal and state officials, environmental health professionals, builders, realtors and homeowners. I formed this task force to come with me to hear Sweden's story.

Let me tell you what we learned in Sweden:

-- Radon is associated with a range of identifiable geologic structures, and its impact varies in predictable ways with home construction.

-- Radon can kill. As much as 40 percent of Sweden's lung cancers may be traced to radon.

-- An aggressive public education program initiated on the federal level, but implemented locally, can make a vast difference.

-- Buildings codes to reduce radon in existing and new homes are feasible and can be effective in lessening the dangers of radon.

-- Educating the business community can stave off economic harm to communities by reducing the fear of locating in these areas.

In short, radon is a treatable hazard. But, it requires government attention.

WHAT DO WE HAVE A RIGHT TO EXPECT FROM OUR FEDERAL GOVERNMENT?

The federal government has resources which make it best suited for assessing the extent of the radon problem. EPA is now designing a national radon survey, which will be started in several months.

The federal government employs skilled environmental, health, and housing experts. It has a leadership role to play in developing low-cost radon mitigation methods, developing model building codes, and establishing adequate health guidelines which the states can use and which will permit homeowners to assess risks.

The federal government can assist in protecting consumers from fraud. EPA has already established a voluntary program to calibrate testing equipment to ensure that detectors work properly and that readings are accurate. States should take advantage of this resource as they set up consumer protection programs.

Providing financial assistance to needy homeowners is a difficult issue. Unfortunately, the federal budget crisis dictates a limited federal role.

I have pursued two options on the federal role for assisting homeowners. One is an agreement I was able to secure from the Federal Emergency Management Agency. FEMA is the federal agency that responds to natural disasters like floods or fires. At my urging, FEMA agreed to classify radon as a natural disaster. This agreement opens the door for Governor Kean to seek federal relief for New Jersey through a presidential declaration of disaster.

No one wants to think of his or here home as a disaster. But, if contamination is pervasive enough in a community, a disaster declaration would allow federal funds to be used to mitigate radon in homes.

Another potential means of assisting homeowners is for a taxpayer to claim a deduction for expenses incurred for the protection of health. I am seeking to clarify the definition of the medical deduction to include radon mitigation. No family should be forced to choose between its health and financial stability.

THE STATE ROLE

States and localities have a vital role to play in an effective radon program.

States with budgets running in the black, like New Jersey, should consider financial assistance to low and moderate income homeowners. Pennsylvania is providing radon detectors and low interest loans to residents in high risk areas.

States need to reach out to the business community. Builders, realtors, bankers, and firms looking to locate within high risk areas must be targeted. They need to know that investing in a radon-prone area is not risky business.

New Jersey's builders and realtors are interested in helping to fight the radon problem in our state. These professions have enormous expertise to offer. With the cooperation of these industries, we can aggressively develop radon-proof homes and buildings.

These industries have been diligent and constructive in their approach to the threat of radon. They decided very early to get involved in this environmental problem, rather than run away from it, minimize it, or just complain about it.

Representatives from the New Jersey Builders Association and the New Jersey Realtors Association acted immediately on the request to join our radon task force in Sweden last winter, and they are playing a pivotal role in determining the mitigation techniques appropriate for New Jersey.

States need to identify radon prone areas within their boundaries. Such assessments can complement EPA's national survey and pinpoint hot spots. Community planners can work around hot spots when siting developments.

Schools should be tested for radon problems. Parents and teachers need to know if school buildings pose health threats. States and municipalities have an obligation to see that the buildings under their jurisdiction are safe.

High radon levels in water supplies must be identified. They can lead to elevated indoor levels from showers, sinks, and dishwashers.

CONCLUSION

New Jersey is not alone in this challenge. Radon has been found in 45 states. When EPA's national assessment is done, radon contamination will command national attention.

But we can't wait for that. Every year wasted could mean more lung cancer cases nationwide.

In New Jersey, we are not waiting. We have a head start on most of the country. Together, we can lick this problem. I pledge myself to work with you.

THE RADON PROBLEM: A FEDERAL PERSPECTIVE

Christopher J. Daggett*

Good evening and thank you for the opportunity to take part in this important conference -- to lend my voice to the debate about the radon problem, and to explain EPA's program to deal with it.

Will Rogers once told a fellow newspaperman at the 1930 world disarmament conference, "We won't really hear what was done at this conference 'til we read one of the delegates' memoirs after the next war." Well, I think this conference is going to be more effective than that one, and it has already achieved a great deal simply by bringing together so many experts and decision-makers working on the radon problem.

I particularly value issue-oriented forums like this one because, through the informational and rhetorical give-and-take they provide, our body of knowledge grows and our policies inevitably become more solidly grounded. This is especially vital with the radon issue because there are still important questions to be answered about the scope of the danger throughout the country.

We know it's a big problem, and we've seen ample direct evidence that exposure to the products of radon gas can cause lung cancer. We also know that a large percentage of dwellings over the Reading Prong and other radium-bearing geological formations that stretch from Pennsylvania through New Jersey and into New York have high levels of radon. And we've found high levels of radon in other parts of the country.

But we do not know how pervasive the problem is nationwide because, as yet, no statistically significant screening of the entire country has been undertaken.

We have established health effects criteria on the basis of reliable studies of the incidence of lung cancer in underground miners. We have not had sufficient information, though, on radon exposure among the general population to make any definitive findings. But based on the results of the studies dealing with miners, various scientists and research organizations suggest that between 5,000 and 20,000

---

*Regional Administrator, U.S. Environmental Protection Agency, Region 2. The conference paper was delivered by William Muszynski, Director, Water Management Division, United States Environmental Protection Agency, Region 2.

cases of lung cancer a year in the United States -- up to a fifth of the total -- may be related to radon exposure in the home. It is worth noting that these estimated risk levels are significantly greater than the risks we deal with at nearly every Superfund site involving toxic chemical exposure.

THE EPA NATIONAL RADON ACTION PROGRAM

It was principally to address this paucity of information, and to take advantage of some of our research tools and technical expertise, that EPA crafted its national radon action program.

The program is aimed primarily at helping homeowners understand and deal with any radon problems they may have by giving them access, through the states, to the best scientific information and technical know-how. Toward that end, the Agency is focusing now on developing and disseminating the necessary information in the shortest possible time. We are also encouraging the development of state programs and private sector capabilities to meet radon assessment and mitigation needs. We are focusing our limited resources heavily on those states where we know that a strong concentration of radon sources exist. However, our national action program is meant to lay the groundwork for finding and dealing with similar problems elsewhere in the country.

The national strategy addresses four key issues:

1. The assessment of the overall problem;

2. Mitigation of the existing dangers and prevention of future ones;

3. Institution-building to make sure that local governments and industry groups have the guidance and information they need to develop programs; and

4. Getting information out to the general public through the states, so homeowners and businesses can make informed decisions.

It is important to understand that this is a multi-year effort that, of necessity, will be focused initially on the development of technical knowledge and of the needed institutional mechanisms. Once these areas have been dealt with, the program will shift as quickly as possible into state-by-state problem area identification and mitigation. The initiatives I will describe today, which are an extension of the four main components of the action program, should be understood in the context of those time considerations.

In the area of problem assessment, EPA has several initiatives either already in motion or soon to be.

1. We've published seven protocols for appropriate methods of measuring indoor radon.

2. We've got a quality assurance program to evaluate the proficiency of private firms and other organizations, including government labs, that measure indoor radon levels. Participation in this program is entirely voluntary, but states may use the evaluations as a basis for certification or some other action geared toward assuring competency. This is important because, as one New Jersey resident pointed out at a recent public meeting, what are homeowners to think about Joe-the-Handyman, who last week was a professional roofer and this week is a radon measurement specialist? We're going to periodically issue lists of the organizations that have demonstrated adequate proficiency, and we've got a toll-free number for people to get that information [800-334-8571, Ext. 7131].

3. To improve our own bank of statistical information, EPA is planning a national survey of radon levels in a random sample of homes around the country. Included in the survey will be studies designed to evaluate techniques for predicting indoor radon levels based on geological and other factors. We're also developing generic designs that states can use in conducting their own surveys. And we've developed the capacity to analyze several thousand charcoal canister detectors each month, so that states with the appropriate plans can request EPA help in supplementing their analyses. Thus, interested states can make their own assessments of radon problems sooner than they could through participation in the national survey. Of course, New Jersey, New York, and Pennsylvania already have surveys underway. States can request EPA assistance in designing radon surveys later this fiscal year.

4. A final element of the Agency's problem assessment initiative involves the use of geological data to help identify areas where high radon levels can be expected. We're exploring the potential applications of this data right now with the U.S. Geological Survey, as well as evaluating alternative soil gas measurement techniques.

In the area of mitigation and prevention, EPA's office of Research and Development is supporting demonstrations and evaluations of selected mitigation techniques in homes where high indoor radon levels have been found. We've already begun a project in the Boyertown area of Pennsylvania, where 18 homes underwent remedial demonstration work. The idea is to expand this program into both New York and New Jersey, and, accordingly, we recently announced our plans to work with New Jersey's Department of Environmental Protection to develop a mitigation demonstration project in Clinton. The Clinton Project will demonstrate low-cost mitigation techniques in single-family houses with elevated indoor radon levels.

Ten houses in the Clinton area have been selected for the project from a list of 70 candidate homes that was supplied by New Jersey's DEP. These houses will be fitted with radon mitigation techniques designed on a case-by-case basis to provide the maximum reduction of radon and radon-daughter levels. Up to another 20 houses will also be diagnosed as part of the project. Because our resources are limited, this second group of homes will not be subject to mitigation demonstration techniques by EPA. But the homeowners will get technical assistance to help them install mitigation systems using private contractors selected and paid for by the homeowner.

Additionally, the information from our remediation techniques applied to the ten homes will be made available to all other homeowners. Thus far, our EPA team has finished its survey phase in Clinton and is working with the state to make arrangements with the homeowners of the ten homes that have been selected. Fortunately, the affected homes in the area break down into three basic construction types. This will facilitate the transfer of the remediation techniques to the various homeowners. We are also providing information as to immediate actions individuals could take in the meantime to begin to lower their radon levels.

In the longer term, EPA expects to have demonstration projects like this in other areas of the country, to develop a wide variety of low-cost mitigation approaches that can apply to the broad gamut of housing construction and to the variations in climate found in different regions around the U.S. States that want to participate in these projects will have to supply EPA with a list of candidate homes, and we'll make our decision based on the criteria we applied in the Clinton case. We are looking, first of all, for homes with significant levels of radon contamination. But, for the purposes of our research, we also need homes that are in a concentrated geographic area and that have types of construction that fit our project designs. Other factors we consider include the type of heating system present, the age of the house, and of course, the owner's willingness to participate. The selection process should normally take about three to four weeks.

Right now, we are evaluating a proposal just received from the new York State Energy Research and Development Authority. The Authority proposes to work with the New York State Department of Health to identify a number of existing homes in different areas of the state as candidates for demonstration remediation efforts and also to develop a demonstration program for radon resistant techniques to be used in new construction.

EPA'S national Office of Radiation Programs will be complementing regional efforts to mitigate radon in existing homes with evaluations of new home design and construction techniques that would prevent or at least minimize future radon problems. Some of the information we come up with here could be incorporated into model building codes throughout the country.

Perhaps the most important component of EPA's radon action program in terms of future radon problems is the insitution-building among state and local governments, and in the private sector, to make possible local, flexible, but informed responses to regional radon problems.

Our national Office of Radiation Programs -- in conjunction with Regions II and III, and the states of Pennsylvania, New Jersey, and New York -- is designing a training program for federal and state employees to measure and diagnose radon problems and to give homeowners the available information on mitigation alternatives. The courses are planned to begin in the coming weeks. We're going to work on an aggressive effort to share all our information with the states and also to help them evaluate their survey data.

It's important, as well, for us to cultivate expertise in the private sector -- in national regional groups like home builders, realtors, and mortgage lenders. As many of you may know, when Sweden experienced its radon problem some years ago, there was, for a time, something of a real estate panic in areas with high radon levels. I think such a panic can be avoided in our case if we get sufficient and accurate information and get people qualified to assess local problems.

We're also going to work with manufacturers of radon measurement devices to encourage them to enter the residential market, as well as with manufacturers of heat exchangers and air cleaners to encourage proper testing of their products.

Now as far as getting information to the general public, EPA has prepared two pamphlets providing guidance on assessing and mitigating indoor radon. They explain what radon is, how it threatens health, how it can be measured, and what can be done to reduce radon levels in homes. The pamphlets have been out for review by state officials and are coming back to us with comments. We're also working on a much more detailed technical manual on radon mitigation techniques. These should be available to the states in final form for distribution by June.

Meanwhile, EPA staff will continue to refer public inquiries about radon to the appropriate state officials. The staffs, in general, will also be accelerating their participation in various technical and general conferences and workshops like this one.

So that's our national program for dealing with the radon problem. It's flexible and it's going to unfold over a two- or three-year period, and it will be open to any needed revisions as we learn more about the problem. This is a new one for us, after all, and we're on a significant learning curve. So new information could have considerable impact on our national strategy. Of course, the French say that "only that which is provisional endures," so maybe it won't change at all.

A NON-REGULATORY APPROACH

But it's obvious from what I've said today that ours is not a regulatory approach to this problem. And I don't think that whatever new information arises, you're going to see us move in that direction. There are several reasons for this, and I want to go into them because they explain a lot about EPA's current policies and the presently evolving relationship among the three tiers of government in environmental protection in general.

Our radon policy is not regulatory first of all because there are no easy culprits to go out and regulate. Although some people can see villains in any scenario, there really aren't any in this one, unless you consider the environment the villain. And I guess you might say in this case the environment is exploiting us. Mother Nature has unleashed her radon daughters, so to speak. This is not a man-made problem, except to the extent that homeowners unwittingly helped radon do its damage by sealing up their homes for insulation.

And, although a serious problem, its inherent variability from state, to state, and even from home to home, makes it an essentially decentralized one. This is an area where we must take maximum advantage of the unique flexibility of local institutions, and the proximity of these institutions to the individual problems.

EPA's policy is also not a massive funding campaign to underwrite remediation at homes around the country. Our mitigation demonstration projects, like that in Clinton, are not unlike many we have undertaken in other areas to help states find the appropriate technology to deal with a problem. They are not pilot experiments for an all-encompassing federally-funded mitigation program.

In its conception, EPA's radon program is an extension of what we have come to see as the most effective role the federal government can play in solving most environmental problems. That role has evolved over the years, and is still evolving, as is the division of labor among the three tiers of government. I believe the division of labor is beginning to reflect more accurately our original national vision for an environmental program.

I've been doing a lot of thinking -- and talking -- lately about this original vision, as EPA celebrated in December the fifteenth anniversary of its founding. I guess the sentiment I saw, that sums up the vision best, was expressed in the early years by the scientist Rene Dubos, who said that the key to solving environmental problems was to "think globally but act locally."

That philosophy guides what I feel is, when possible, the most effective scenario -- one that has the federal government providing the necessary expertise and information, the state providing the infrastructure and funding, and the local government exploiting its proximity and flexibility. Indeed, in the case of radon, I believe this issue will evolve to the point where the individual builder, or homeowner, will deal with radon as a routine matter, just as he would deal with a wet or sandy building lot.

This scenario doesn't work with all environmental problems. Some are so vast and complex that they require a huge and extensive federal commitment. The Superfund program is the best example. But, in general, what I call a federated approach to environmental problems works best.

This approach not only more fully embodies the vision that spawned EPA, but it incorporates some of the lessons we've learned over the past fifteen years. It reflects a greater appreciation of the power and efficacy of local institutions. It also reflects a greater awareness of the difficulty Washington bureaucrats have when they are removed, in many cases, hundreds of miles from the source of a problem. Even Eugene McCarthy, a man not exactly known for his abhorrence of an active federal government, said, "The only thing that saves us from the bureaucracy is its inefficiency." While, as a public official, I take exception to the breadth of that statement, I understand where it's coming from. And, in my capacity, I try to put EPA'S resources and expertise to work where they are most effective.

We're going to work to get the information on radon out to those who need it and help states and private groups take advantage of our technical capabilities. We're going to make it as easy as possible for homeowners to get the facts they need to make the appropriate decisions for their particular situations. And we'll keep our policy open to any new initiatives that prove necessary.

THE EPA RADON ACTION PROGRAM

Daniel Egan*

It is vitally important that we learn to understand what the role of the federal government and various state and local governments should be in dealing with indoor radon problems. We certainly don't have this sorted out yet. All the governments involved are still feeling their way through dimensions of the problem. I suspect that different answers will emerge as we go along. There needs to be continuous dialogue about the role of government.

This paper presents an overview of what EPA sees as its current role and the diversity of issues we are facing. Rather than being entirely descriptive, some of the issues behind each of the parts of the program will be highlighted, specifically, "Why we are doing some of the things that we are doing, and why are we not doing some other things that some people think we should be doing?" Finally, the paper gives a flavor of how EPA's response fits into a coordinated effort with the various state and local governments addressing the radon issue.

EPA's Radon Action Program has evolved since early 1985 when we first learned of the discoveries of very high radon levels in homes in eastern Pennsylvania. Our initial strategy was designed in concert with the role of other federal agencies. At about the same time, the states with the immediate crises -- Pennsylvania first and, subsequently, New Jersey were evolving their response.

REASONS FOR A NON-REGULATORY APPROACH TO RADON

One of the first questions that has to be answered is whether EPA or the federal government should set regulations, rules or some other proscriptive guidelines for what should be done about radon. There are a number of different perspectives with which to look at this question. An economist will tell you things like, "There are no externalities in the problem." This means that there is no external party causing the elevated radon concentrations in affected homes. There is no industrial firm or polluter who is imposing a social cost from which they are reaping a benefit. One of the basic tenets of regulation is lacking

---

*Daniel Egan is Project Leader for Radon Action Program, United States Environmental Protection Agency, Office of Radiation Programs, Washington, D.C.

here: nobody is externally causing the problem. If you are a political scientist, you may ask, "Does a government have a right to dictate to individual homeowners what level of risk they should tolerate in their own dwelling?" Stated this way, the government role becomes very controversial. There are many complications in that homes are visited or occupied, from time to time, by other people who don't own them. That raises a whole series of other questions. These all will need to be answered, but it is too early in the program to claim that we have the answers. Therefore, it is far too early to create a regulatory program, at least from a federal perspective.

A second issue in considering the appropriateness of a regulatory program is that there are far, far too many unknowns right now. Even if it were appropriate to set regulations in the face of the above issues, there is a great deal of necessary information that we don't know. We don't understand the extent of the problem, how expensive it will be to fix homes, and even how one approaches fixing some of the homes we discover. There is a tremendous lack of information as well. When you meld all those things together, it is fairly clear to us in the federal government that it is inappropriate at this point for the federal government to take a lead in setting requirements relative to indoor radon. It really is a problem that is most effectively handled at local levels by individual homeowners in many cases, or by local or city governments as they feel appropriate. There is nothing in the program that envisions the federal government coming in and setting rules for how the situation should be dealt with. On the other hand, we have created a very effective program to help state and local governments and the private sector in coming up with responses to some of the imminent hazards that are being found, to best help those people that have problems, and also to help find where similar problems might exist elsewhere. We are working very closely to try to help characterize or define the size of the problem, and at the same time, rapidly pursuing technology development programs to try to find effective and fairly low-cost methods to fix homes that have levels that clearly call for action on the part of the homeowner. We are also working in the longer term to find ways that we can help to insure that we don't just continue the problem by building new homes that have radon problems from the start, if this can be prevented to begin with.

DETAILS OF THE EPA RADON ACTION PROGRAM

There are four different categories into which the EPA program can be divided. In providing details on each of these four areas, I will continue to draw some distinctions about what we feel our role appropriately is at the moment and what we feel it is not.

(1) Problem Assessment involves trying to define how many places have elevated radon levels and how high they are. One of the first tasks we identified, which we are still working quite hard on, is to design a survey of the country to determine average radon levels, how those radon levels change in different parts of the country, and generally what the scope of the problem is nationwide. We also want to look at what causes different levels of radon in different areas. In other words, are there geographic associations or correlations, are there geologic correlations? Do certain types of house structures seem to cause elevated radon levels more often than others? There are probably other variables that will emerge as good predictive techniques in discovering where elevated radon levels are and what types of programs across the country should be put in place to address these problems. This is one of the baseline parts of our program. It is also a program that will take several years to develop. We have to design and work out all the statistical problems of survey design, actually go out and make the measurements, and then analyze the results. In the interim, we must address some other pressing problems that, in fact, can't wait that long.

Another dimension of defining the problem is the attempt to identify areas that have particularly high risk, such as those discovered all through 1985 in the Reading Prong in eastern Pennsylvania and in Spring, 1986 in Clinton, New Jersey. We plan to develop this only in conjunction with the states who wish to do surveys within their own borders to find out if they have similar problems. There are good reasons for that. For one, the state will know far better than we the local geographic and geological factors necessary to predict where there may be a problem. If a problem is identified, they will provide the primary line of communication to the homeowners regarding what to do to fix the problems. We must closely coordinate our efforts with the states. We have been offering the states a program of technical assistance to make it relatively easy for them to do screening surveys. The program uses charcoal canister devices which are the most cost-effective and rapid means of discovering problem areas. There is no reason to think that the elevated radon levels we see in Pennsylvania and New Jersey are limited to those two states. In fact, we have reason to think there may be many similar but as yet undiscovered problems in other parts of the country, and it is particularly to those states that we address this program.

We are also working closely with the United States Geological Survey and the Department of Energy to identify predictive clues as to where high radon levels might be. This is a very inexact science right now. We are still learning, and we are only at the very rudimentary stages of being able to predict, based on geologic facts, where elevated radon levels may occur. At the same time, there is probably a wealth of information out there that we need to understand better. The intent of this program is to see *if* it is possible to use this geologic knowledge to focus surveys in areas where problems are likely to be found.

The third EPA effort is to establish a quality assurance program for radon measurements. This involves two steps. We established and published in late February a series of specific protocols for seven specific instruments used to measure indoor radon levels. To match those protocols, we set up a measurement proficiency testing program for the benefit of firms involved in radon testing. Firms wishing to participate send us either their passive or active radon or radon decay product measurement devices. We then expose the devices to an undisclosed concentration of radon and then ask them to determine the results. In this way, we can see whether they have basic and adequate capabilities to perform radon measurements. We plan to make a list available of those firms who successfully participated in the program to the states and members of the public who request it from us. From our point of view, of course, it is entirely voluntary. But at least one of the states, New Jersey, has made participation in this program a mandatory requirement for making radon measurements in this state. This was our intent in establishing the program. The federal government has the opportunity to establish a technical baseline upon which state governments can draw, as they wish, to enforce the conduct of an industry within their state. This is uniquely a federal role because many of the radon measurement companies operate on a national scale. It is much easier for the federal government to provide the technical foundation for this type of program than it would be for any particular state.

(2) Efforts to mitigate or prevent radon exposure are necessary once you have determined that a house or an area of the country has elevated radon levels. There are three different efforts in this area. First, we have a research and development program that is just completing 18 homes in the Reading Prong area of eastern Pennsylvania. The federal government, with the cooperation of the state, identified these 18 homes. We went into them and diagnosed where the elevated radon levels were coming from and then performed and paid for highly successful mitigation efforts. The fixing of those homes at no expense to the homeowners was a beneficial side effect of a program to learn from the scientists and engineers what approaches are most effective to lower the radon levels in a variety of types of construction. We are now just in the stages of finalizing the early plans of the expansion of this program in Clinton, New Jersey, where we will similarly do ten homes in that area. We are also looking at other areas in other parts of the country which will have different types of housing characteristics or different geologic situations so that we can expand our basic knowledge of how to fix homes. It should be clear that this program does not imply a responsibility on the part of the federal government to pay

for mitigation. Our selection criteria for choosing houses are entirely based on what type of technical knowledge we and the private sector's need to further the state of the art of radon mitigation.

In a similar program, to begin shortly in Pennsylvania, called the "House Evaluation Program," we will work with state officials in some higher level homes diagnosing radon levels for those homeowners. We, as the federal agency, will not pay for the mitigation. We will prepare a menu of approaches that homeowners may use to lower radon levels, defining the effectiveness of mitigation at different costs. If the homeowner chooses to pay for that mitigation, we will then offer him free follow-up testing to see if it has been effective or not.

This program goes beyond the objectives of the first. In addition to helping people fix their homes while we get more data on what are the effective ways to lower radon levels in the homes, this particular program has a very strong emphasis on training both federal employees and state officials in radon diagnosis. They will then develop a base of knowledge to identify highly elevated radon levels in homes, where the radon is coming from, and ways to stop radon entry. We will involve state officials from all parts of the country so that government personnel can give homeowners advice as problems are discovered in the future.

(3) The third program in this area gets to the issue of new home construction. We are working on a grant with the National Association of Homebuilders to develop the organization's radon expertise, prepare it to advise its builders and help its builders advise us on steps that can be taken in new construction to prevent the occurrence of radon problems. We also plan to expand our demonstration program to look at some new houses, assuming we can find builders to cooperate. We hope to test the effectiveness of various techniques in new houses as they are built. Again, we hope to build up simultaneously the state of knowledge of radon mitigation for application by the private sector and to encourage government officials to obtain the knowledge necessary to advise private citizens on mitigation. As part of this institution building, we are also planning an intensive three-day course in radon mitigation specifically addressed to state officials. We are also providing technical assistance to states and a fairly aggressive program of outreach to private sector organizations, such as the National Association of Homebuilders and the National Association of Realtors, to increase the awareness and preparedness of groups that have particular concerns with radon so that they can help their sector of the economy come up with more constructive answers to the problem.

(4) As a natural outcropping of this effort, a fourth area that we are working on is public information. At speeches or conferences like this one, we participate across the country to try to help people to understand, and other technical experts to explain, the technical dimensions of

the problem to the public. We are also working on a document called the Citizen's Guide to Radon, which we expect to have out some time in June. It will explain the basic nature of radon and why people should be concerned about it, give some indication of the levels at which people should become concerned, to help distinguish between situations where radon levels may be only slightly elevated and some of the very serious problems that have been discovered. In some press reports, it is difficult to distinguish the fact that say, a .03 WL (6 pCi/l) or .04 WL (8 pCi/l) home represents a far different, and a much lower risk, than say, a 1 WL (200 pCi/l) or 10 WL (2000 pCi/l) home. Those types of situations call for rather different responses by the homeowner involved. Part of the purpose of this document will be to try to give people an appreciation of the different risks associated with varied exposures, the responses that they demand, and the distinctions between the responses that should occur in widely different situations. This guide will also provide a rudimentary overview of home mitigation and will make reference to a companion document that summarizes what we've learned in the ongoing demonstration program in Pennsylvania and from some of the private sector companies that have been doing radon mitigation. The document will also refer readers to sources of information in their respective states.

CONCLUSION

What is the characteristic role that the EPA sees for itself, as distinguished from those roles we think are most appropriate for the states we work with? We see ourselves as providing an important research and development dimension to the problem. We have resources to learn more about radon mitigation and health effects. We don't mean this to be a laboratory exercise; we really plan to go out in the field, as we are already doing, and test things so we can demonstrate what works and doesn't work. We also plan, beyond that, to respond to specific, and perhaps unique, needs of different states as we go along. We also have a requirement, to some extent, to coordinate with the national organizations in the private sector and also to work fairly closely with other federal agencies to make sure that there is a fairly unified response from the federal government. We work pretty closely with the Department of Energy, which is emphasizing basic research needs both in the area of health effects and characterizing soil, geographical factors, geological structures and the phenomenology of radon migration in homes. EPA's primary concentration has been in the transfer of knowledge to actual application in the private sector and by the states, so there is a fairly smooth coordination of effort between the two departments. We expect the states to take the lead in assessing the distribution of radon levels and the identification of particularly high risk areas within their respective areas.

They will be the primary vehicle for actually transferring information to the public, because it is to the states that the people will ultimately have to turn for the most assistance and help as they discover the seriousness of the issue. At the same time, we expect that the states will be the appropriate organizations to coordinate and interact with the private sector in terms of local constituency and local structure. There is a very careful coordination that has to go on between our interaction with the private sector associations on a national level and the states' interaction with them at the local levels. Similarly, we expect the states will be the source of advice to citizens on what types of mitigation techniques should be applied to particular situations, either directly or through the private sector.

In closing, we seek to encourage the reduction of radon risk as much as possible in existing and new structures, not directly through a federal effort, but only in concert with the respective states working very closely with the private sector, because that, in fact, is where the radon problem is going to have to be addressed.

A REVIEW OF THE RADON PROGRAMS UNDERWAY AT THE UNITED STATES DEPARTMENT OF ENERGY

Wayne Lowder*

INTRODUCTION

In the Department of Energy, we have emphasized the evaluation of the research needs in terms of developing the scientific information base to really understand this whole radon problem. The department, over many decades, has been the main sponsor of radiation research in the federal government, and it was quite natural when the radon issue emerged that we would mobilize whatever resources we could within our budgetary limitations and try to address the scientific issues. To understand what we are trying to do currently, and what we intend to do in the future, it is probably useful to emphasize some of the essential facts about radon.

First, we have identified pressure differentials and convective flow of soil gas as being the dominant influences on radon entry rates and on radon concentrations in houses under most circumstances, particularly in the houses that are at the higher levels of exposure. This is a very important finding because it suggests that the most effective remedial actions involve reducing this entry rate. Second is the fact that as ventilation decreases, houses are increasingly radon collectors. This is a particular interest to the Department of Energy because of the question of the effect of energy conservation and the application of conservation technology on future radon levels. Third, the distribution of sizes of particulates to which radon daughters attach has a very important influence on where the daughters eventually get deposited in the respiratory tract. There has been a great deal of information very recently developed on the particle size distribution under various environmental conditions. This has helped us to understand how critically important it is to really determine what fraction of the radon daughters are on very tiny particles, tiny liquid drops or molecular clusters, and what fraction

*Physicist and Radon Program Coordinator, Office of Health and Environmental Research, Environmental Measurements Laboratory, Department of Energy, as well as co-chair of the Radon Work Group of the Federal Committee on Indoor Air Quality.

are deposited on the larger particles that go to different places in the respiratory system. One of the reasons I list these essential facts is because I think they describe a chain of events that starts at radon production and ends with the irradiation of the critical cells and the development of cancer. To understand the health risk and to understand what we might do to reduce exposures, we have to understand this whole process that goes from production to cancer induction.

With the infusion of some new money next year, we are developing a more comprehensive research program on radon. In doing so, we are looking at what the key public health and public policy questions are, because we are not just doing research to answer some interesting scientific questions, we are trying to provide the needed input data to respond to the questions that the public is asking, such as: "What is the risk associated with the exposure in my home or the workplace I'm in?" and "If this risk is unacceptable, what can be done to reduce it in a cost-effective manner?" However, there are some other questions that may not be quite so much in the public consciousness right now, but which are certainly of concern to public agencies. Namely, "How can we avoid undue risk in future housing?" Also, what you see in the newspapers, in the media, and most people's concerns is the issue of high individual risk. However, there is an overall public health issue as well. This stems from the fact that most people's exposure is at much lower levels. If there is a significant risk at normal levels, or somewhat elevated levels, then clearly a lot of the lung cancers that are appearing in the U.S. population due to radon may be occurring in people that are exposed to much lower levels, simply because there are so many more of them. In terms of the public health risk, we have to look at that question too. Again, this is very relevant to the question of the health effect associated with energy conservation, because it is conceivable that energy conservation efforts may shift the whole frequency distribution of radon exposure to higher levels.

OVERVIEW OF DOE RADON PROGRAM

Within the Department of Energy there are many radon-related activities going on. I will just very quickly go through some of these. Under the Assistant Secretary for Nuclear Energy, there is a major remedial action program going on having to do with "contaminated sites" where you can trace the cause to human activities. In another office of the department, there is an aerial measuring assisting capability, which has been applied to surveys of areas in the Reading Prong to see where uraniferous areas are where the high houses might be located. There is also an information synthesis handbook on radon which was produced last year.

The Bonneville Power Administration (BPA) in the northwest is doing a very extensive radon survey as part of its residential weatherization program, and we are shortly expecting a great deal of useful information on radon in homes in that region. Also, BPA has done some studies on mitigation effectiveness in cooperation with Lawrence Berkeley Laboratories.

Now, to get to a discussion of the research programs sponsored by the DOE. Within the Office of Conservation and Renewable Energy programs, there are current investigations of the features of structural design and operation that affect radon entry rate and exposure. There is also an effort to develop design techniques for construction of low-radon homes, identify and evaluate energy-efficient control and mitigation strategies and develop a predictive capability for identifying areas with the potential for indoor radon exposure. Some of the expertise that has been developed in the mitigation-related programs is now being applied to field demonstrations of mitigation methods in collaboration with EPA.

Finally, I can speak to the program that is presently in place in my own office, the Office of Health and Environmental Research within the Office of Energy Research. I will emphasize our future plans.

Within the Office of Energy Research, the overall program goal is to provide a firm scientific basis for the assessment of radon exposure and associated risk under all environmental conditions. I should emphasize that a key phrase is "under all environmental conditions." We want to understand how actual conditions in the home affect radon exposure and resulting risk. The question of particle size properties is a very good example of a situation in which different kinds of conditions may be critical, such as whether you have a smoker in the house, whether you are cooking, and whether you are using a kerosene heater. These all affect, very much, the particle size distribution, and the aerosol concentrations which have a very significant effect on the exposure and on the dose to the critical cells of the lung. I will highlight several goals of our research effort.

(1) Our first goal is to develop a scientifically-based model of the geological influences on radon entry rates starting with radon production, how it transports through the ground, how it gets into houses, what the driving forces are and what are the factors that influence it. In this area we are collaborating very closely with the Geological Survey and the Environmental Protection Agency. Whatever model we develop, whatever its parameters, we need good extensive data on all the relative factors in a reasonable sample of homes. Of course, we would also like to have this data initially, before we do the final model development, so that we understand which factors are the important ones, and define the parameters of the model adequately.

(2) We are also very concerned with how we determine radon exposure accurately, and in particular the dose to the critical cells which determines what amount of cancer is going to be induced. So the elements of this program are actually quite extensive. In a sense we want to develop a complete model of the indoor atmosphere that's perhaps as extensive and detailed as our knowledge of the outdoor atmosphere. There is a great deal that needs to be done. We are also working with EPA on developing improved measurement methods, evaluating all existing techniques, and developing an appropriate quality assurance program. I don't know how often thoron has been mentioned during this conference, but there is an isotope of radon which comes from the thorium series which has been very much neglected in radon studies. We are not quite sure if this is really appropriate, since in some cases, the thoron daughters can make a very significant contribution to lung dose, and we have to look into this question.

Another area involves lung dosimetry. The effectiveness of the alpha particles emitted by radon decay products in inducing lung cancers depends on where the daughters go in the respiratory tract relative to the critical cells. Quite frankly, we are in a bind here because we are not really quite sure what the critical cells are. It is hard to hit a target when you don't know what the target is, and a great deal more needs to be learned before we can really give a solid quantitative estimate of the doses that really matter in terms of lung cancer induction.

(3) The third goal has to do with appropriate epidemiology, which has been much discussed here. The key word here is appropriate. Epidemiology is rather expensive. We want to find situations where epidemiological studies will have enough power to really provide us with useful information. That's not easy because there are many problems associated with such studies: finding control populations, having high enough exposures so you can get out the noise of conflicting factors and so on. We are sponsoring now a study in eastern Pennsylvania which is quite extensive. Hopefully, we can get some information on the risk for environmental radon exposures at least at the higher levels. Other agencies of the federal government, notably the National Cancer Institute, the Centers For Disease Control, and EPA are either sponsoring or considering sponsoring related studies on environmental radon epidemiology.

(4) Finally, and probably most importantly, we want to develop a quantitative model of lung cancer induction from radon exposure. This should be done by a multi-disciplinary approach involving physics, chemistry, and biology to try to characterize the progression of events from energy deposition to neoplasia. So we are really refocusing a lot of our radiobiological research to look specifically at lung cancer induction. This now appears to be probably the most significant radiation hazard in our experience. This is

going to be a big job, and certainly a lot of the work being done in other agencies, notably, the National Cancer Institute and EPA, are very relevant here. But in a sense, it doesn't appear that any alternative approach will give us the answers we want. For example, the public is asking, "What is the risk in my house?" Now the data we've gotten from miners, and some suggestive environmental data, gives us a pretty good handle on the risk somewhere between a factor of three and ten. That is a pretty good degree of accuracy when you look at most environmental pollutants. And in terms of what public policy makers need to make policy decisions, it may well be that the risk factors we have now are adequate for making these judgments. However, factors of from three to ten are not adequate in response to what the public is asking because, if you are a factor of ten off, a homeowner may be spending money that, if he actually knew what the real risk was, he may not want to spend. So it would seem to me that there is a need for much better accuracy in our risk assessments, particularly down near the 0.02 WL (4 pCi/l) area where, in fact, our existing data is extremely weak. This is where the kind of work I'm describing comes in, because I think it's the only way we are going to effectively get these kinds of answers. Epidemiology will not give us a good handle on the risk at the lower exposure levels.

CONCLUSION

Finally, I just want to mention what we might get as a payoff from these kinds of programs. The last one is really the big payoff. If we can develop a good model of the process of lung cancer induction, we can really look at how smoking and radon interact. I think the jury is certainly still out on whether they are acting independently or whether they are acting in a complex way in the induction of lung cancer. I get very nervous when I hear people talking about the lung cancer risk from radon and the lung cancer risk from smoking independently, when animal experiments indicate that they are not acting independently here. In fact, there is one school of thought that says that radon is primarily an initiator of the process of carcinogenesis and that smoking is primarily a promoter. In a very elementary way this suggests that if you didn't have radon to start it, smoking wouldn't be very effective in completing the process.

Well, these are questions that have to be answered, and, in a sense, it's only going to be when we have a good model of the whole process of carcinogenesis that we are going to really answer these questions.

HIGHLIGHTS FROM THE DISCUSSION: THE ROLE OF GOVERNMENT—THE FEDERAL RESPONSE

THE IMPORTANCE OF TESTING

P. Sandman: Senator, would you advise single-family homeowners who live in north Jersey to monitor or have their homes monitored now, or would you rather see them wait a year or two until there is more research done?

Senator Lautenberg: I think it is a good idea to do it sooner rather than later. The costs have come down to where I think that most people in a home can afford the testing without having a serious difference in their financial condition. As one person who is being forced to sell her house in the Reading Prong because her husband's had a job transfer said to me, "I don't want to test my house and find out. The heck with it, I'd rather have someone else find out." And I said to her, "You are not going to get by with that; the issue is raised to the forefront and the question is going to be asked. If the question isn't asked, I don't think you have a reliable buyer on the other side that is going to fulfill the contractual obligations." I think it is better sooner than later, and one thing that we are all concerned about now is making sure that the contractor or the agency that offers you the test is a reliable contractor. There is some panic selling going on, and there are some underhanded operators out there. But I think it's a good idea to test.

RADON AND FAMILY STRESS

Q: My concern as a school psychologist has been that we have done nothing in the schools to inform children. Children's stress levels, especially in the areas of high radon concentration, have increased remarkably. It's like the kids are the last to know when there is a health threat of such magnitude. No effort whatsoever has been made in the schools.

Senator Lautenberg: We know certainly that children do scare as a result of circumstances, sometimes even those that are removed from them. The Challenger explosion, for instance; the nuclear explosion at Chernobyl. You do know that kids respond to those things. Radon seems to be somewhat removed, but I appreciate your mentioning it. We are beginning to study this problem in far greater detail. Hopefully, when conferences such as these have finished their work, we will be able to present a fuller picture of what ought to be done, including an alert to the children that this building that they are in isn't injurious to their health, or if it is, they ought to be able to send the message back home and have something done about it.

K. Varady: The stress element is very, very prevalent in the children of the homes that have been found to have high levels of radon. The stress comes with the change of their lifestyle and the fact that they do understand the lung cancer evidence that is there.

H. Sachs: I would like to comment on the stress element for families. Until the first of the EPA courses this week, the mitigation work has been almost exclusively in a research and development environment. I personally know almost everyone in the country that has been heavily involved in mitigation. I believe that it is fair to say that just about everyone who has been involved -- there are only a handful of groups -- has taken it as one of the highest and most important ethical concerns to sit with the homeowners to discuss the problem to the extent possible for them as building professionals rather than health professionals. They have tried to help the homeowner understand, and in turn, to understand the homeowner's concern. It will be a challenge to the EPA to maintain and build upon this tradition as we transfer mitigation skills to other groups in the private sector. Radon is a problem of the home, but it is also a problem with the homeowner. The house doctoring approach in energy conservation pioneered at Princeton University is a model for programs which involve the homeowner as the active participant in the diagnostics and the evaluation and remediation.

RADON AND CONSERVATION OF ENERGY

Senator Lautenberg: As we conserved on energy, we created the larger problem with radon; if the houses weren't so tight...

H. Sachs: I beg to differ. As a public health matter, the small changes in ventilation rate that have occurred in tightening up the building stock are important. This point has been made by other speakers. But, in terms of an individual house, and an individual risk, it is not important. When we are talking about the problems in Clinton, in Boyertown, and generally in eastern Pennsylvania, not only the Reading Prong, we are talking about the problems of large sources. When the sources of radon are as large as they are in these areas, the amount of ventilation is entirely a secondary matter. I am sure that Rich Sextro, who is here from Lawrence Berkeley Laboratory, will concur in that; much of their data has supported this. It is a problem of the source rather than the minor modulations that come from ventilation rate. It is a public health issue, not a mitigation issue.

FEMA AS A MEANS OF AIDING VICTIMS

K. Varady: We at People Against Radon are concerned about getting immediate help for the citizens in these high level homes that need immediate attention, and you have mentioned the possibility of going to FEMA.* What is FEMA's role? Are they strictly coordinators or do they give grants to the homeowners, and how immediate is their response?

Senator Lautenberg: If the governor of a state makes the request, FEMA action is then triggered. FEMA is a quick response agency. FEMA grants are directly associated with the individual's income level. Additionally, the Small Business Administration can give low-cost loans to help people.

EQUITY AND RADON RESPONSE

H. Horowitz: This is a question really that all three panelists may be interested in commenting on. I am profoundly distressed, in one respect, in the decision to back off from regulation in radon management. We have a situation now in which the great majority of the people testing for radon are in the upper income bracket. We have very little testing going on in the lower and middle income portions of the population, and there seems to be very little ability right now for the majority of the public to be able to mitigate readily. So I think there is a fundamental equity problem that I would like to see addressed by the various federal people.

Senator Lautenberg: There is no doubt about it; a thousand dollars in a home that costs $400,000 or $500,000 is not the same as a thousand dollars in a home that is worth $50,000 or $60,000, if you can find them in New Jersey. It's a very serious problem. People are entitled to good health, whether they are poor or rich. We are trying to link assistance from the federal government to some form of testing, because it becomes a financial tragedy as well as an emotional one for those who don't have the means to do the mitigation. We believe that in most cases, mitigation will involve a relatively modest cost. But then again, modest cost depends on your ability to pay in the first place.

D. Egan: I know some of the states are in fact considering questions about what should be done to help lower income homeowners deal with mitigation questions, and they may also be able to provide some answers to that.

---

*FEMA is the Federal Emergency Management Agency.

THE OFFICE OF MANAGEMENT AND BUDGET AND RADON

M. Edelstein: Given the Gramm-Rudman bill and given the response of the Office of Management and Budget in general under this administration, have we in fact moved as definitively on radon as we might have? That may be an easier question for the Senator to answer than our other guests.

Senator Lautenberg: The answer is no. What we are doing right now in our country, I find difficult to understand. There is kind of an ideological struggle underway over where we put our money. Do we protect the public health best when we build MX missiles, Star War defenses, and B-1 bombers? Do we make the country stronger as a result of those investments, or do we make the country stronger when we build a healthy society, an educated society, a nourished society, where we can send our children to school without fear of damaging their health from asbestos fibers that have been used to insulate pipes, where we know that the water supply is clean, where we know that there is no risk from radon gas? Do we build a stronger society when we put our money into health research? Do we build a stronger society when we clean up toxic wastes? The Gramm-Rudman Budget Program, in my view, is an act of cowardice because what we have said is that we as your elected representatives don't really want to tell you what we've done about cutting the budget in ways that may have affected you. We have turned on the automatic pilot so that there is a formula out there that is going to tell us where we cut and where we spend, and we have literally almost no voice in how those decisions are made once Gramm-Rudman is in place. The result is that programs like this, new programs where you discover a new health hazard, where you discover a new problem, can't really get the financial resources for thorough investigation. The budgetary constraints are very real. I sit on the budget committee; I sit on the appropriations committee; I sit on the environment committee. The budget committee is where many of the philosophical decisions are made as well as the financial decisions. They become almost one in the same, in fact. OMB, which has the responsibility for managing the budget, has invaded the province of the decision-makers by deciding which programs are worth funding and which programs are not worth funding. Well, that's not their job anymore than the accountant in a company should make those decisions. It ought to be up to the management. Here's the problem: how do you make those decisions? So it does affect the radon gas problem as well as some of the other health problems that we are faced with. Fortunately, there is enough support in the Senate because of the needs back home, to say, "OK, we are going to put some of that stuff behind, and we are going to focus on some of these problems, like radon, and deal with them."

H. Sachs: I very much appreciate your comments, particularly your comments on the awkward role of the OMB. I would like to point out that there has been an incredible amount of action taken by the present administration, beginning in 1981, to take apart the established mechanism for addressing risks such as radon, particularly the interagency committee on indoor air pollution. Members were banished to bureaucratic Siberia; some left the government. The ideology of OMB has been very, very important. That's a particular case. The more important point, I think is that somehow within our nation, we have to develop better mechanisms by which the civil servants can accept political direction without our losing the benefit of their technical knowledge. These people were literally silenced for four years. That to me is the much larger issue than the indoor air pollution issue and its radon ramifications, one which I hope you will be able to bring some influence on.

Senator Lautenberg: I agree with you. Thanks very much for your comment. It's almost as though this gentleman and I had kind of conspired before, because I see it that way. There are a lot of very qualified people who work for government. These are people with often terrific training, who believe that they can contribute something to society; they can make their contribution in ways that aren't seen, or in most instances aren't even rewarded, but want to do it for the sake of doing the right thing. As more and more we in government contract and squeeze these budgets, and as we make short-term decisions, we resist difference; we resist challenge.

D. Egan: As one of the civil servants who have managed to stick around the federal government, I'd like to give a little of the silver lining on the financial side. Within the constraints and the dimensions of the federal role I talked about in my talk, we at EPA and DOE and the other federal agencies are operating at the moment as aggressively as we believe appropriate to implement the types of efforts we have described. Within the fiscal year 1986 budget, there was no explicit appropriation for radon other than the research and development Senator Lautenberg was very instrumental in assuring to fund our demonstration program. However, Lee Thomas, EPA's administrator, has gotten us whatever resources we feel are appropriate. I am not constrained by a lack of contractual resource dollars to proceed. In terms of the 1987 budget, the President's budget has in it all that we were smart enough at the time to ask for. So there is no part of the agency request for radon funding in 1987 that was turned down. Now, it is certainly true in a program like this, we are learning a lot more about what needs to be done, and those needs will change and perhaps expand as time goes on. In fact, I am quite impressed with the efforts that Pennsylvania and New

Jersey, and increasingly New York, are putting forth in dealing with radon levels. It is certainly fair to say that the government, although perhaps getting a late start as governments are wont to do, is now proceeding fairly rapidly. That isn't to say that there aren't lots of arguments and discussions to have in the future about what the role and funding should be, but I don't think it's entirely accurate at this point to imply, at least at the federal executive level, that there is a reluctance to proceed now that the dimensions of the problem are becoming known.

Senator Lautenberg: Lee Thomas, while no doubt well qualified, is a soldier of the ranks. And, his boss, the President of the United States, feels just a little bit differently about radon than perhaps someone like myself. I don't think that there is sufficient funding or commitment for the projects. I think what has happened, in fairness to the agency, is that the radon problem, the indoor air pollution problem, is considered in the context of all the environmental problems. And, if we don't have enough money to deal with all the environmental problems, the pressure sweeps us down. So, somewhere between Dan's optimism and my pessimism lies, I think, the position that we find ourselves in. It is true that the agency gives radon its functional support, but it really hasn't put out the request for the kind of funding that we need to get assessment done, to get the studies moving. Since I am neither a scientist nor an engineer, I would rather take the hasty way with this thing and not wait until all the parts of the puzzle are in.

W. Lowder: My position is kind of in between these two gentlemen. In my talk, I tried to indicate where we in the Department of Energy saw the basic research needs to provide the scientific information that is much needed to really understand the problem, to find the most effective control and mitigation techniques and to provide the public with very firm estimates of risk at all levels of exposure. When you look at the sorts of things I was talking about, some of you may have actually done a little mental arithmetic on how much it might cost to achieve all these wonderful purposes. There is clearly nothing on the immediate horizon in the way of money to accomplish all these things. In a sense, we are beginning to understand what needs to be done, we've developed a good qualitative picture of the situation, but somehow we are going to have to make better efforts in the future to redirect existing programs, to have interagency collaboration in applying resources to all of these questions, and hopefully, within one or two more years, to develop additional funding to respond to the really high-priority needs for research.

The Pennsylvania Response

PENNSYLVANIA: A LEGISLATIVE PERSPECTIVE ON RADON

Senator Michael A. O'Pake*

In December 1984, when Stanley Watras set off the radiation detectors at the Limerick Nuclear Plant in eastern Pennsylvania, he literally sounded an alarm that has been heard around the world. For, as many of you know, the radiation detectors at this facility were sounded, not because of radiation exposure at the plant, but due amazingly to radon contamination in his home, which is in my Senatorial district.

Radon gas is a colorless, odorless, tasteless radioactive gas emitted from natural uranium deposits and similar rock types which exist throughout Pennsylvania and the United States, and extended exposures to high levels of radon gas can lead to an increased risk of lung cancer. The federal Environmental Protection Agency now declares that radon gas represents the most dangerous indoor air pollutant, and they estimate that between 5,000 and 20,000 people die of lung cancer each year due to their exposure to high levels of radon gas.

As many of you know, the radon level in Stanley Watras' home was detected at over 14 Working Levels (2,800 pCi/l), with the EPA determining that 1 Working Level was the cancer-causing equivalent of smoking 10 packs of cigarettes a day.

With the discovery that Mr. Watras' home contained radon contamination at levels equivalent to smoking 140 packs of cigarettes a day, we, in Pennsylvania, embarked on a year-long struggle to find solutions at the state and federal level for people exposed to this cancer-causing environmental hazard.

Shortly after learning of the severity of the problem in early 1985, I introduced legislation in the Pennsylvania Senate to appropriate $900,000 in state funding to our Department of Environmental Resources for a massive radon testing program. To my knowledge, this was the first radon legislation ever introduced in the eastern United States. At that time, on behalf of a bi-partisan coalition of legislators from Berks, Lehigh, and Northampton counties, I also wrote to each member of the Pennsylvania Congressional delegation urging federal assistance for innocent victims of radon contamination.

---

*Michael A. O'Pake, is a state senator from Pennsylvania.

On April 29, 1985, we, in the Pennsylvania Senate, introduced a Resolution, which subsequently passed, memorializing Congress to provide financial assistance in the federal budget for victims who are put to the expense of buying equipment and materials to reduce high levels of radon gas in their homes.

As a result of these measures and the mounting pressure on the state government to respond to the radon problem, $1 million was appropriated in the 1985-86 Pennsylvania Budget for the largest radon testing program known at the time. As a result of this appropriation, the Department of Environmental Resources embarked on a massive testing program, sending free Terradex radon detectors to over 20,000 homes along the "Reading Prong." This was the first state-sponsored radon testing program in the country and led to the gathering of a substantial amount of data which has given impetus to action by other states.

The preliminary results from that testing program indicated that over 40 percent of the homes along the so-called "Reading Prong" had levels of contamination that were above the stated "safe" level of 0.02 Working Levels (4 pCi/l). In addition, it was discovered that it could cost homeowners anywhere from $100 to $30,000 to remediate their homes. At a cost of over $32,000, the Watras home was eventually remediated by the Philadelphia Electric Company.

In August of 1985, the National Centers for Disease Control wrote to the Pennsylvania Department of Health stating that: "Sampling is not complete; however, current data indicate that the radon decay product levels in a large number of these houses constitute a major public health problem that requires attention. Risk estimates extrapolated from epidemiological studies of uranium miners show a substantially increased risk of lung cancer at these levels of radon daughters. From a health risk standpoint, it is essential that the exposure to these residents be reduced."

Unfortunately, while the CDC was sending major health warning signals and our Department of Environmental Resources was discovering many homes with high levels of radon gas contamination, the national government was slow to act. The EPA had not set a national radon standard. There existed no reliable low-cost remediation technique, and Congress had not provided direct federal assistance to homeowners despite the hard work of a few federal representatives, such as Senator Lautenberg.

For my constituents and many others along the "Reading Prong," the uncertain danger posed by high levels of radon gas contamination combined with no reliable low-cost remediation technique led to anger, frustration, and anxiety. These emotions eventually coalesced into the formation of a citizens' group named Pennsylvanians Against Radon. Under the able leadership of Kay Jones and Kathy Varady, this grass-roots organization became an effective lobbying group for increased state and federal assistance to affected homeowners and continues to play an important role in shaping public policy.

As a result of the CDC findings and in an effort to provide state assistance to those families who found high levels of radon contamination, I introduced two bills on September 25, 1985 regarding the radon problem. The first bill would appropriate $5 million to the DER for the reimbursement of "costs actually incurred" by homeowners to reduce radon gas levels in their residences. The second bill would appropriate $500,000 to the Department of Health, similar to a bill which was recently signed into law by Governor Kean of New Jersey, to conduct a radon epidemiological study to determine the effects of long-term exposure to radon.

Unfortunately, Pennsylvania Governor Dick Thornburgh chose not to endorse my bills and opted instead for the establishment of a low-interest loan program, which still has not gotten off the ground. However, I am happy to report that a compromise proposal, which I authored, to establish a $1 million radon remediation program, passed the Pennsylvania General Assembly last week and is now before the Governor for his signature. The state funding would be allocated to the Department of Environmental Resources to work with private industry and the federal government in developing a low-cost remediation technique that will work to successfully lower the radon levels in homes. These state resources should substantially augment the federal involvement in remediation and hopefully lead to a prompt and low-cost solution to this problem for use by all affected citizens in the United States.

This legislation also grants the Pennsylvania Housing Finance Agency immunity from liability in administering the Governor's proposed $3 million low-interest radon loan program. As a result, Pennsylvania will soon become the first state to give direct assistance, in the form of low-interest loans, to the victims of high-levels of radon contamination to enable them to afford remediation work. The PHFA will have the ability to authorize loans up to $7,000 at rates of either 2 percent or 9 percent, depending on a family's income.

Of course, more needs to be done. In Pennsylvania, a more thorough testing program of the entire state should be undertaken, and I am happy to report that the proposed 1986-87 state budget contains an additional $1.2 million for further radon testing. In addition, I am studying possible state legislation to establish regulations for new home construction to safeguard future homeowners from the dangers of radon contamination.

At the federal level, I strongly believe that Congress needs to appropriate more money for radon testing and mitigation work. The EPA needs to take a more active and leading role in the area of radon contamination, and it can only do this with adequate funding from Congress. With the current estimate that high levels of radon gas could exist in between 5-10 pcercent of all U.S. homes, I believe that it is a disgrace that current radon expenditures by some individual states are larger than the entire EPA budget for radon. The EPA needs to conduct a national survey as soon as possible and find the "high" areas around the United States so that innocent victims do not continue to die. I challenge the EPA to move quicker and more responsibly in designing a testing program in as short a time period as possible.

I also believe that EPA needs to establish national standards for "safe" levels of residential radon gas contamination. It has become increasingly difficult to act without adequate knowledge of what is and what is not safe residential radon exposure.

Finally, I also believe that the federal government should assist families with high levels of radon. it seems only logical that if the federal government can grant energy tax credits for tightening a home, they should be able to provide assistance for homeowners who need to "untighten" their homes.

However, while these steps will hopefully be taken in the future, I sincerely hope that the valuable information being developed from the Pennsylvania experience will enable other states to more quickly and responsibly assist their many affected citizens. History will judge us by our response to this very real threat to the health and safety of countless Americans, and now is the time to act.

COMMONWEALTH OF PENNSYLVANIA RADON MONITORING PROGRAM

Jason Gaertner*

BACKGROUND

In December of 1984, an engineer for the nuclear Limerick Generating Station set off radiation detection monitors. It was discovered that the source of his contamination was radon gas (the product of decaying uranium and radium in rock and soil) emanating from the ground beneath his home in Colebrookdale Township, Berks County. Air samples taken in the home showed radon daughters (particles emitting high energy alpha radiation which are the result of decaying radon gas) at levels as high as 13.5 Working Levels (2,700 pCi/l). Working Levels is the standard measurement term for radon daughter concentrations. The Commonwealth has adopted 0.02 WL (4 pCi/l) as the acceptable threshold for average annual exposure; 0.02 WL is assumed to carry a 1 percent lifetime risk of lung cancer.

Since this discovery, this initial household was evacuated under a Department of Environmental Resources advisory and an extensive and successful remedial project was completed at the expense of the Philadelphia Electric Company. The family has since moved back into the home. Other homes in the immediate vicinity of this initial home also were found to contain high concentrations of radon daughters. Many of these homes are participating in the U.S. Environmental Protection Agency's (EPA) low-cost remediation demonstration project. These homeowners form the nucleus of the People Against Radon (PAR) citizens' group (formerly Pennsylvanians Against Radon).

Following the initial discovery of radon in homes, the DER established the Radon Monitoring Program in nearby Gilbertsville, Montgomery County, and undertook an intensive door-to-door campaign in Colebrookdale Township, Boyertown Borough and surrounding municipalities offering a five-minute air sample to detect radon daughters. Approximately 2,800 households (30%) accepted the offer. Of these homes, approximately 45% had radon daughter concentrations above the 0.02 WL (4 pCi/l) guideline.

*Jason Gaertner, a Community Relations Officer with the Department of Environmental Resources, State of Pennsylvania, works with the Radon Monitoring Program in Gilbertsville, PA.

The municipalities canvassed are within the boundaries of a uranium and radium bearing geologic formation known as the Reading Prong. The Prong runs in a northeasterly direction from east of Reading through portions of Berks, Bucks, Lehigh and Northampton counties. It continues through New Jersey and New York, and into Connecticut.

Because the door-to-door survey was a slow and labor intensive process, the DER developed a radon screening strategy that would allow it to determine the scope and seriousness of radon in Reading Prong homes much sooner. Through a contractor, Terradex, Inc., the DER offered in-home radon detection kits (Track Etch) free to residents of the Prong. At least 26,000 requests for kits were received. Of these, at least 19,000 were eligible residents of the Prong and were sent kits. Early results of this mass screening show close correlation with the door-to-door results. The kit program was later expanded into the area around the City of Easton due to high readings being discovered there.

Radon Monitoring Program staff continue to conduct air sampling in homes where test kit results indicate radon daughter concentrations of 0.1 WL (20 pCi/l) and higher, in homes where remedial action has been taken and in schools.

After a home is tested, interpretation of the results are important since many variables are involved. A Community Relations Coordinator at the Gilbertsville office interprets the data and provides other health and remedial action information. Community Relations focuses on providing accurate information without unduly alarming the public.

The public has been appreciative of the testing program and the home and school data it provides, but concern is repeatedly expressed that specific corrective measures with predictable results need to be developed. The Department's Radon Remedial Action Booklet provides suggestions, but the predictability of their effectiveness is low due to the many variables involved in the radon phenomena. The EPA has completed remediation demonstration projects in 18 Reading Prong homes, and is scheduling approximately 80 more projects in the state to compile data to increase predictability for corrective measures.

The DER anticipates a similar $1 million project and remedial training courses for contractors, both at the state level, will be conducted in the future. The project is included in a bill now before the Governor, a bill which will also assist the start of a $3 million low-interest loan program proposed by the Governor. The bill grants immunity from liability lawsuits that could result from the Commonwealth's involvement in the loan program. The training courses are scheduled to be a cooperative effort between the DER and EPA.

PUBLIC CONCERNS

Pennsylvanians living outside of the Reading Prong who cannot receive free in-home radon detection devices or other state services believe it is unfair to offer these services only to residents of the Reading Prong.

The concerns raised about the Commonwealth's response to the radon discovery that have received the most publicity have come from the PAR group. Below are a list of concerns which have been brought to the attention of the Radon Monitoring Staff:

> They feel there is a lack of Administrative concern because residents with continuing high concentrations of radon are not being offered relocation assistance.
>
> They feel that the low interest loan program should be a grant program and that it is too early to encourage remedial measures which have not been fully proven.
>
> They feel that the DER should be more involved with EPA's remedial demonstration projects so that it will be better prepared to advise others.
>
> They feel that something should be done to determine what damage has been done to them and their children by their radon exposure.
>
> Those that are part of the EPA demonstration project are resentful that EPA is experimenting with less effective low-cost measures rather than the extensive remediation that was contracted by PECO (Pennsylvania Electric Company) for the initial house.

The Pennsylvania Assocation of Realtors has expressed fear that home values and resales in the Reading Prong will drop. They reported that buyers have begun to request radon detection tests before buying.

The realtors also reported that radon is becoming a factor in prospective employee decisions to accept job offers and move into the area.

PROGRAM STATUS

Radon Monitoring Program Office

The DER has established the Radon Monitoring Program office in Gilbertsville, Montgomery County. The staff is comprised of five administrative and 17 technical personnel. Fifteen of these 22 positions have been filled. At least six additional seasonal student interns will be added to the office workforce this summer.

Home Air Sampling

The initial DER response to the radon discovery in Colebrookdale Township was to go door-to-door there and in surrounding municipalities offering five-minute air sample tests (Kusnetz Method). Approximately 30%, or 2,800 of the homeowners contacted in person, accepted the air sampling offer.

Home Sampling Retests

The Radon Monitoring Program Office has committed significant resources to resampling the air in homes in the Reading Prong where remedial action has been taken. The homeowner is responsible for advising the DER when the work is complete and for requesting this service.

School Air Sampling and Diagnostic Tests

The air in all schools in the Reading Prong has been sampled for radon. Where samples indicated radon daughter concentrations above the 0.02 WL (4 pCi/l) guideline, DER provides contracted diagnostic studies and advises the school districts on possible remedial actions. At least fifteen school buildings are currently under contract for these services.

Now that we have sampled air in all of the schools inside the Prong boundaries, the DER is currently complying with requests from some to sample school buildings which lie outside the Prong. A few of these schools outside of the Prong have been found to contain radon daughter levels slightly above the 0.02 WL (4 pCi/l) guideline in some student use areas. Contracted diagnostic services will also be made available for these buildings.

Home Radon Detection Kits

To more quickly determine the scope and degree of the radon phenomenon in the Reading Prong, the DER offered free home radon detection kits to residents of the Prong. Of the more than 26,000 kit requests received, more than 19,000 were from residents of the Prong. Requests from homeowners living outside the Prong received rejection letters. Those eligible are sent home detection kits by the DER contractor. They are installed in the basement of the home and left there for approximately three months and then returned to the DER contractor for analysis. So far, the concentration distribution has closely paralleled the door-to-door effort results.

The results of the more than 11,000 homes tested by the DER are as follows: 0 to 0.02 WL (0-4 pCi/l), about 4,400 homes; 0.02 to 0.1 WL (4-20 pCi/l), about 4,700 homes; 0.1 to 1.0 WL (20-200 pCi/l), about 1,400 homes; and 1.0 to 13.5 WL (200-2,700 pCi/l), less than one percent, or about 70 homes.

Pennsylvania Representative Donald Snyder and Congressman Ritter have been offering nominally priced radon detectors to their constituents outside the Reading Prong. Detectors are also available from the University of Pittsburgh and the Northeastern Pennsylvania Environmental Council, among others. The DER regularly refers residents living outside the Prong to these sources.

Community Relations

Since the creation of the Gilbertsville office, the program has included a contracted or staff community relations professional. The individual responds to public, media, and government inquiries, provides health and remedial construction information, and informs homeowners and school officials of the results of tests in their buildings. Outreach efforts, in the form of speaking engagements, have been increased in order to provide accurate, timely information to organizations and concerned citizens.

Hispanic Community Outreach

The DER has contracted with the Council of Spanish Speaking Organizations of Lehigh Valley, Inc., to solicit participation of Hispanic Reading Prong residents in the home detection kit campaign and to inform the Hispanic Community of the radon situation. The contract runs through June, 1986.

Public Information Materials

The DER has developed a comprehensive fact sheet on radon describing radon gas, radon daughters, measurement terms and methods, exposure and health risks and other related topics. Tens of thousands of fact sheets have been distributed to individuals and through groups.

Community Relations also has developed new information materials. Interpretation letters for homeowners, lot owners and others have been written to provide important perspectives on the radon situation. Supplier lists also have been compiled to assist the public locate supplemental testing kits and testing and remedial services.

Remedial Action Booklet and Course

The DER contracted for the development of a remedial action booklet and course for homeowners and contractors. Given what is expected to be learned from EPA's demonstration projects, the booklet and course are considered to be early and preliminary efforts. Tens of thousands of these booklets have been distributed. Two sessions of the course based on the booklet have been taught at a Berks County vocational school.

Program Expansion

An inordinately high number of homes, at least 16, in the Paxinosa Ridge area of Forks Township and the City of Easton were found to have very high radon concentrations. This discovery caused an expansion of the DER's administrative Reading Prong Program to include additional parts of the City of Easton, Forks Township, and adjacent Palmer Township. The DER is in the process of sending letters to at least 2,000 homeowners in the expanded area, encouraging them to utilize the free kit program.

Low Interest Loan Program

A stateside Low Interest Loan Program is anticipated for the summer of 1986. On the advice of local banks which have shown an interest in originating the loans, the Pennsylvania Housing Finance Agency (PHFA) is looking at ways to reduce loan fees. PHFA will not begin the program until sovereign immunity legislation is enacted. DER responsibilites under this loan program are:

* To certify that applicants homes were tested and show concentrations of radon daughters above 0.02 WL (4 pCi/l)

* To provide homeowners with advice on possible routes of radon gas entry and what methods might reduce concentrations

* To provide the homeowners with a list of potentially qualified contractors to perform remedial work

* To approve the homeowners remedial plan

* To certify that work has been completed satisfactorily

* To retest air in homes to determine the effectiveness of the remedial work

Contractor payment is not conditioned on the effectiveness of the work, only the quality and completeness.

Lung Cancer Mortality Survey

The Pennsylvania Department of Health has conducted a limited lung cancer mortality survey in and around the Colebrookdale/Boyertown area and found the rate of lung cancer deaths to be consistent with the rest of the state.

The DER Gilbertsville office is also assisting the U.S. Department of Energy with an in-depth lung cancer-radon health study that will encompass about one-quarter of the state in scope.

Proposed Pennslyvania Legislation

Proposed legislation for a $1 million DER Remedial Demonstration Program has passed both Pennsylvania houses and is before the Governer. The bill would require the DER to determine low-cost radon reduction procedures appropriate for Pennsylvania homes. The bill also grants immunity to the Pennsylvania Housing Finance Authority for any liability lawsuits filed as a result of the $3 million low interest loan program.

HIGHLIGHTS FROM THE DISCUSSION: THE ROLE OF GOVERNMENT— THE PENNSYLVANIA RESPONSE

ADDRESSING RADON'S VICTIMS

Kathy Varady* (Discussant): You people are very fortunate coming on the scene at this point in time because the market has been flooded with all kinds of radon information. When it was first discovered in our homes, we did not have that benefit. In an attempt to not scare us to death, we were not presented with some of the statistics that you were given this week.

It is important to understand that a public awareness is vital. If a neighbor's home burned down this week, I'm sure you would feel it was important to help that family. You wouldn't feel that just because it didn't happen to you, it wasn't important. There are families that are living with extreme radon levels, and you shouldn't feel "that's a shame, but there is nothing I can do." As Senator O'Pake said, with public support, you can get a legislative response.

We are extremely concerned about people now being identified with extremely hot homes. There are homes now being identified in the Easton area outside of Allentown where people are living with extreme working levels of between 7 and 10 WL (1,400 - 2,000 pCi/l). These people should be addressed immediately. An EPA official recently told me, "Time is running out for these families." This is a pretty extreme statement for an EPA official to make. And he wasn't kidding. At these levels, the families have suffered extreme exposures.

PAR is there for everybody. In New Jersey there is a need to go after assistance for the people with high levels. There is a potential for 200,000 homes on the Reading Prong in New Jersey to be affected, and the problem is not confined to this area. In the future you may find yourself in the same position we are in.

Pennsylvania has a loan program which, if the governor ever gets in a position to sign it, will help those who can not pay an additional amount beyond their current payments. They shouldn't have to make that choice.

Public awareness is important. Support people who have this problem; it is a very stressful problem.

---

*Kathy Varady is Vice President of People Against Radon (PAR).

NATURAL VERSUS CAUSED RADON

Kay Jones** (Discussant): One additional comment. Regarding the EPA issuance of guidelines, they knew what they should be. They already have guidelines for mill tailings and man-made conditions. In our opinion, there is no difference between man-made and naturally-occurring radon. When that radon gets into your home, it doesn't care how it got there, it is going after your lungs. If there is a home in Grand Junction, Colorado that is being cleaned up at a 0.0187 WL (3.74 pCi/l), how can anyone tell us that we don't have a problem at 0.02 WL (4 pCi/l)? In August, 1985 we were told in Washington that guidelines would be easy to put out. They said they would have it in three weeks. Well, now it's May, 1986, and we still don't have them. It just goes on and on.

We were appalled that as much as $800,000 can be spent on man-made radon. I don't want to slight the people in Montclair, New Jersey -- they have a serious radon problem. But some of the levels that we are encountering are 10 to 100 times those levels. It is ludicrous for the EPA to tell us that we have a lot of guts asking them to come in and clean up our homes because our radon is naturally occurring, but then they stand here and admit that they are spending $800,000 in another area.

We must get the help from our representatives. It is only by us acting that they are going to act for us.

THE LACK OF A FEDERAL RADON STANDARD

Question: Would you comment on the lack of a federal radon standard?

Senator O'Pake: That you are concerned about Washington's failure to act on a timely basis in setting standards is interesting. I think the problem is that nobody knows what the standards should be. What should be set? What is safe? What isn't safe? You know you can't create that knowledge just by snapping your fingers. They can't do, in two months, what it may take a couple of years to establish, or maybe even longer. They've only had a year and a half.

On the other hand, if they were not prepared to set standards in August (1985), why did they tell us they were prepared to do it in the next few weeks. These people (in the high radon area) are living with that kind of concern and frustration and anxiety. A high ranking official in the Environmental Protection Agency told those people at the end of August that they had worked this out and were going to announce it in three weeks at that time. Either they had it and are unwilling to do it, or they were lying to these people and have persisted in that since August, 1985. I

**Kay Jones is President of People Against Radon (PAR).

think we are entitled to some action, some results. When you say, "Nobody knows," I'm not a technical person and I don't know who knows what. But certainly these are supposed to be the best brains that we have in radon in this country, and somebody better come up with some definition pretty soon, or what are we all doing here? I think some people know, and my question is, why aren't they sharing that information, and why aren't they doing something? I think they aren't sharing information because they are not prepared to follow through and make the kind of commitment that is needed in light of the seriousness of the problem. But that is only my opinion.

I can understand the problem at the federal level, because it was difficult enough to get my fellow senators and my fellow legislators and the governor and the administration of Pennsylvania to understand what we are talking about. Unless you live there, unless somebody comes and says, "We have a 14 WL reading in your home, or we have a 7 or we have a 10," it is really somebody else's problem, it's not your problem yet. And that is the mentality. And that's the problem that we have got to overcome. And it is an effort. Believe me, it is not going to be easy, but it is worth it.

WAS THERE A RADON COVERUP IN PENNSYLVANIA?

H. Sachs: My name is Harvey Sachs. I did the original interpretation of the Pennsylvania Power & Light data, and a fair amount of study after that. Many of us tried very hard to help the state. We were very, very frustrated. Do you have any sense of how bad a job DER did from 1979 to 1985?

Senator O'Pake: That's a loaded question, but let me try to respond this way. Until December of 1984, I don't think there were more than a handful of people in Pennsylvania who knew there was any kind of cause for concern about radon in that area. Ironically, if Mr. Watras had not worked in a nuclear generating plant which had the equipment to detect radiation, we might not yet know about this. Now, you point the finger at DER. DER will have to respond to that. All I can say is that we in the elected realm of government were not aware that radon was that kind of problem in our area. Believe me, had we known that, something would have been done sooner. Why don't you direct that to the man from DER?

H. Sachs: Mr. Gaertner, I'm delighted to be here with you today, and I'm particularly delighted that you have come into this work recently, and have been working so actively to help, because I've got nothing but bad news to say about the state DER in Pennsylvania. I want to review very, very briefly, the history of the Department of Environmental Resources' inaction for a period of five years. The inaction has lead to excess cancer deaths which are probably ranging from 50 to 100 per year. The excuse that has been given is that Three Mile Island diverted all of the resources of the DER. Let it be stated unequivocally that the best estimates that have ever been given of the excess cancer deaths total over the next forty years from Three Mile Island are either zero or one. The real alarm of significance was never heard. It was sounded by the DER's own geologist, Robert Smith after he returned in 1979 from the Reading Prong where he had discovered very high levels of background radiation. He communicated these findings to the health physicists in the DER. These findings were buried and no action was ever taken. Incidentally, I am grateful for an article by Steve Drachler in the Allentown Call a few months ago.

In 1981, Pennsylvania Power & Light and our group in Princeton independently presented to DER the results of an initial study north of the Reading Prong indicating more severe problems with radon than had ever been seen in the United States under natural conditions. There was no action. In 1982 and 1983, I sent no fewer than three letters, which I have with me today, to DER, outlining the problem and calling for action by DER. I suggested ways of using Terradex cups provided by the U.S EPA with which a research program could be conducted at no cost to the state. At my own expense, I went to Harrisburg to discuss this with the DER and present my findings. I am pleased to have done so. I am somewhat disappointed that, to this day, I have never heard from them. On the other hand, the homeowners who were contacted by the geologists were told, "Your radiation problem is so severe that we are sure that DER will get back to you immediately." This was in 1979; they never heard from DER either. In 1983, Health Physics, the most respected journal in this field, published an entire volume on indoor radon as a problem.* That's important, because after the Watras' house was discovered, one of the responses from the state DER was, "How could we have known anything about it? There has never been anything published." I would only point out that a member of their professional staff is on the board of editors of this journal. There has been a little bit of inconsistency.

There are also statements that there had been no work done. I would like to list quickly a few things. Beyond the 1983 Health Physics, there was a request for proposal from the U.S EPA in 1984, which generated a great deal of response and summarized much of the knowledge about radon. Also in 1984, the Air Pollution Control Association had a symposium on radon mitigation. In 1980-82, Pennsylvania Power & Light and Princeton University were active in eastern Pennsylvania studying radon distributions and looking at the problems. From 1982 to 1984, the National Indoor Environmental Institute, with which I was associated after Princeton, also carried out surveys in the area. From

*Health Physics. Vol. 45, No. 2, August, 1983.

1983 on, the Argonne National Laboratories studied the health physics of radon distribution in eastern Pennsylvania. Both at Princeton and Argonne, readings up to 200 pCi/l were discovered at that time. In 1984, the University of Pittsburgh began its work in Cumberland County. In 1985, a significant study was undertaken of approximately 100 houses by Channel 16 in Scranton. This is the base of knowledge which the state claimed did not exist. I want to state, unequivocally, that I am disappointed in the previous actions by DER, and I believe that they have been unprofessional. What was not stated, even today, is that 13 percent of the houses in the study of 11,000 houses have concentrations greater than 0.1 WL (20 pCi/l). If you extrapolate the ratio of exposure in these houses to account for the fact that people tend to be in their houses much more than on the job, you are saying that something like 13 percent of these houses have exposures greater than would be allowed in monitoring a uranium mine. Thirteen percent! That is a pretty high number, a shockingly high number. And they haven't even surveyed some of the high valued areas as yet. Rocks related to those that are a problem in Clinton run through the north side of Easton and Allentown. We pointed out the problems there in 1982. It is my understanding from members of Pennsylvanians Against Radon that the state initially was unwilling to tell people anything at all about what their figures meant. When asked, "Should we leave our houses?", all that the state would say was, "We can't tell you that; we can't help you understand that. You can live there because there is no regulation." Under the mine safety standards, if you employed somebody, and this were a mine, that person couldn't work more than 2 weeks a year. In the face of this evidence, the DER responded, "We can't tell you whether you should leave or not, we can't help you understand your risks."

I am pleased that the State of Pennsylvania seems to be trying very, very hard. I am pleased that their programs have finally started. But let us not forget that a lot of people, statistically, have died because there were six years of inaction. I was not the only one who tried; many others tried, but only Stanley Watras' alarm was heard. Thank you.

J. Gaertner: I deal with the public every day. I have faith in the public, assuming most of you are here as private homeowners, private citizens. I was up here a few minutes ago as a professional representative of the State Department of Environmental Resources, for which I am proud to work for. I am no less proud after being in this radon program for several months. I have also worked for the U.S. Environmental Protection Agency, the State of New Jersey, and the State of California.

Rather than respond to some of the comments that were made, I would like to ask a few questions, and I'm in trouble up here if I can't get a little bit of assistance.

They are very simple questions. Does anyone out there recall my saying that many things in life do not go as quickly as we would like them to? Does anyone recall my saying that? I believe I also said, extremely strongly and pointedly, that there are many disputes about many things, including radon. And there is uncertainty with radon, probably a lot more than with other situations, such as asbestos that you can see, you can pick it up in your hands or water contamination, you can pick up a jar of chemicals. I just want to double check, if I could get a hand, does anyone out there recall my saying that many things are difficult to pin down? "It is difficult to come up with definite answers," does anyone remember that? One further question, and then I'll make a compliment. Does anyone remember, when was the date or month that the Stanley Watras' situation was discovered publicly, or governmentally, or what have you? And that discovery is why we are here today. Does anyone recall that date? December, 1984. Before the very day when that happened, in the state budget of Pennsylvania, there was a proposal to study radon. Before that date. Why? Because of the concerns expressed by Dr. Sachs; because of the concerns expressed in the scientific community, it was in the budget before the Watras' home was discovered.

The New York Response

NEW YORK STATE DEPARTMENT OF HEALTH RADON PROGRAM

William Condon*

The number of radon measurements in homes in New York State has been rapidly increasing as the interest in the question of indoor radon has grown. Several early surveys showed that a measurable fraction of the homes had indoor levels above a few picoCuries per liter, and there was no obvious relationship to location or house construction.

With the emergence of the radon problem in Pennsylvania and New Jersey, interest in New York State has also increased. The New York State Energy Research and Development Authority (NYSERDA) is presently doing a year-long study of radon levels in about 2,300 homes across New York State. As a part of this study, scientists from the Department of Health are doing measurements of surficial geological characteristics in an effort to understand how this influences radon levels in homes and the usefulness of such data in predicting whether homes in a specific location will have elevated radon levels.

Several county health departments have also begun making radon measurements in homes, particularly in areas impacted by the Reading Prong formation. Limited data to date indicates that the fraction of homes in New York State above some arbitrarily chosen radon level is less than in Pennsylvania. However, this fraction may vary from one area of the state to another as it does everywhere. Upon completion of the NYSERDA study we will have a much better feeling for the distribution of indoor radon levels, and also which areas of the state may have the largest fraction of homes with elevated levels.

Several new programs will be starting shortly. Two are being funded by the U.S. EPA and one by the state. The EPA funded programs include:

1. A further study of 1,000 homes in areas where radon levels are expected to be highest.

2. Location of two areas where remedial action can be done on 8 homes per area to demonstrate techniques for remediation and to train contractors in these methods.

---

*William Condon is with the Radon Program, New York State Department of Health, Albany, N.Y.

The state funding extends over a three-year period and will include: (Laws of 1986)

1. Making radon detectors available to homeowners at reasonable cost.

2. Provisions for information via a radon hotline.

3. Providing training in proper radon diagnosis and mitigation methods.

4. Develop criteria for dealing with the radon problem and develop appropriate recommendations to the Governor for dealing with the various aspects of this problem.

5. Investigate areas of the State where elevated levels of radon might be expected.

6. Assess the impact of indoor radon levels on energy conservation efforts.

NEW YORK STATE ENERGY RESEARCH AND DEVELOPMENT AUTHORITY RADON PROGRAM

Joseph E. Rizzuto*

INTRODUCTION

This paper presents the New York State Energy Research and Development Authority's (Energy Authority) radon program. Both past and present efforts as well as plans for future programs will be described.

The Energy Authority's radon program is part of a larger effort on indoor air quality. The Energy Authority is a public benefit corporation established to foster new energy production and conservation technologies. The Energy Authority's involvement in indoor air quality is based on a concern that State conservation efforts might have a negative effect on indoor air quality. Its strategy is to implement conservation in a manner consistent with acceptable indoor air quality.

The Energy Authority has been involved in indoor air quality for approximately five years. It has conducted its programs in cooperation with several other State agencies including the Department of Health, Energy Office, Department of Public Service and the Department of State. The Department of Health has direct responsibility for all health-related aspects of indoor air quality. The energy Office, the Department of Public Service and the Department of State are responsible for the management of major conservation programs involving the weatherization of homes. The Energy Authority also has conducted its indoor air quality programs in conjunction with the State's electric and gas utilities which conduct home conservation programs in accordance with the State's Home Insulation and Energy Conservation Act.

COMPLETED PROJECTS

The Energy Authority's first indoor air quality project was conducted in cooperation with Lawrence Berkeley Laboratory and the Rochester Gas and Electric Corporation. The purpose of the project was to explore the relationship between infiltration, construction practices and indoor air

*Joseph Rizzuto, an engineer, directs the radon research effort for the New York State Energy Research and Development Authority (NYSERDA).

quality. The performance of air-to-air heat exchangers in mitigating high concentrations of indoor air pollutants also was examined.

As part of this project, the radon concentrations in nine tight Rochester homes with air infiltration rates between 0.2 and 0.5 air changes per hour were monitored continuously for periods of one week. The results of the monitoring indicated all nine homes had radon concentrations of less than four pCi/l. Although the levels found were not considered very high, air-to-air heat exchangers were installed for evaluation purposes and found to be an effective means of reducing radon levels in tight homes. With regard to radon and several other indoor air pollutants that were monitored, it was found that if source strengths are low, air infiltration can be significantly reduced while maintaining acceptable indoor air quality.

New York has not yet adopted a state position on acceptable radon levels for homes or established recommendations on remedial action guidelines. The approach that the Energy Authority is utilizing in conjunction with its radon monitoring programs is to inform homeowners of the radon levels measured in their homes and also to provide information on the guidelines established by both the United States Environmental Protection Agency (EPA) and the National Council on Radiation Protection and Measurements (NCRP). These guidelines are as follows:

1. Environmental Protection Agency Guideline: 0.02 Working Levels = 4 pCi/l.

2. National Council on Radiation Protection and Measurements Guideline: 0.04 Working Levels = 8 pCi/l.

A second project on indoor air quality was conducted by the Energy Authority in cooperation with the Niagara Mohawk Power Corporation. The project examined the indoor air quality in 61 homes in the Niagara Mohawk franchise territory. The homes were divided into two construction categories: 34 of the homes utilized special energy efficiency construction techniques; the remaining 27 homes were of conventional construction. Approximately half of the homes were selected based on having known indoor pollutant sources such as kerosene heaters, wood stoves or fireplaces, gas cooking or smokers. The other half of the homes had no known pollutant sources. Most of these homes were total electric homes. No preselection of homes was made with regard to geology or other external factors that could affect radon source strength. Thus, the potential for a high radon source strength was essentially a random choice.

The radon levels in each of the 61 homes were monitored for one or more periods of at least two months each. The radon concentration levels for these homes are shown in Table 8.3. The values for both the basement and first floor are provided. The number of homes in each category falls short of the total sample size for two reasons: 1) All houses did not have basements, and 2) in some cases the monitors were lost. The results of the monitoring indicated that a total of 14 homes had basement radon levels exceeding 4 pCi/l and eight of these homes exceeded this concentration on the first floor. Furthermore, the distribution of radon concentrations in energy-efficient homes appeared similar to that in conventional homes. This was not unanticipated since radon source strength can vary by three or four orders of magnitude while energy efficient construction can reduce air exchange rates by only one order of magnitude. This finding does not imply that air exchange rate does not affect indoor radon concentration since, for any house with a given source strength, reducing the air exchange rate will increase the radon concentration, and vice versa.

In the 14 houses in which the radon concentration in basements exceeded 4 pCi/l, a demonstration of mitigation techniques was performed. The techniques utilized included:

1. sealing cracks in basement floor and walls, perimeter joint and the top of concrete block walls;

2. venting sump and subslab areas, and unpaved crawl spaces; and

3. using air-to-air heat exchangers.

The results of the radon mitigation demonstration were that all radon concentrations were reduced to below 4 pCi/l. The cost of the radon mitigation, including labor and materials, ranged from $15 to $1,245 per house. The average cost per house was $640.

CURRENT STATEWIDE PROGRAM

In December, 1984, the Energy Authority initiated a major Statewide investigation of infiltration and indoor air quality in New York State homes. The cost of the effort is approximately $1,600,000 and is co-sponsored by the Energy Authority, New York State electric and gas utilities, the U.S. EPA, the Gas Research Institute, and the Electric Power Research Institute. The overall program has three primary tasks:

1. to assess the impact of conservation measures including caulking and weatherstripping of windows and doors and the use of storm windows and doors on house leakage and air exchange rates;

Table 8.3

Radon Concentration Levels Found in Niagara Mohawk Territory

	Number of Homes					
	Basement			First Floor		
Range pCi/l	Energy Efficient	Typical	Total	Energy Efficient	Typical	Total
0-2	10	13	23	27	20	47
2-4	8	4	12	0	2	2
4-8	5	0	5	3	2	5
8-16	2	1	3	1	2	3
16	1	5	6	0	0	0
	26	23	49	31	26	57

Energy-efficient homes: 34
Typical homes: 27
Total homes: 61

Table 8.4

Radon Survey of Four Geologic Areas

County	Geology	Mean RA-226 In Soil (pCi/g)	Mean RN-222 At 2 ft.(pCi/l)	Indoor Radon 13-15 Homes pCi/l Basement	Indoor Radon 13-15 Homes pCi/l First Fl.
Onondaga	Black Carbonaceous Shale (Marcellus)	2.6	2700	0.9-23	1.4-19.3
Erie	Black Shale	2.4	1060	0.6-12.3	0.3-8.4
Orange	Metamorphosed	1.3	450	1.4-29.7	0.1-11.8
Long Island (Nassau/ Suffolk)	Sandy Deposits	0.5	160	0.4-4.4	0.2-1.4

2. to perform a Statewide radon study in New York State; and

3. to determine the impact of combustion sources and air infiltration rates on indoor air quality.

Task 2 of the program focuses on obtaining a better understanding of the radon problem in New York State. The specific objectives of the radon study are as follows:

1. to make a statistical determination of the extent to which radon concentrations constitute an indoor air quality problem in New York State;

2. to determine the extent to which home and site characteristics (age, construction characteristics, geographic location/geology) can be correlated to high radon concentrations;

3. to develop low-cost radon mitigation techniques; and

4. to develop public information on radon risk and remedial action.

The first and major subtask of the radon effort is to conduct a Statewide radon survey of New York State homes. The purpose of this subtask is to determine the statistical distribution of indoor radon concentrations in New York State homes.

Four constraints are placed on the survey population. The homes must be:

1. year-round residences;
2. a single-family unit;
3. owner occupied; and
4. non-moving for the next 12 months.

A probability sample of 3,100 homes was selected for participation in the survey. All homes conforming to the above four constraints were supposed to have an equal chance of participation in the survey. The houses were selected by random-digit telephone dialing (hence a minor additional constraint was that the home must have a telephone although having it listed is not a requirement).

The survey population was divided into seven strata on a geological/geographical basis. A pilot survey of 50 homes was conducted on Staten Island to test the experimental protocols and to get a preliminary handle on the response rates. Based on this pilot survey, it was determined that it would be necessary to enlist approximately 3,100 homeowners in order to obtain the desired sample size of 2,000 measured homes.

Homeowners were interviewed with a telephone questionnaire upon initial contact. The purpose of the questionnaire was to determine:

1. eligibility;
2. house characteristics;
3. weatherization treatment;
4. type of heating system; and
5. basement construction characteristics.

Each of the 3,100 homeowners agreeing to participate in the radon survey typically was mailed three alpha-track monitors. For some of the homes a fourth monitor was provided for quality assurance purposes.

The three alpha-track monitors were to be utilized as follows:

- One winter living space monitor; two months exposure - 1 (pCi/l)-month sensitivity;

- One annual living space monitor; twelve months exposure - 4 (pCi/l)-month sensitivity;

- One annual basement monitor; twelve months exposure - 4 (pCi/l)-month sensitivity.

Additional monitors were provided for quality assurance purposes. These monitors consisted of both duplicates and blanks. A total of 453 duplicate monitors were provided. These were divided approximately equally between the above three monitored parameters. A total of 453 blanks also were provided. These blanks were subjected to the same handling and mailing procedures but were not exposed in the home; hence, they should indicate no radon exposure.

A third quality assurance provision was to perform site visits on 154 homes. The purpose of the visits is to ensure that homeowners responded accurately to the radon questionnaire and properly deployed the radon monitors.

The second subtask in the Energy Authority's radon effort is to determine the extent to which home and site characteristics (age, construction characteristics, geographic location/geology) can be correlated to high radon concentrations. Four areas of the State were selected for study. Three of the areas were selected because, based on geology, the areas were believed to have the potential for high radon concentrations. The three areas selected on this basis were in Onondaga, Erie and Orange counties. A fourth area in Nassau and Suffolk counties was chosen for comparison because it was believed to have little potential for high radon concentrations.

In each of the areas, radium content of the soil and radon content of the soil gas were measured in addition to taking radon measurements in fifteen homes. Data obtained

Table 8.5

Radon Concentration Levels
Onondaga County Marcellus Shale Area

Range pCi/1	Basement (2 Month Ave)*	Number of Houses First Floor 2 Month Average*
0-2	1	2
2-4	3	7
4-8	5	1
8-12	1	2
12-20	-	1
20-30	3	-
30-40	-	-
	13	13

*January-March 1986

thus far are shown in Table 8.4. The data indicate a somewhat greater potential for high radon concentrations in Onondaga County than in the other counties. This potential appears to be related to the high radium and radon content of the black shales, and particularly of the Marcellus shale.

The radon concentrations in thirteen homes located in the Onondaga County Marcellus shale area are shown in Table 8.5. Eight of the thirteen homes had basement radon concentrations exceeding 4 pCi/l. Four of the homes exceeded this concentration at the first floor level. The results obtained thus far on this program are very limited. A complete report of project results is expected to be published by June of 1987.

FUTURE PROGRAMS

Several additional radon-related programs to be conducted with the U.S. EPA are now in the planning stage. These include:

1. obtaining additional radon measurements in the Reading Prong area of New York State and other areas of the State where survey results show homes with high radon concentrations to be located;

2. demonstration of radon mitigation techniques for existing homes;

3. demonstration of radon-resistant construction techniques for new homes; and

4. development of a training program for radon mitigation.

The first three tasks are to be conducted in cooperation with the New York State Department of Health. The fourth task will be performed by the New York State Energy Office.

NEW YORK STATE ENERGY OFFICE RADON UPDATE

John Paul Reese*

In 1982 the New York State Energy Office initiated a program entitled "Building and Remodeling for Energy Efficiency." This series of two-day workshops was designed to address critical issues in energy-efficient construction and remodeling. Because of a focus on tightening homes and providing adequate ventilation, indoor air quality was a focus from the program's inception.

One of the critical indoor air quality issues addressed was radon. Radon is a colorless, odorless, tasteless gas produced by the natural decay of uranium and radium in the earth. Radon can enter a house through a variety of entry routes and is thought by many researchers to be responsible for causing a significant proportion of all lung cancers. Information on radon entry and means of reducing it was presented in the workshop series. With the discovery of several homes in Pennsylvania where radon had obtained alarming levels, awareness of radon as a national health issue began to grow.

In the fall of 1985, the U.S. Environmental Protection Agency (EPA) expressed interest in providing technical assistance and training to the public and private sector because of a pressing need.

In early 1986 thousands of homeowners in New Jersey, New York and Pennsylvania began to receive results of government-sponsored radon testing. As homes with elevated levels were identified, it became apparent that these individuals needed more information and assistance. It was essential that a support network exist for those homeowners whose homes had high radon concentration levels and needed to take expeditious action. The necessary support network was not in place.

The need was to create radon diagnosticians, people who could determine the level and sources of threat posed by radon in a structure and who recommend and direct appropriate mitigation methods. Diagnosticians also work with the contractor to ensure quality control and evaluate the success of mitigation efforts.

The State Energy Office submitted a proposal to help develop a support network by providing a series of training and remodeling sessions which would expand upon the radon

*Radon Project Director, New York State Energy Office.

portion of the Building for Energy Efficiency workshop. A $200,000 grant was awarded to SEO to develop and conduct a series of 20 three-day workshops on radon, its health effects, the types of radon monitoring devices available and their proper application, how to identify radon entry points into homes and how to effectively reduce radon levels. The program was to be a model which could be passed on to other states.

The program as designed brought together for the first time all of the current research on radon. Information was drawn from around the country and around the world. Continually, through the development and conduct of the workshops, materials were updated to include new information.

The workshops were developed and conducted by the State Energy Office with the assistance of Solaplexus and Camroden Associates, which are building research design and education firms. The workshop uses a combination of slides, videotapes and hands-on demonstration.

The persons trained by the Energy Office fell into two categories:

> Public-sector representatives, including public decision-makers, local building inspectors and individuals who respond directly to public requests for information and guidance. These typically are the first persons contacted by homeowners seeking help with radon problems, and it is essential that they have the necessary information to respond quickly and accurately.
>
> Private-sector "diagnosticians." Drawn from a number of sources including private-sector building inspection services. They included professional engineers, architects and other individuals with a firm understanding of the structural components and operation of residential buildings. Their role will be to actually provide diagnostic services.

The series started off in Washington, D.C. in May, 1986. Since the first workshop, over 600 individuals have been trained in the 15 workshops in New York, New Jersey, Pennsylvania and the additional 5 workshops in Atlanta, Denver, Boston and Washington. The audience included staff of the U.S. Environmental Protection Agency, the U.S. Department of Housing and Urban Development and U.S. Department of Energy as well as the public health officials from over 20 states. The primary audience, however, were the private sector representatives, the engineers, architects and building professionals who will actually be radon diagnosticians.

As a model program, the ultimate goal of the program is to transfer these materials and teaching expertise to other states so that they may carry out similar types of technology transfer programs to create radon diagnosticians. In December, the three-day workshops will be professionally video taped and distributed to all 50 states to assist in their adoption of the program. Due to the success of the program, EPA is considering a grant extension to conduct additional workshops for the 1987 federal fiscal year.

The SEO/EPA radon program has helped the state of New York to be in the forefront in addressing this issue. Because of the concern over radon levels in New York State, the Governor and the Legislature passed legislation as part of the allocation of Exxon petroleum overcharge dollars, which will allocate $2.25 million dollars for a joint Energy Office, Department of Health testing and education program. The Energy Office will have the lead responsibility in providing technical assistance and training to targeted audiences as well as assisting in providing information to the general public. The Department of Health will have the lead in conducting a radon testing program for individuals involved in energy conservation programs.

Additionally, during the last year, the Energy Research and Development Authority and Department of Health have also been actively investigating the scope of the radon problem in New York with:

A statewide radon survey of 2,300 homes;

An investigation of soil factors that affect radon levels;

Research into the impact of air infiltration reduction conservation measures on radon levels;

An assessment of the effectiveness of transferring information to residents using different presentation styles; and

Two new projects to survey additional homes in areas of New York with potentially high radon risks and a small scale demonstration of radon mitigative techniques in houses.

The state's radon activities in 1986 are among the most extensive in the country, but there is still more that needs to be done before the true scope of the problem within the state is known and can be adequately addressed.

A GRASS ROOTS MODEL FOR RADON RESPONSE

Michael R. Edelstein and Liana Hoodes*

INTRODUCTION

It is quite likely that few people in the Mid-Hudson region of New York state had thought about radon gas before Spring, 1985. While national attention had highlighted the radon problems suffered by residents of eastern Pennsylvania, little connection had been made between their misfortunes and the Mid-Hudson region. On Sunday, May 19, a story in The New York Times apprised people of radioactive gases affecting Pennsylvania, New Jersey and New York "west of Suffern and Peekskill."[1] Subsequently, on Sunday, June 1, the major county-wide newspaper again alerted people to the geological connection between Pennsylvania and New York.[2] This connection, a granite-gneiss formation sometimes called "the Reading Prong," quickly became the focal point for local concerns over radon.

While there is no firm measure for how much public concern was aroused by media attention, in the Town of Warwick, in southern Orange County, the Conservation Board received numerous calls from worried homeowners and several realtors reported stigma due to the association of Warwick with the Reading Prong. It is probable that many residents, having been alerted to a possible threat, were actively seeking additional information in order to guide their response.

When the Mayor of Tuxedo, New York sought to clarify the threat to his town by holding a public forum on the radon threat in July, 1985, the first public clarification of the local threat occurred. He invited representatives of the New York State Department of Health (DOH) and New York University to speak. Based upon first hand observation and media coverage, it was evident that this panel sought to minimize public concern about the issue. While little actual incidence data was reported, the impression was given that area sampling showed very low levels of radon and that radon itself was not particularly dangerous. Slides were

*Michael Edelstein is President of Orange Environment, Inc. as well as Associate professor of psychology at Ramapo College and conference co-organizer. Liana Hoodes is Vice President of Orange Environment, Inc. She coordinated the Orange Environment Radon Program during 1985-86.

even shown of attractive Austrian women in a spa sought out for its high radon level. Overall, the panel discounted the threat of high radon concentrations in the region and likened the radon risk to accepted personal risk. The Mayor was able to conclude the meeting with reference to the potential for land development in his town, implying that the biggest threat from radon in his community would be a decrease in land values from unwarranted panic about a non-existent radon problem.

The regional newspaper gave this meeting thorough coverage. A resulting article was headlined: "Scientists Discount Radon Risk." A state health official was quoted as saying,[3]

> ...it does not look like this area has a big problem...It's doubtful that we have found anything that would require further testing...The samples are terribly dull from a scientific point of view.

Additionally, a New York University professor was paraphrased as saying of radon that

> ... as a cancer risk factor, researchers believe it causes disease in less than one percent of the cases. Cigarette smoking is the largest factor for lung cancer.

The researcher was then quoted as saying,

> What we are really talking about here is a very small risk.

Subsequent articles repeated these themes. For example, in a piece entitled, "Scientists Skeptical About Radon Threat," the reporter led the story with this parpagraph:[4]

> Although no homes in the region been (sic) sampled, experts are beginning to discount the possibility that many are threatened by radon contamination in southeastern Orange County.

As these quotes suggest, the public's concerns about radon were mollified by two distorted propositions. The first asserted that the threat in Orange County was minimal. The basis for this assertion was a preliminary soil sampling survey. While the state health department expert admitted that, "We don't know everything there is to know about radon or there would be nothing to study,"[5] the message was clearly conveyed that radon was not a major problem.

The second proposition was equally distorted. A risk of less than one in a hundred was discounted as minimal, a specious argument at best considering that risks from toxic substances are considered to be serious even if smaller by a factor of ten thousand. Secondly, the greater risk from smoking was used to discount the radon risk. Such risk comparisons are by nature misleading. In this case, it is particularly misleading because smoking is associated with so dramatically high a risk. Thus, the radon risk may be extremely high in comparison with other life risks even if it is less than the risk from smoking.

In short, the public was misled by the Tuxedo meeting into believing that radon is a relatively minor problem. This view was further reinforced when shortly thereafter the Orange County Commissioner of Health wrote a letter to the editor in which he reiterated the Tuxedo conclusions. Additionally, to the extent the problem was recognized, it was clearly associated with one particular geographic area, the Reading Prong.

The effects of the Tuxedo meeting with its ample media coverage appeared to quiet much of the initial radon concern in Orange county. For Board members of a local non-profit, tax-exempt organization, Orange Environment, Inc., the meeting had an opposite effect. Several of the Board members sit on the Town of Warwick Conservation Board. Because of the focus upon Warwick in much of the early publicity about the Reading Prong, Conservation Board members had devoted considerable time and effort to investigating the issue and providing information to the public. Based upon their research, they neither agreed with the risk evaluation given at the Tuxedo meeting nor were they satisfied with the state Health Department's predictions that radon was not a widespread issue. While the Conservation Board received many requests from homeowners to assist them in conducting radon tests, the Warwick Town Board did not support their continuing efforts because addressing a health threat was seen as outside the role of local government. As a result, the Conservation Board turned to Orange Environment, Inc. to address the need for a local program of testing. After the Tuxedo meeting, Orange Environment became particularly concerned that a testing program go beyond meeting the needs of concerned citizens. Such a program would also need to test the state health department's assertions that area residents were not likely to be exposed to significant levels of radon. If this conclusion was not supported, then Orange Environment would have a body of data to use to convince the state to take the radon question seriously.

It may be helpful to briefly describe Orange Environment, Inc. (OE), a non-profit, tax-exempt organization. The organization was formed in 1983 by local environmental activists and professionals who were concerned that research on environmental issues serve as the basis for policy decisions, that compliance with environmental regulations be enforced and that victims of environmental hazards be assisted in their dealings with government agencies. The corporation is also deeply concerned with the impacts of growth dynamics on agriculture and naturally sensitive areas and with the social as well as environmental impacts of projects. Orange Environment contains a public interest law center and a land conservancy. Although it is moving toward becoming a membership organization, OE has relied chiefly on some 15 members of its Board of Directors to carry out the organization's activities.

The Board of OE was ideally suited to the radon gas project. Three members had several months experience working in Warwick on the issue. Expertise in nuclear physics, earth science, environmental law, environmental toxicology, and environmental and community psychology was available within the board. One officer of the board volunteered to coordinate the radon program and to serve as a one-person hotline and information service while another board member served as the project's technical consultant. Numerous members helped in various ways to execute the program.

THE DESIGN OF THE PROGRAM

In fall, 1985, OE began to design its radon response program. Three goals were highlighted in the planning. First, the program sought to provide accurate information to the public despite what was seen as a reticence on the part of New York State to address the issue. Second, the program needed to collect sufficient data to provide a basis for assessing the extent of the problem. Finally, the program sought to serve regional residents who wanted to acquire reasonably priced radon tests and to have assistance in interpreting and acting on the results.

The DOH approach in downplaying the issue appeared to be dominated by fear of public panic. In contrast, OE operated on a presumption of informed consent; only where citizens had accurate information could they reasonably judge what their response to an issue should be. Accordingly, we believed that the threat due to the radon should not be suppressed because residents would not test for radon if they had no reason to be concerned. At the same time, concern could be channeled into vigilance rather than panic if residents felt that they were receiving open information and if the opportunity to do something about the threat was available to them.

A variety of considerations led to the specific design of the program. These were:

(1) Cost and Participation -- OE began the radon program with virtually no resources. As a result, we were not in a position to provide services at less than the real cost. At the same time, we were aware that testing can be expensive enough to discourage residents from doing measurements of their homes. Fortunately, the Terradex Corporation offered a bulk rate for purchase of its Track Etch Detector. As a result, we were able to offer the Track Etch detector for half price while including a small

handling charge to defray our expenses. In this way, the resident would bear the costs of the program while benefitting from a substantial savings by participation.

Deriving the program funding from the participants was more than a utilitarian arrangement. We felt that the active involvement of participants in the testing process was desirable. Beyond their paying for the tests, residents were trained in the installation of the detection units at our pre-installation seminars. The resident thus made key decisions and was directly involved in installing the detector.

(2) Choice of Testing Device -- From a technical standpoint, we opted to utilize the alpha track detector because of the superior ability of this testing device for averaging across radon conditions over time. The other primary device for resident-controlled home monitoring is the charcoal canister. While even less expensive to use, the canister provides a reading which averages over less than a one-week time frame. This technique may lead to inaccurate home radon evaluations because radon gas levels in the home have been found to vary dramatically according to season and weather conditions. The alpha track detector is normally placed in location for a minimum period of at least a month, and its accuracy increases greatly as it reaches and surpasses three months in situ.

We planned for a three-month test period to commence in November, 1985, timed to correlate with the colder months when homes are maximally closed. We decided to recommend that homes be tested with two units if basement spaces existed. In this way, the basement detector would measure radon gas at the point where it was most likely to be at its peak while another unit located in an upstairs location would provide for a measure of living space exposure. The testing plan developed proved to be consistent with Environmental Protection Agency protocols issued after the OE testing program commenced.

(3) Confidentiality and legal responsibility -- Given the early instances of stigma associated with radon exposure, as reported to us by realtors and others, we were concerned that adequate protection of participant identity be provided. Furthermore, it became clear that some potential participants would become reticent if they felt that their identities or location might be released. This was of particular concern because the regional newspaper had filed a Freedom of Information Act request seeking the identity of a Warwick resident whose home had been identified as having a high level of contamination. The paper subsequently threatened to sue the Town of Warwick for denying the request. As a private, non-profit organization, we learned that OE could avoid such challenges. The greatest confidentiality issue arose with regard to our desire to publicly release a map which correlated geololgy and soil conditions with radon levels. Because too great a

specificity would identify individual homes, the mapping was delayed in favor of a simple summary of findings by Town.

Clear statements of agreement were written for use as a contract with participants. These alerted the participant to our research aims, clearly indicated the limits of confidentiality and also provided limits to our accountability for errors by Terradex. Participant signatures were required prior to the ordering of testing devices.

(4) Research -- The basic research objectives of the project were straightforward. Because we were confined to a convenience sample, consisting of people who volunteered to participate, we knew that it would not be possible to construct a carefully controlled study. Simply, we wanted to assess the extent to which homes in the Orange County area would evidence levels of radon in excess of the generally accepted guideline of 4 pCi/l. If higher radon levels were in fact, evident, then we would use our data to pressure government to respond to the threat. A second concern was to test the assumption, firmly entrenched from the onset by media coverage of the issue, that radon exposure was associated with and limited to the Reading Prong. We sought a sufficiently large sample to allow for at least a crude test of both of these questions.

An additional research effort was also undertaken in conjunction with the OE program by board members of the organization. This involves an ongoing study of the perception of radon gas and the impacts of radon exposure. All participants in the OE program were asked to fill out a questionnaire at the time that they picked up their testing devices. A follow-up questionnaire was sent several months later along with the results of the testing program. This study was intended to provide a reading of the level of stress and concern engendered by a radon testing program. A comparison group was also sampled which consisted of residents who had not tested for radon. It was felt that this data would help OE develop means of managing stress associated with radon testing programs while also providing valuable research data (see Gibbs, et al., Section 6).

(5) Publicity -- In order to attract participants to the study, OE contacted a list of people who had approached the Warwick Conservation Board about radon testing. A news release, additionally sent to all local media, received good coverage. This release included the phone number of the OE hotline, which immediately began to receive inquiries.

IMPLEMENTATION

From those sent radon testing agreements, OE received actual orders for 96 detectors to be placed in 56 houses. These were distributed in November, 1985 at the home of the OE radon coordinator. As people arrived to collect their detectors, they were given a brief installation seminar by OE's radon technical expert. Any questions were answered, and the risk perception questionnaire was distributed. A postcard was completed with the address of the participant for purposes of notification of the date by which the testing kits were to be returned.

During the three-month period in which these tests were conducted, OE continued to provide information to callers. In response to public demand, a second radon testing group was begun in February, 1986. This group involved 138 test kits used in 81 homes. Because of its late start, we did not initially intend to run this testing group for more than two months. However, a cold spring made it possible to extend the testing period to just short of three months.

A number of issues arose as we implemented the radon response program. First, frustrating delays occurred in the return of detectors. Because detectors from each of the testing groups had to be batched and returned for evaluation collectively, the program depended upon participant cooperation to return the detectors to us in a timely fashion. However, particularly in the second group, individuals within the group caused extensive delay, during which time care had to be taken to assure that the detectors already collected would not receive additional exposure. A major effort was required in tracking down the delinquent radon detectors. These delays were compounded by events at the testing lab. When we first planned the program, Terradex was returning measurements within three weeks. However, the national concern over radon caused the lab to more than double its turn-over time. As a result, the November group did not receive its findings until April and the February group until July.

We were careful with our management of the detectors. For a small non-profit group, the loss of $2,000 in detectors would have been catastrophic. Thus, all mailings were insured. In one case, in the first test group, however, a detector opened at the lab was found to be ruined. The lab claimed that the damage could not have stemmed from their handling; the participant made a similar claim. The result was that OE ended up bearing the cost of the detector.

The other major issue involved the development of a feedback system for participants. Because we hoped to manage the program to minimize stress, we wanted to be in the position to provide clear information about the meaning of the results. We were also aware that once residents had tested their homes, those finding elevated radon levels would want to know how to deal with the problem. It was not sufficient to say to them, "You are at risk. Goodbye."

We discussed at great length the type of feedback to be provided to homeowners. OE was now in the position that regulators find themselves in. We faced the difficult task of accurately characterizing findings in a way that was not unnecessarily alarming. We eventually settled upon a strategy that recognized that the task of setting standards belonged to government, as did the responsibility for the legitimacy of those standards. It was not our place to create such standards. Our role was to present available information clearly and to help answer questions as best as we could given the levels of uncertainty present with an issue such as radon gas.

Several OE Board members had attended the New York Governor's Conference on Radon held in Albany in March, 1986. At that conference, radon expert Anthony Nero had presented a proposal for the evaluation of risk for varying radon levels which related the level of radon to the needed response. The result was a surprisingly lucid and useful feedback tool that far surpassed the practice for presenting standards for toxic chemicals. Risk was expressed in an ordinal rather than nominal way. Residents were given a basis for judging their comparative risk and a means of weighing what level of action they should take. We were aware that the Nero proposal was similar to the format utilized in a draft EPA radon brochure for homeowners. However, the EPA experienced repeated delays in issuing the brochure, and despite repeated requests, we were never shown a draft of the brochure. As a result, we cited Nero in our own feedback letter to participants in our testing program.

Upon receipt from the laboratory, OE transferred the test results to a personalized letter for each participant in the testing program. The Nero format was given as a basis for judging the significance of results. We also alerted recipients to the additive nature of radon exposure, suggesting in a low-key manner that even the 4 pCi/l guideline carried risks, the acceptability of which they would have to evaluate.

We further decided that some type of public program was needed to answer questions. To present official responses about risk evaluation and public programs, it was important to involve the Department of Health in this program. Under New York State law, public health issues are clearly their responsibility. Furthermore, we wanted to address the potential desire of the audience to learn about radon mitigation. Consequently, a leading radon mitigator, Terry Brennan, was invited. Participants in both phases of the OE program were informed of the program. Public notice of the program was given, although the major county newspaper delayed publicity about the event until the morning of the meeting. Despite this, some 90 people attended the session held on the evening of April 30. The panel consisted of the head of the New York Department of Health radon program, OE's technical expert, an OE Board member expert in health risk assessment, and Terry Brennan. OE's president chaired the meeting. The meeting was opened with a brief overview of OE's first phase findings. More than an hour of questions were fielded from an active audience. Finally, Terry Brennan lectured and presented slides on radon mitigation, taking an additional set of questions. The meeting succeeded in addressing concerns by participants in the testing program in a thorough manner.

RESULTS OF THE TESTING PROGRAM

In all, 137 buildings were tested with 236 test kits. For this test group based largely in Orange County, New York, results in excess of 4 pCi/l guideline were found in 12 communities. Within this test group, 28 tests in 24 buildings were found to have levels between 4-10 pCi/l; 10 tests in 7 buildings were found to be in excess of 10 pCi/l with the highest test being 63 pCi/l.

We have broken results down by municipalities for the purpose of a crude comparison (Table 8.6). Note that in some cases, our reliance upon mailing addresses and zip codes may have caused us to inaccurately identify the town. In a few cases, villages have been delineated as well.

TABLE 8.6: BREAKDOWN OF COMMUNITIES WHERE TESTS EXCEEDED 4 pCi/l (Note: listing is for number of tests; the number of buildings is indicated in parentheses).

	<4 pCi/l	4-10 pCi/l	>10 pCi/l	Total
Warwick	64 (38)	11 (8)	4 (1)	79 (47)
Greenwood Lake	2 (2)	2 (2)	2 (2)	6 (4)
Goshen	4 (3)		2 (2)	6 (3)
Pine Island	2 (2)	1 (1)	1 (1)	4 (3)
Mahwah, NJ	12 (3)	1 (1)	1 (1)	14 (5)
Chester	5 (4)	2 (2)		7 (4)
Florida	5 (4)	4 (3)		9 (5)
Otisville	1 (1)	1 (1)		2 (1)
Monroe	45 (29)	3 (3)		48 (29)
Newburgh	2 (2)	1 (1)		3 (2)
Highland Falls		1 (1)		1 (1)
Harriman	3 (3)	1 (1)		4 (3)
Other	51 (30)			51 (30)
	196 (121)	28 (24)	10 (7)	236 (137)

Other communities where tests were completed had results which fell below the 4 pCi/l guideline. This, of course, does not mean that more comprehensive tests would not identify buildings with higher levels in these communities. The community is listed with the number of tests (and number of building): Tuxedo 6(4), Cornwall 4(2), Bloomingburg 2(1), Hewitt, N.J. 1(1), New Milford 2(1), Middletown 5(3), Campbell Hall 3(2), Highland Mills 9(5), West Milford, N.J. 2(1), New Windsor 2(1), Woodcliff Lake, N.J. 1(1), Wykoff, N.J. 4(2) and Irvington, N.Y. 2(1).

DISCUSSION OF TEST RESULTS

As a comprehensive survey of the region, this data has severe limitations. It represents a convenience sample based upon people's voluntary participation. The number of tests is not great enough to serve as the basis for firm regional conclusions. The Orange Environment, Inc. radon survey, however, represents the best available data on the Orange County area at this time. The results are suggestive of a widespread problem which demands active attention from individual building occupants as well as from government.

Some additional conclusions can be drawn from the results:

*Where upstairs and basement tests were both obtained, the general outcome is to find higher results in the basement. However, a number of exceptions to this were found, in one case where outcomes exceded 10 pCi/l. These findings suggest that testing at two levels of a dwelling is a desirable practice.

*Although accurate geological plotting of the test sites was not possible, approximate plotting generally supports emerging understanding of some of the geologic correlates of radon gas exposure. Thus, while some of the higher scores were found in Reading Prong tests, many of the test results over 4 pCi/l are found in other bedrock formations which allow for geological passageways to the soil, from which gases then can reach buildings. In some cases, higher scores correlate with geological fault lines. In other cases, limestone and dolomite (calcium carbonate) formations, such as the karst formation which crosses southern Orange County, are implicated. This bedrock represents a highly soluble material where cavities allow ready passage of gases and water. The karst topography has been previously identified as a prime site for potential pollution problems in Orange County.[8] This region should now be thoroughly studied for radon gas exposure as well. The widespread distribution of findings over 4 pCi/l, however, suggests that radon exposure cannot be conveniently isolated as belonging to one or several bedrock conditions. Orange Environment, Inc., therefore, recommends that all area homeowners conduct tests regardless of their location.

*There are no major differences for the sample as a whole and for just the Orange County portion. In Orange County, readings of ten or greater were found in 4 percent of the tests and 5 percent of the buildings. Readings between 4-10 pCi/l were found in 13 percent of the tests and 18 percent of the homes in Orange County. Thus, in total, 17 percent of the tests and <u>23 percent</u> of the buildings exceeded 4 pCi/l. While this rate and the levels of exposure are not excessive when compared to some areas of Pennsylvania and New Jersey, a widespread radon problem of moderate severity is suggested. Given the high risk of lung cancer associated with exposure at the 4 pCi/l guideline, this suggests a major public health threat in the survey area.

EDUCATIONAL AND INFORMATION EFFORTS

The OE program has placed a major emphasis upon public education regarding the radon issue. Towards this end, we have carried out several activities beyond those described earlier. These can be briefly summarized:

-- Three OE Board members organized a three-day conference entitled Radon and the Environment under the auspices of the Institute for Environmental Studies at Ramapo College of New Jersey. Six directors of OE participated in the conference along with radon experts from across the country. The objective was to present the most current information on radon and to create an event in which the public from northern New Jersey as well as southern New York would have direct access to the top experts on the topic. The publication of the conference proceedings was a further outgrowth of the concern to disseminate this information to the public.

-- A packet of current readings on radon was placed in four Orange County libraries for public review.

-- OE officials participated in numerous radio and television interviews in an effort to disseminate the results of our research. Press releases were sent to all regional media.

Barriers to Correct Information -- It would be inappropriate to ignore a major source of frustration which has limited the effectiveness of the OE effort. This involves the seeming inability of the major regional newspaper, The Middletown Times-Herald Record, to provide correct and adequate coverage of the radon issue. The problem was first noted in the failure of the paper to counterbalance the coverage of the Tuxedo meeting. Despite editorial and feature efforts to the contrary, the news department of the paper has acted systematically to limit information on the issue. Wrong information has been repeated dozens of times. In part, the problem stems from the continual changes in reporters assigned to cover radon. Thus, reporters are often forced to rely heavily upon prior stories for information, compounding the confusion. One reporter admitted to never having heard of radon prior to his attempting to cover the OE public meeting with Terry Brennan. However, there is reason to believe that much of the problem stems from editorial interference and distortion. This is evidenced by the complaint of reporters that their field stories have been changed, by the refusal of the paper to give adequate notice of the Brennan meeting despite its news value, and by the refusal of the paper to send a reporter to the Ramapo conference because, as the news editor explained, "there is never anything new at conferences." Possibly most destructive is the paper's continual use of misinformation and refusal to correct its mistakes. For example, virtually every article filed contains a reference to the Reading Prong, indicating that this is the source of radon gas. On a number of occasions, brief stories have been carried which contradict this, but immediately thereafter reference to the Reading Prong is returned. When OE released its first radon results, it strongly stressed that the findings supported what radon experts now generally recognize, that the Reading Prong is not the only radon source. The paper's blind dependence upon the Reading Prong notion, however, is revealed in this quotation from the article:[6]

> Edelstein said that one of his goals was to counter the myth of the Reading Prong, an underground belt of radium stretching from northeastern Pennsylvania into parts of Orange County that scientists say emits radon -- a colorless, odorless gas -- even when it decays.

Even in reporting the inaccuracy of the Reading Prong concept, the paper reintroduced the idea in the same paragraph. Later, in reporting OE's second round of testing, the paper ran a map of the Reading Prong alongside the findings of radon outside of the formation.

Other examples of distortion in this paper have occurred. When the first set of OE findings indicated that about 29 percent of Orange County buildings appeared to exceed the 4 pCi/l guideline, the banner headline on the front page read "Tests downgrade county radon problem."[7] When this and other misleading aspects of the coverage were reported to the news editor, he refused to correct them. When it was suggested that an analysis piece be written which would set the record straight, the editor indicated that radon had received too much coverage already and did not warrant further attention. When OE's second set of results were released, the paper failed to print them in a timely manner. They then nearly decided not to provide coverage, ignoring a filed story and printing instead a story on a state testing program in one Orange County town. This story asserted that no data existed for radon testing in Orange County other than limited state data. This was despite the fact that the paper had already covered OE's first set of results and had a filed story on the second set. To finalize the distortion, the paper then forced the reporter to rewrite the story covering the OE findings as an attack on the state program so that it would have more of a "yellow press" news appeal.

The point of this is not merely to vent frustration over the response of the newspaper in question. Rather, the paper has acted so as to prevent area residents from having a clear understanding of the OE findings. With nearly one quarter of the area homes potentially at risk, if our results can be generalized, we believe that a major public health issue has been identified which requires immediate and firm action. The paper has confused and hidden this issue. Furthermore, by repeating the myth of the Reading Prong at every turn, the paper has continually reinforced the belief that people living off of the Prong are not at risk. This is simply untrue. Yet, the paper is unresponsive to feedback on its errors. We suspect that such horror stories regarding the press are not uncommon, yet, as this instance indicates, an unresponsive newspaper holding a near monopoly on regional news can have a large negative impact on the public response to a major issue.

CONCLUSIONS

We have learned a great deal from our efforts about conducting a radon testing program. Some of our experience would lead us to modify aspects of the program were it to be repeated. For example, prompt return of radon test kits at the end of the testing program must be clearly made the responsibility of the homeowner.

Rather than dwell upon such specifics here, however, it may be useful to consider whether there should be a continuing role for OE in the radon issue. To consider this, it may be helpful to identify several clear policy objectives for the Mid-Hudson region.

(1) Given the health threat associated with radon exposure, a comprehensive radon testing program should be in place for the winter of 1986-87. The objective of this program follows directly from everything that we have learned about the threat from radon: AN ATTEMPT MUST BE MADE TO URGE ALL RESIDENTS TO TEST THEIR HOMES PROMPTLY. Confirmatory testing for marginally high results measured previously should also be encouraged.

Based upon our experience, we recommend a program in which the homeowner able to bear the expense undertakes the initial testing effort. While government prefers the charcoal canister as a cost-effective screening device, we believe that, from the homeowner's perspective, the alpha track detector left in place for at least three months gives a more accurate basis for deciding whether remediation is required. Where basements are present, we recommend separate testing in the basement and in an upstairs living space. Such testing should be centrally coordinated so that results can be monitored to update the data reported here. At this time, it is not clear what efforts will be made by the New York State Health Department in meeting this need. Despite recent funding, it may be some time before a state testing program comes on line. Furthermore, the level of funding suggests that only a small fraction of New Yorkers will be reached through the program. The OE effort will be aimed at filling the gap until a state program arrives and then acting to extend the scope of the state program to reach the greatest number of residents. In this regard, we are seeking a means to provide radon detection to those who cannot directly afford it so that low income people are not excluded from testing programs.

(2) There is a need for an aggressive program for training contractors in radon mitigation techniques and in radon-free construction. Fortunately, the New York State Energy Office with support from the EPA is moving to develop such a program. Several OE members took the course during the summer of 1986. We view it as a good first step toward training mitigators. We have begun planning for ways to extend the EPA program by creating an apprentice mitigation workshop complete with a co-op equipment pool.

(3) Building codes for new construction are required in the high growth mid-Hudson counties if we are to avert constructing new buildings which will only need to be subsequently remediated. Such codes should be put in place prior to the 1987 building season. Likewise, policy for real estate transfers is needed.

(4) With increasing reports of radon in water within the mid-Hudson region, we intend to devote more of our attention to this source.

The role of a non-profit, tax-exempt organization in achieving these objectives needs to evolve. OE is actively seeking funding to offer a continuing testing program and to work on building code development. At the same time, a comprehensive New York State program may finally be emerging. If this program is far reaching, then there may be little need for a duplicate effort on the part of OE. Our role would then be to continue to monitor the program and advocate a firm response.

The success of the OE radon testing program suggests a different scenario. The use of a community-based organization in dealing with response to environmental hazards may in fact be a preferable model in many respects to that of direct government action. The community-based organization can serve as an advocate for resident needs and can help to bridge the communication gaps which frequently occur between government and the public. Furthermore, the use of such organizations may be a cost-effective way to stretch government resources in addressing issues such as the threat from radon gas.

We wish to acknowledge the roles played by many of the OE Board members in developing our radon program. William Makofske served as technical consultant, Pat McConnell and David Church helped to gather information and provide advice, Margaret Gibbs assisted in the assessment of psychological impacts, Robert Michaels advised us regarding health risk assessment, Nancy Schneider provided legal advice and the use of her office and staff in distributing and collecting materials, Maureen Quatrini provided support assistance, as did Marissa Biegel. The local press aided in informing the public of the program. The First Presbyterian Church of Goshen provided space for our public radon program.

FOOTNOTES

[1]Shabecoff, Philip, "Radioactive Gas in Soil Raises Concern in Three-State Area," New York Times, Sunday, May 19, 1985.

[2]Boice, Ruth and Alvis Meehan, "Tests to Determine Radon Gas Safety," Sunday Record, June 2, 1985, pp. 3, 8.

[3]Boice, Ruth, "Scientists Discount Radon Risk," The Times-Herald Record, July 10, 1985, p.3.

[4]Boice, Ruth, "Scientists Skeptical About Radon Threat," Times-Herald Record, August 15, 1985.

[5]Boice, R. and A. Meehan, Ibid.

[6]Zurier, Steven, "29% of Tested Homes Show Radon Risk.", The Times-Herald Record, May 1, 1986.

[7]Boice, Ruth, August 15, 1985, Ibid.

[8]Edelstein, Michael R. and William Makofske, "The Karst Topography of Southern Orange County: An Analysis of the Relationship between Soluble Bedrock, Land Use, and Land Use Policy." New York Land Report. Vol. 5, Issue 10, 1985.

HIGHLIGHTS FROM THE DISCUSSION: THE ROLE OF GOVERNMENT–THE NEW YORK RESPONSE

Liana Hoodes* (Discussant): Mainly I would like to say that the problem in New York State is not with the programs that you have just heard about. The problem is with what is not being done in New York State. There is no comprehensive radon program.

I am with Orange Environment, Inc., a non-profit environmental research firm. We got involved in this issue about a year ago, when our region, in Orange County, became increasingly sensitized. We saw an avoidance of the problem by local and state officials. Playing down the issue did not convince all local homeowners; there was a public demand for testing and for some help in understanding what kind of testing is available. We initiated a radon testing program which was paid for by anyone who could afford testing. We offered discounted alpha track detectors by batching groups of 100 or more tests together. We have had about 250 tests in our program. We advertised my phone number in the paper, and I became the New York "hotline". There was no other place for people to call. They called me day and night, asking, "What is going on with radon? How do I test?" People who had already tested asked, "What does it mean, and how do I mitigate this problem?" Beyond a $5 surcharge per test, we had no funding for this program. This was a labor intensive program that our group had to subsidize, but we felt that it was needed. We realize that a part of the problem in New York comes from partisan political fighting over limited legislation which only begins to address this issue. One wonders that if Orange Environment could set up a largely self-funded program, why can't the state do it also? If you can get batch reductions from various labs, why can't the state make these tests available? I would also like to know, does NYSERDA have aresponse component, and does DOH have a research component, and are they intending to get together?

W. Condon: One of the legislative proposals is to do just what you mentioned, to make money available to the state health department to buy detectors in bulk to provide to citizens in New York. I think your group is to be commended in the absence of a service for getting together to do this. I know that I have referred a number of people who lived in that area and wanted to get testing done to your

---

*Liana Hoodes is Vice President of Orange Environment, Inc., Goshen, New York. She coordinated the Orange Environment Radon Program during 1985-1986.

organization. They called, and I had to say, "I'm sorry, I don't have anything to offer you, but you can get some help."

J. Rizzuto: The Energy Authority is a research organization. Continuous services are more appropriately offered by the NYDOH. There has been legislation written to do this, but unfortunately it is moving very slowly. The agencies are working closely together to bring results into play in new programs that the DOH might be running in the future.

L. Hoodes: I would like to add that despite their lack of funding, the DOH has been very helpful to us in coming to informational meetings for citizens.

M. Edelstein: I might add that Orange Environment is trying to play a role that goes beyond what the government can do, particularly in moving quickly to address a need. We are also doing research to legitimate that there is an issue. As a non-government agency, we can act as an advocate for a government response in our area. This suggests that another potential actor in the radon issue includes a way of brokering between what government does and the citizens who may not even know what their needs are. In the future, I expect that you will find more of this type of development with non-profits becoming active regarding radon. PAR represents another model of grassroots involvement. I think that both are important developments in the radon issue.

S. Conroy (from the audience): I'm a town official with a long history of community involvement. I urge you to push for funding. I live in a community on the outer edge of the Prong where no radon research has been done. I got my testing done through Orange Environment, and I'm grateful that after floundering around for a long time, I found someone to help me and to answer my concerns. However, out of 2,500 in my community, only 10 or so homes have been tested. There are people in my community who are becoming concerned and don't know where to turn and even who to trust given the conflicting reports. It is important that the state funds your program.

The New Jersey Response

THE NEW JERSEY RADON PROGRAM: A LEGISLATIVE PERSPECTIVE

Senator John Dorsey*

I will try to be very brief, because a great many areas have been covered. It is a pleasure to be here, and it is interesting to hear the comments, particularly from the other state representatives, relative to the New Jersey program. I think, frankly, at least at this point, and I hope you won't think this too self-serving on my part, that New Jersey has perhaps come up with the most responsive program of any of the other states. It is due, in a very large part, to the fact that the New Jersey program was put together based upon very close coordination between myself and three professionals whom I hold in very high regard, Drs. Gerald Nicholls and Donald Deieso from the Department of Environmental Protection and Dr. Judith Klotz from the Department of Health. It is important to keep in mind that the New Jersey program was designed as part of a very specific piece of legislation. The legislation spells out a great deal of the program, and originally was to make an appropriation of 4.2 million dollars. My Democratic colleagues cut a million dollars out which we are in the process of restoring. We hope that within a month, the additional million dollars will be appropriated so that in an eighteen-month period, the Department of Health and the Department of Environmental Protection will be able to expend 4.2 million dollars in connection with the overall program. That is a significant amount of money.

You have heard Senator O'Pake speak of the money spent in Pennsylvania. At a recent indoor air pollution conference in Boston, I learned that the EPA's appropriation for study of all indoor air pollutants is only about 2 million dollars. This is a substantial amount of money, but, of course, it is not enough to deal with the entire problem. Going into the New Jersey legislative program, a number of factors were recognized that have turned out to be both true and important in consideration of what will happen in the future. First, we were not going to receive any enormous sum of money or any very direct assistance from EPA in connection with the program that we were undertaking. We, therefore, approached our legislative program on the

---

*Senator John Dorsey was the author, in conjunction with State Assemblyman Zimmer, of New Jersey's radon legislation. He is Assistant Senate Minority Leader.

basis that the basic surveys and studies that we were to undertake would be the state's responsibility and would not be paid for by the EPA, but that we would hope to hold in reserve EPA assistance to help pay for the mitigation that might arise. Secondly, we observed that the radon problem is not limited simply to the Reading Prong, and that has proven to be particularly important now that the Clinton situation is known; Clinton not lying on the Reading Prong. So it is a problem which was recognized at the beginning to be much wider than simply the Reading Prong. Thirdly, we noted from the outset that there was no way that the State of New Jersey could come up with sufficient funds and sufficient employees to test every home requiring it. There was a need for a certain amount of cooperation between the private sector and the public sector.

With those facts in mind, we then designed a program to raise an initial 4.2 million dollars in connection with two basic studies. The one that is simplest to describe is one to which $600,000 was allocated. It is the epidemiological study which is now being conducted by the Department of Health of the State of New Jersey. Its purpose, of course, is to try and relate what the incidence is between exposure to radon and lung cancer. When it is completed, it will be, I believe, based upon my trip to Sweden with Senator Lautenberg, the most thorough epidemiological survey ever conducted in connection with this issue. The Swedes have had only very cursory epidemiological surveys completed even though they have been dealing with the problem for a longer period of time than we have in this country. The balance of the money is dedicated to the program that Gerry Nicholls is in charge of. It is a program which is in the process of performing a comprehensive review and analysis of available geographic, geological, demographic and radiometric data; developing a model to predict radon exposure in New Jersey residents; and testing for accuracy of that model as a predictor of radon. We are to conduct, I think, 6,000 tests by the state alone. The state is in the process of conducting some 3,000 or more confirmatory tests; it is in the process of analyzing the risk associated with the exposure of radon in New Jersey, developing recommendations for future study and also in the process of developing mitigation techniques. The program has been designed very specifically to interface with the public and has sought to develop the ability to utilize the other resources which are out there. From the very inception of the program, money was set aside to educate, to brief, and, in some instances, to finance operations by local health officers in the state. The program has also utilized a system of briefings. Early on in the program, for example, we held a briefing with all of the superintendents of schools in one county in which there is known to be a significant problem. The DEP is receiving calls from the public at the rate of 1,200 a month. Fifty-five percent of the calls are for general information, 30 percent of the calls are from people with questions after initial testing, 10 percent of the calls are regarding health risks. So we have interfaced, as much as we could under this program, with the public.

We have also attempted to interface with two industries in this state which are directly affected. The first one is the realtors. The realtors have been very responsive in the sense that they have recognized at the outset that there was a problem that they could not conceal; one with which they wished to deal in an intelligent manner. I think because of the interfacing between our Department of Environmental Protection and the real estate community, we have been able to avoid what might otherwise be a situation in which areas are red-lined or black-marked in terms of the real estate market. We have avoided that because of the briefings which we have held. We have attempted to deal closely with the New Jersey Builders' Association because they, of course, are intimately involved. We are now working on a process by which the DEP, the EPA, and the New Jersey Builders' Association will, in a jointly funded project, attempt to develop appropriate mitigation measures. Gerry Nicholls tells me that that program is fine as long as it doesn't take any money from his personal budget, and therefore, he has laid upon me the responsibility to fund the program. We have attempted to work with these two industry groups, and I think the results of that are being seen.

Because we also recognized at the outset that there was no way that we could physically go out and test all the potentially affected homes, we realized that we would have to depend on individuals to have a great deal of testing done on their own. That has occurred. In the process, we have attempted to protect our citizens against certain fraudulent practices that sometimes arise. From the very beginning, Dr. Nicholls' department has been arranging a list of so-called "approved" testers for radon, and we are now in the process of waiting for legislation which will specifically authorize the DEP to license those who are found to be qualified to make those tests. We also have been looking forward to the point in time when we must deal with the issue of mitigation. We will require, again, that only those certified by the DEP will be able to be gainfully employed in the mitigation business, an area in which we know that great sums of money will have to be expended. We are approaching the problem of mitigation with the thought that there will have to be some kind of relationship between the private homeowner and the bank that holds the mortgage on his home. We are trying to develop a program by which there will be subsidized loans in which the state, private industry and the homeowner will all have to share in the cost of mitigation.

What we have found, more than anything else, is that the radon problem is an evolving problem. We have not as yet uncovered all of the problems and all of the nuances that are involved in the radon problem. We have to deal with such concerns as the confidentiality of the private results received of radon testing by an individual homeowner. In New Jersey, we are going to require every individual homeowner to report the results of his testing to the Department of Environmental Protection in order to expand our data base. But we are going to grant, on the other hand, confidentiality of those records. The New Jersey program is not static. It began with certain knowledge of what had occurred in Pennsylvania and with a sense of what was reasonable to expect in terms of assistance from EPA. The EPA, particularly in Clinton, has been of significant assistance to the state. Basically, the states have been called upon to address the radon problem by default on the part of the federal government. It is a problem that is continually evolving, raising new issues that are going to have to be dealt with in the context in which they arise. Based upon the quality of our professionals and the bipartisan support that we have generally had in New Jersey in connection with the radon problem, I am confident that we will be able to deal effectively with each of these issues as they arise.

AN OVERVIEW OF THE RADON ISSUE IN NEW JERSEY

Donald A. Deieso*

When Stanley Watras walked through the radiation monitors at the Limerick Nuclear Power Plant in December of 1984, he sounded an alarm that has been felt across the country, and especially here in New Jersey. At first this issue was limited to those living atop uranium or phosphate mill tailings. Now we know that structures built above naturally-occurring sources of radium can also exhibit unacceptably high levels of indoor radon.

By all measures, the dimensions of this issue are staggering. First, the human health risks posed by high levels of radon and radon progeny are among the highest of any environmental exposures known to us. As an example, consider that dioxin levels found in the flue gas from a resource recovery facility are judged to be acceptable with associated risks of one excess cancer in a population of one million, or one in a hundred thousand. In comparison, radon, even at a 4 pCi/l concentration, poses a risk of three excess cancers in a population of a hundred. At elevated levels of radon, models indicate risks exceeding one in two. Clearly, this is one of the most serious issues environmental health officials have had to face.

Second, the geographical extent over which structures could be affected may include entire states or large portions of states. In New Jersey alone, we estimate that 1.6 million structures are candidates for high radon levels. Over 35 states have documented cases of high indoor radon. Without question, this issue has assumed national and even international significance.

Third, we do not adequately understand the scientific principles governing radon. Pathways into the home, geological factors, and remedial techniques remain in large part a mystery. Each day more is learned and each day more gaps in our knowledge are revealed.

During this conference, you will have an opportunity to hear many stimulating and provocative scientific and policy concepts about radon. The conference agenda is outstanding, including many investigators who have contributed nationally to our present-day knowledge. You will hear in the next day of the approach developed by New Jersey in addressing the

---

*Assistant Commissioner, New Jersey Department of Environmental Protection. This was the opening address to the conference.

radon issue. If our program is to be judged successful, it will be because our Governor, Legislature, the state Departments of Environmental Protection and Health, as well as our county and local health officials, have a strong common purpose: to eliminate the unacceptable risks to our residents from radon. As any state representatives here today can attest, the most dedicated of public health officials can only be marginally effective without a committed administration and an informed legislature. On January 10, 1986, Governor Thomas Kean enacted our state's first radon law, providing 3.2 million dollars to our Department of Environmental Protection (DEP) and Department of Health for radon activities. This is nearly twice the budget of the entire EPA national radon program. An additional one million dollars is now pending before our legislature which would expand DEP's activities into the certification of commercial radon testing and remediation firms. A group of extremely dedicated professionals becomes the next ingredient necessary for an effective program. Under the direction of Dr. Gerald Nicholls, the assistant director of DEP's radiation program, a group of twenty scientists have already made outstanding contributions in addressing this issue.

I would like to discuss now several of the most significant public policy issues we face with radon. Unlike hazardous waste or air toxics in which the role of the federal or state government is clear, indoor radon presents many unique differences. For one, the source of the contamination is not an offending industrial facility or landfill; it is nature. Next, the properties that are affected are primarily homes with no discernible geographical or geological pattern of occurrence. The sheer number alone, over 1.6 million in New Jersey, suggests that conventional approaches to environmental analyses will not be appropriate or effective. Also, the scale of program costs is devastatingly high; a quick calculation reveals that in our state, radon testing of potentially affected properties will cost over 200 million dollars with remediation costs exceeding 750 million dollars in the long term. Putting aside for a moment the philosophical issue of using public dollars to remediate private properties affected by a natural phenomenon like radon, we are left with the prospect of the state government administering a one billion dollar program in New Jersey. Considering the urgent remediation time frames associated with high radon levels, it became clear that the proper role of state government must be to establish a regulatory framework within which the private sector could fill the need.

In May of 1985, during the early stages of this issue, our residents, responding to press accounts describing the health impacts of radon, telephoned in record numbers requesting that the state sample their homes. The combined resources of our state government could not have possibly satisfied their requests within an acceptable time. Faced with this fact, we decided to encourage commercial testing firms to enter the radon measurement business.

One year ago, only two commercial firms offered radon measurements; today over seventy firms seek certification in our state. During the past year, these firms have tested approximately 10,000 homes.

I might add that little real encouragement was necessary in New Jersey. Business opportunity provided the real incentive. Through a well-organized certification program administered by the DEP, the residents of our state can be assured that these commercial firms have staff with appropriate educational backgrounds, are using DEP-approved sampling methods, and in all other ways are conducting themselves in a responsible manner. We are proceeding with precisely the same strategy for those firms offering remediation services.

Let us consider next the public expectation that government will provide the funds for the corrective actions necessary in their homes. After all, the public argues, this is precisely the relief Superfund provides to those affected by hazardous waste sites or the relief the Federal Emergency Management Agency (FEMA) provides to those affected by floods or hurricanes. Why not radon? Several notable differences exist, not the least of which is the widespread nature of the radon problem. Asking that the federal or state government underwrite the initial correction action and the follow-up actions at millions of homes is not workable. An entirely new bureaucracy would be necessary. Further, we are finding that the average cost of remediating a radon home will be on the order of $2,000, a reasonable sum for a homeowner to bear. In short, it appears that low interest loans or even prevailing rate loans that are available to homeowners within a week or two of application would provide the necessary relief. We are pursuing just such a plan in our state.

Of course, the list of policy questions goes on and on and many, no doubt, will be the subject of discussion during this conference. There is one final thought I would like to offer. All of us have watched during the past months as exaggerated statements concerning the health effects of radon have caused the public to become truly alarmed. While I would be the first to say that this issue is serious and deserves our full attention, I must also say that we must foster a calm, well-informed citizenry who fully understand the issue and associated risks. Each irresponsible and self-serving statement which distorts the truth and causes the public to panic makes it difficult for us to communicate the facts. Whether the statements are attributed to charlatans in the scientific community, avaricious radon testers, or short-sighted elected officials, the consequences are the same; public confusion and distrust. I would ask that all of you work aggressively to find answers to the many uncertainties about radon but work no less aggressively to see that only the most accurate of facts are shared with the public. It is our responsibility as scientists and public health officials to see that the public is informed and not ignorant, concerned and not panicked over this issue.

RADON IN NEW JERSEY: PRELIMINARY CHARACTERIZATION AND STRATEGY

Gerald P. Nicholls and Mary K. Cahill*

INTRODUCTION

Early in 1985, officials of Pennsylvania Department of Environmental Resources (DER) informed their counter parts in the New Jersey Department of Environmental Protection (DEP) of the December, 1984 discovery of very high levels of radon decay products in certain Pennsylvania homes located atop the Reading Prong, a geological structure running from within northeastern Pennsylvania, through northern New Jersey, and into New York State. As the significance of the Pennsylvania findings became evident, various personnel within NJDEP became involved in attempting to characterize the nature of New Jersey's potential problem and developing a strategy to address it. By May, 1985, the following general characteristics of the potential problem in New Jersey were reasonably well defined:

1. In Pennsylvania, approximately 22,000 homes were located on the Reading Prong; in New Jersey the number of homes on the Reading Prong was approximately 250,000.

2. Preliminary analysis of the National Uranium Resource Evaluation (NURE) Data[1] demonstrated that significant concentrations of uranium were present in the surficial layers of northern New Jersey soil outside the area of the Reading Prong. Further, a significant but relatively unsupported body of data existed indicating that various geological studies had at least tentatively identified specific locations in New Jersey where radioactive minerals are present.[2] Geographically then, the area of potential risk extended both north and south of the Reading Prong and might encompass the approximately 1.6 million homes in New Jersey north of a line drawn roughly from

---

*Gerald Nicholls and Mary Cahill are, respectively, Acting Director and Staff Scientist with the New Jersey Department of Environmental Protection, Bureau of Radiation Protection, Trenton, N.J. This paper was revised, November 19, 1986.

Florence, New Jersey on the Delaware River to Asbury Park, New Jersey on the Atlantic Coast.

3. Many New Jersey residents were already sensitized to the potential health problems posed by radon gas due to the industrially related radon contamination incident in Essex County which DEP and the United States Environmental Protection Agency (EPA) had discovered and are in the process of addressing at a cost exceeding $8 million.[3] Also, a radon decay product concentration of 0.02 WL (4 pCi/l) had been used by EPA, DEP, and the Centers for Disease Control as a level which triggered remedial action for these Essex County properties.

4. With the exception of the Essex County sites, only meager and inconclusive residential radon level data existed.[4]

Attempts to evaluate a rational strategy to address the potential radon problem were given added impetus by the publication of an article in the New York Times in late May of 1985 alerting New Jersey residents to the nature of the radon problem in Pennsylvania and the potential for a radon problem in New Jersey. During the week following the New York Times article, more than 200 calls were received by NJDEP, primarily from New Jersey residents directly concerned about their own health or that of their loved ones.

RADON STRATEGY

A simplistic approach to New Jersey's radon problem would call for a screening program for those homes at potential risk, a second round of testing for "high" homes identified in the screening process, and testing of remediated homes on an as-needed-basis to confirm the effectiveness of remedial actions. Pennsylvania DER had indicated to us that approximately 40 percent of the roughly 1,200 homes they had tested as of June, 1985 exceeded the 0.02 WL (4 pCi/l) remedial action indicators which had been used with regard to our Essex County sites. Depending on whether one selected the 250,000 homes on the Reading Prong in New Jersey or the 1.6 million homes in northern New Jersey as the population at potential risk, our estimates for the cost of the simplistic solution ranged from $20 million for a program involving at least two years of work with 40 employees to $200 million for a program involving a five year effort and 250 employees. Funding for such programs was clearly outside the normal DEP allocation. Just as clearly, hard scientific evidence, which we did not have, would have to be presented to the legislature to justify such expenditures of resources.

What finally emerged from numerous formal and informal meetings on the radon problem was a commitment to obtain the necessary scientific evidence to define the radon problem and, once defined, prioritize the eventual testing of homes so that the residents at greatest potential risk would receive attention first. Additionally, an informational program for the public would be developed, voluntary confirmatory monitoring would be conducted in homes which tested at or above 4 pCi/l as indicated by results from a commercial testing service, and support would be provided for an epidemiological study of lung cancer in a residential setting by the New Jersey Department of Health (DOH).

Senator John Dorsey of Morris County and Assemblyman Richard Zimmer who represents portions of Hunterdon and Morris counties introduced legislation in the summer of 1985 to provide the necessary mandate and funding for the DEP and DOH to proceed. An initial budget of $4.2 million was developed but later cut to $3.2 million in legislative committee. This legislation was signed by Governor Kean on January 10, 1986.

The statewide scientific study for radon consists of four major components:

1. Mapping of the NURE data along with relevant geologic and demographic data. Radiation anomalies mapped will be ground-truthed by survey teams.

2. Development of a predictive model for radon exposure in New Jersey. This model will include parameters such as location, soil type, type of housing, type of heating system, degree of weatherization, etc.

3. Development of a sampling program to test the validity of the predictive model. It is anticipated that approximately 6,000 structures will be tested.

4. A risk analysis for lung cancer in New Jersey due to exposure to radon and its decay products utilizing existing major data based models.

Parallel to the statewide scientific study of radon, the New Jersey Department of Health will conduct an epidemiological study of the link between lung cancer and radon exposure in a residential setting. It is currently anticipated that the study will involved approximately 1,000 lung cancer cases drawn from DOH's cancer registry and 1,000 controls. Radon testing of houses where cases lived and controls lived will be undertaken to attempt to reconstruct radon expsoure retrospectively.

Under our confirmatory monitoring program, we perform confirmatory testing upon request in homes where a first test obtained from a commercial vendor yields a result of 4 pCi/l (0.02 WL) or greater. We have utilized both Kusnetz and charcoal canister methods in this testing. We will also test again if remediation is performed so as to determine its effectiveness.

The radon information line is the centerpiece of our public information effort. It is active during normal business hours and has just been converted to an 800 number. Calls are received by clerical personnel who send out informational literature and refer technical questions to radiation physicists on our staff. We are in the process of retaining a public relations firm to improve the readability of informational literature. We also provide speakers for groups as diverse as Chambers of Commerce, citizen's awareness groups, bankers, realtors, appraisers, and school officials.

RESULTS

For the purposes of this publication, the results provided have been updated to the latest available at the time of submission of the final manuscript for publication.

Figure 8.2 gives the distribution of radon levels in basements of houses as determined by either charcoal canisters or alpha track detectors. We are indebted to eight commercial testing firms which have voluntarily provided this data to us. Note that approximately 28 percent of the houses tested were at or above the 4 pCi/l action level recommended by EPA.

Figure 8.3 gives the distribution of radon levels measured by DEP as part of its confirmatory monitoring program. Note that in 80 percent of the cases, DEP's charcoal canister measurements identified a level of 4 pCi/l or higher. This is in good agreement with the original testing firm data. Much of the 20 percent of the measurements where DEP did not corroborate the original measurement of 4 pCi/l can be attributed to natural fluctuations in radon levels.

Figure 8.4 gives the results of radon-in-water testing conducted by DEP in homes where confirmatory monitoring was conducted due to a resident reporting 4 pCi/l or higher in the air of the home. Note that approximately 14 percent of the homes have a radon-in-water level at or above the 10,000 pCi/l level which EPA has indicated can produce a concentration in the ambient air of the home of about 1 pCi/l.

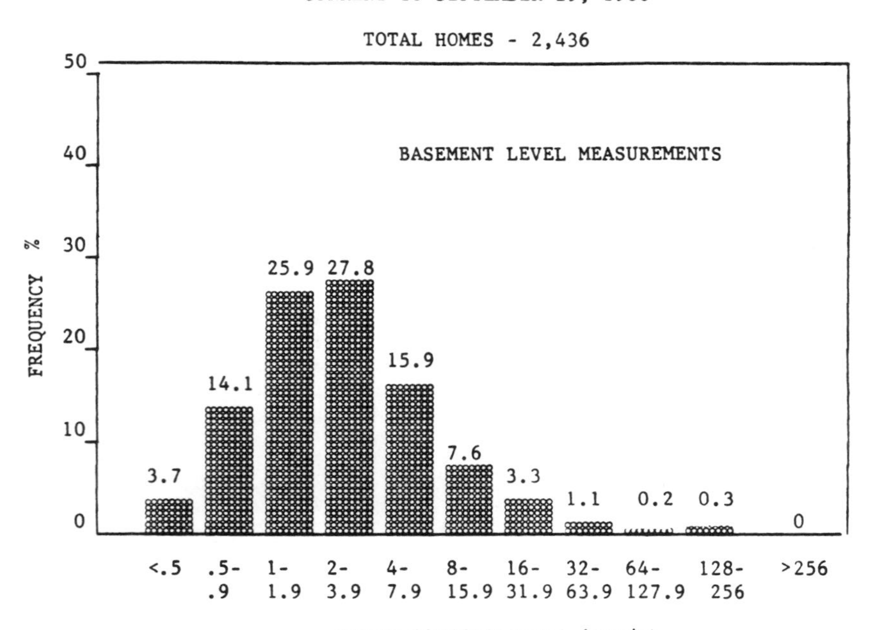

FIGURE 8.2
FIRM RADON TEST DATA DISTRIBUTION
CURRENT TO SEPTEMBER 29, 1986
TOTAL HOMES - 2,436
BASEMENT LEVEL MEASUREMENTS
FREQUENCY %
50
40
30
20
10
0
3.7
14.1
25.9
27.8
15.9
7.6
3.3
1.1
0.2
0.3
0
<.5
.5- .9
1- 1.9
2- 3.9
4- 7.9
8- 15.9
16- 31.9
32- 63.9
64- 127.9
128- 256
>256
RADON CONCENTRATION (PCI/L)

FIGURE 8.3

DISTRIBUTION OF
RADON CONCENTRATIONS* IN HOMES
TESTED BY DEP OCTOBER, 1985 TO SEPTEMBER 12, 1986

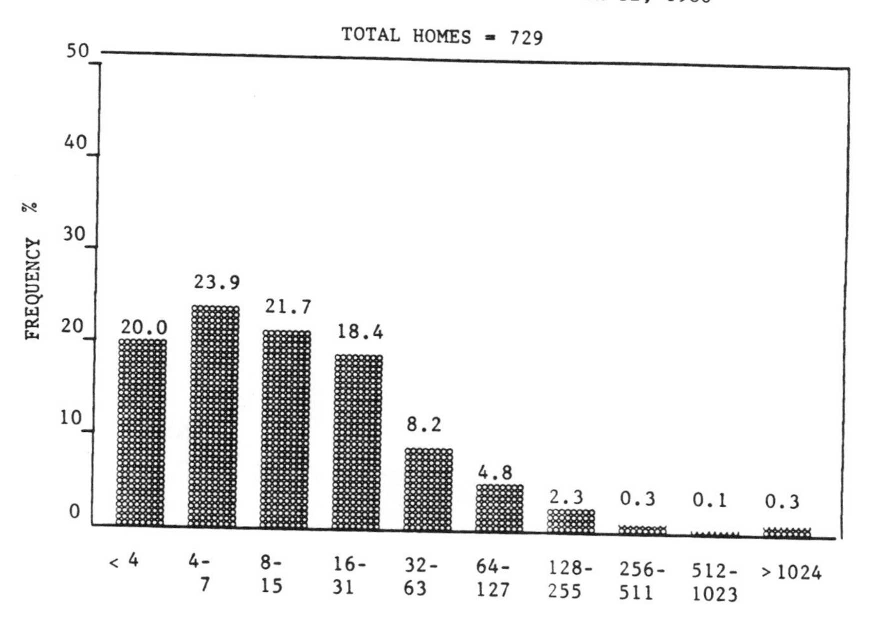

*THE EPA AND CDC HAVE RECOMMENDED REMEDIAL ACTION AT 0.02 WL (0.02 WORKING LEVELS IS EQUIVALENT TO A RADON CONCENTRATION OF 4 PICOCURIES PER LITER AT 50 PERCENT EQUILIBRIUM).

FIGURE 8.4

DISTRIBUTION OF RADON LEVELS
IN WATER SUPPLIES TESTED BY DEP
FEBRUARY, 1986 TO SEPTEMBER 12, 1986

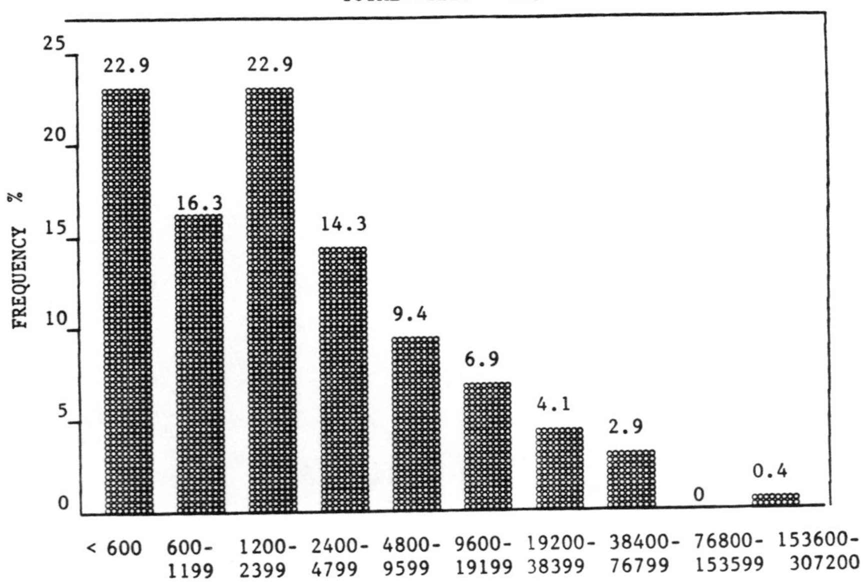

(10,000 PCI/L OF RADON IN WATER CONTRIBUTES APPROXIMATELY 1 PCI/L TO RADON IN AIR)

FOOTNOTES

[1]Popper, G.H.P. and T.S. Martin, "Natural Uranium Resource Evaluation, Newark Quadrangle, Pennsylvania and New Jersey", Bendix Field Engineering Corp. Report P.G.T./F-123(83) for the U.S. Dept. of Energy (1983).

[2]Bell, Christy, Radioactive Mineral Occurrences in New Jersey, N.J. Geological Survey Open File Report # 83-5, 1983.

[3]Eng, Jeanette, "Investigation of a Former Radium Processing Site", NJDEP Report, December, 1980.

[4]George, A.C. and J. Eng, "Indoor Radon Measurements in New Jersey, New York and Pennsylvania," Health Physics 42 (2) August, 1983. pp. 397-400.

ROLE OF THE STATE DEPARTMENT OF HEALTH IN THE RADON PROGRAM FOR NEW JERSEY

Judith B. Klotz*

In close coordination with the lead agency, the Bureau of Radiation Protection of the New Jersey Department of Environmental Protection (DEP), the Department of Health is conducting a multi-faceted program focusing on health aspects of indoor radon. Our activities are mandated and funded in part by the state legislation enacted early in 1986. The components of the program are described briefly below.

1. Epidemiologic Study of Radon Exposure and Lung Cancer

A statewide case-control study of lung cancer in females in relation to former radon exposure is being conducted under the direction of Janet Schoenberg, Chief of the Cancer Epidemiology Program of our Department. The Environmental Disease Prevention Service is also involved, as is the National Cancer Institute and the N.J. DEP Bureau of Radiation Protection, which will conduct the radon monitoring.

2. Registry of Individuals Previously Exposed

A voluntary registry of persons who have previously accrued high exposures to radon has been instituted by the Department in order to be able to inform people in the future of any clinical tests, treatments, or preventive methods which are developed which could be used to decrease their risks of lung cancer incidence or mortality. Currently, extensive clinical trials of vitamin A analogs are in progress, and the efficacy for prolonging survival of various screening techniques is still being evaluated. At the moment, however, the New Jersey Department of Health agrees with the American Medical Association and the National Cancer Institute in <u>not</u> recommending chest X-rays or other screening procedures for asymptomatic people.

Staff of the Department continually evaluate current research which could be applicable to those in the Registry, and we maintain at least annual contact with all registrants.

---

*Judith B. Klotz is Coordinator of Radon Programs for the New Jersey Department of Health.

3. Public and Professional Education

In response to individuals' requests and in anticipation of community and professional needs, the Department conducts information and educational outreach activities regarding both health risks and radon exposure. To this end, we have a full-time health educator on staff.

4. Coordination with Local Health Departments

County and municipal health departments are often called upon to be the front line in responding to citizens concerned about indoor radon, and local health officers have both proximity to and familiarity with their constituents which state agencies can never duplicate. Therefore, the State Department of Health, through the radon legislation, is funding confirmatory monitoring and related activities by local health departments in areas of New Jersey which have significant radon problems. The Local Health Development Service of the Department is the lead in this function. Pertinent training of health officers has been and will continue to be offered, in cooperation with DEP.

5. Promotion of Further Evaluation and Research

The Department is also engaged in promoting new research projects and evaluation of existing data which could enhance our understanding of risks and chemical techniques potentially useful for prevention. We plan to initiate some projects and serve as resources for others.

6. Recommendations

The Department of Health is also asked by the DEP to formulate recommendations on radon exposure limits in general and in specific instances. We have chosen to issue recommendations very similar to those of the U.S. Environmental Protection Agency and the Centers for Disease Control. An advisory committee of scientists who are leaders in the field of health physics and radiation risk have assisted us in these activities.

Under the New Jersey Public Employees Occupational and Safety Act, we are also recommending procedures for monitoring and limiting the radon exposure of state and local government staff who do testing of residences and other buildings.

The recommendations on exposure limitations are immediately applicable to testing and remediating existing homes. In addition, in cooperation with DEP, we are working with the Department of Community Affairs toward establishing new building codes for future housing in areas with high potential for radon problems.

HIGHLIGHTS FROM THE DISCUSSION: THE ROLE OF GOVERNMENT—THE NEW JERSEY RESPONSE

DEALING WITH EXTREME RADON LEVELS: THE CLINTON EXPERIENCE

MAYOR ROBERT A. NULMAN* (Discussant): Well, if it's a race or a competition, you probably want to be first. But nobody wants to be first in finding a cluster of homes with radon. Unfortunately, it's a dubious distinction that Clinton is the first community in New Jersey with extremely high readings. Unfortunately, there are going to be other communities that are going to be affected in a similar way. They are going to have the benefit of the Clinton experience. Clinton is now a laboratory for remediation of radon. The bad news is that we had to be first; the good news is that we are getting a tremendous amount of attention from federal and state agencies. I am especially appreciative of that kind of attention because we are really getting something accomplished in the town. Let me just give you the benefit of some of the experiences that I have had.

The first thing that we tried to do when we found out that we had a particular problem with a cluster of homes with radon was to urge our homeowners to test. If you know that you are in a potential problem area, with a potential hazard, why wait? Why just put it aside? I feel that we have been somewhat desensitized in our lives to environmental hazards. There are so many things that we hear about that are potentially hazardous to us, whether it is acid rain or salt or saccharin or cyclamates. Along comes something like radon, which is something that we can really do something about if we know that we have a problem, but I continue to hear the response that, "I've lived here for 30 or 40 years, and I've been smoking all that time, and it never got me, so I'm not going to worry about this." But radon is something that you can do something about, sometimes rather simply. It is something that should be addressed.

I am proud to say that another first, as far as Clinton is concerned, is that before we even found out that we had a particular problem in that neighborhood, our Town Council went ahead and planned a bulk purchase of radon detectors which we made available to our citizens at bulk purchase rates. We acted as a facilitator; we set a program up. We were waiting to see if it would be done by the state or the county folks. We kept hearing that somebody was going to set this thing up to get cheap detectors available to us,

*Robert A. Nulman is mayor of Clinton, New Jersey.

and nothing happened. So the Town Council said, "This is ridiculous." We went ahead. I guess the problem that everyone else was worried about was confidentiality. We were concerned about that too, so we put together a little release that a person signs that acknowledges that we will try to protect the confidentiality of anybody who participates in our program, but some people in the office are going to necessarily have to see it. Because it's a bulk purchase, we are getting a listing, but we are doing it by lot and block and trying to protect confidentiality. I wish other municipalities would at least investigate what the possibilities are of trying to come up with a program that makes detectors easily available, especially in those areas where there are potential problems, or where your neighbor tested independently and got some high numbers. I've been telling folks that if you know you are in a high-risk area and you don't get your house tested, then you should get your head tested, because it's just foolish to continue to live under those kinds of circumstances where you have a potential danger for yourself or your family or your kids. Why not just get the peace of mind and know that you've got a situation that is either already acceptable, or if it is unacceptable, find out and fix it.

My role as Mayor has been kind of counseling because we are a very small town. I get many calls from people who are concerned after they got results back on their testing who wanted just to talk to someone and find out what's going on, what is the state doing for us, what's the long term prospect, what's going to happen to my property values, those kinds of things. So I tried to give them assurances and be a good middle man in trying to get the information out.

My job has also been to evoke a healthy response from the citizenry. The worst thing that can happen in a situation like this is to have people panicking. This is not a situation where you grab your children and run into the streets. It is the kind of situation where, if you have a reasonable reponse and a resolved approach, then sanely and soberly we can set about to correct the problem. I hope that's the kind of environment that we have established in Clinton, and I feel that this has done an awful lot to make the job easier for myself and for the agencies that are helping us. We have a good state of mind about how we are addressing the problem.

One of my particular problems was in dealing with the press. Overnight, I became a media celebrity. I found that the press was ready to sensationalize the problem. We heard reports on television that said, "Clinton's sitting on a time bomb, ready to explode; forty families already evacuated." No one was evacuated from homes, and there was no recommendation to evacuate any homes, but we still found those kinds of reports. When the cameras came into town, they wanted to know names. They went knocking on doors asking, "Are you involved in this?, Do you have a reading?"

I thought that this was rather counter-productive. If we have done nothing else, it was to preserve the confidentiality of any of the findings. The state folks have been super in their protection of the identity of individual households, and we've done the same thing. But we've had kind of a rough battle to protect the identify of the individuals involved. I have had threats of restraining orders that were going to be placed against me and the town, and none of them have come to fruition. I don't think it serves any useful purpose to identify whose house is involved. The real problem there is that people don't want to participate in these programs if they know that when they come to pick up a detector, the cameras are right in the room snapping a picture of them and they are going to be on the front page of a daily newspaper tomorrow. So that didn't serve any purpose.

There is a tendency to want to neutralize the sensationalism by underplaying the situation. I've tried to catch myself whenever that has happened and be as factual as I could to make sure that the public information that gets out is at least correct in terms of generally what risks are involved and everything else. We have a very educated public now as far as radon is concerned. We have had meetings with various state agencies.

The biggest concern after the safety and health aspects of this are property values. What's happened to this investment? Their home may be the biggest investment that an individual has in his entire life. And now you see it just kind of going up right before your eyes. That's the initial shock value of having a home with very high radon levels. It's a real concern that people have. I would say that I wouldn't want to have my house on the market this week in Clinton. This isn't the best time to sell. In the very short term it's a bit of a problem. But in the long run, I think we're going to find that this situation is not going to affect property values. It's not going to give a reputation to Clinton. We will get things fixed. We will get on with business and continue our lives safely because the good news is that radon is a temporary problem as far as in-house pollution is concerned. If you go out and test, if you find out what level problem you have, if any, you can address it, remediate it, and get your home to an acceptable level. It's over. You might continue to test every year to make sure nothing has changed, but you can pretty much have the peace of mind of knowing that you have taken care of the problem.

My particular goal, and the goal of the Town of Clinton, is to get a reputation of being the most tested community in the State of New Jersey as far as radon is concerned. And then, beyond that, to be the most remediated town in the State of New Jersey. That is, when we saw we had a problem, we fixed it. We can pretty much certify, at that point, that if you move into Clinton, you are going to move into a house that has been fully checked, has been remediated, and let's get on with business.

One of the things we are also investigating is what we can do as far as changes to our land use ordinance to make sure that new homes that are built have incorporated pre-construction techniques to make sure that if there is a potential problem there, you kind of seal it off before you get started. That's the cheapest way of handling things. So that's one of the aspects that our planning board is now looking into.

I've had daily contacts with my fellow panelists and many of the other people you've seen in the audience; people from the DEP and the Department of Health, the EPA, and the politicians and their representatives, and with the Federal Emergency Management Agency. We have run the gamut as far as our contacts. My particular task has been to try to keep the ball rolling where we saw things slowing down. We are asking the governor to look into the possibility of Federal Emergency Management Agency funds that would be available to Clinton. Senator Lautenberg mentioned earlier that they have already determined that radon is a category that could be included, and we want to see if there are any low-interest loans or any other grants that are available to help us out. So we are looking at every avenue. We have requested some geological studies in our particular area to see exactly what the problems are and if there is any kind of general remediation that can even be done underground to the area.

So we are getting the kind of response that you only dream about. You hear a lot of bad-mouthing about red tape and bureaucracy from state and federal agencies and the response that you get at a local level. I have to say that the response from Dr. Nicholls and Dr. Klotz and the EPA folks has been remarkable. As you know, the EPA has agreed to remediate 10 homes at no cost to Clinton citizens as part of their remediation research project. Another 20 homes will get complete diagnosis. It's a tremendous response, and we certainly appreciate it. So I am trying to spearhead and coordinate that effort and just keep things moving along.

We have made the municipal facilities available so that we have headquarters of the EPA and DEP right in the Town or in the area. People can actually come in and talk to the experts if they have a problem, to allay their fears or to pick up a canister to retest after they've done some remediation on their own. We are expecting the EPA to run homeowner workshops for us on do-it-yourself kinds of projects. You go in, you do something that is recommended and then you pick up a canister from the DEP right in town, set it up again, and retest to see to what extent you have mitigated the problem.

So this is in large part a success story that I am telling you. We started out with a problem in a cluster of homes. Beginning on Monday, the EPA will be actually walking around with their contractor looking at houses and ready to begin the work to remediate the problem. It's a real team effort; everyone is pitching in. I like to think that those of us at the local level have been a real part of that team and have contributed something. But we are a success story. I've asked the members of the press to please give as much attention to the remediation effort when we get this whole deal cleaned up as the sensationalism of what happened at the beginning of this. If any of you have questions about how do handle things at a local level, in New Jersey we are the only ones that really have that experience. I would be glad to share it with you at any time. Just give me a call in the Town of Clinton, and I would be glad to help.

TEAMWORK IN THE NEW JERSEY RADON PROGRAM

DAN JORDAN** (Discussant): I just would like to describe to you a program that got off to a bit of a rocky start but that I think has developed into a program that can be used in any state. Very seldom do you hear anyone say that they are real proud to be from New Jersey; I am. I think the comments you are hearing from the New Jersey experience certainly indicate that I think we did a pretty good job. I am really proud to be part of it. For me, it started in January 1985. I had come back from a meeting on Friday night with a concerned bunch of parents about an AIDS policy - a typical health officer going from one fire to another fire -- and I felt really good about surviving that meeting on Friday. I went home and Saturday I coached my daughter's soccer team. And, then, Sunday I sat down to read my <u>New York Times</u>. There was a rather chilling article in that Sunday's <u>New York Times</u> about radon and about the problem in Pennsylvania, and then lo-and-behold in the last paragraph, there was a listing of towns. Four out of the five towns for which I am health officer were listed in that article. Well, it certainly wrecked my day, to say the least.

I immediately tried to call the DEP hotline; there was no information available for me at that point. I went in Monday. I was really very upset for two reasons: one was that there was a problem in my towns, and the other was that I really didn't have any advance notice of this problem. I was fuming. I was angry with the State Health Department. I was angry with DEP. I was loaded for bear. Well, very little information came through during the first two, maybe three months. The information we had on a local level was from journals that we could dig up and from newspaper articles. In April there was a briefing given by DEP and the State Health Department at Rutgers University, and they really did a very good job of identifying the problem and recommending how to proceed. From that April meeting, to

---

**Dan Jordan is Health Officer for Warren County, New Jersey.

today, a partnership has really developed, and it's miraculous if you look at it. You are looking at two branches of state government, the Department of Health and the Department of Environmental Protection, joining with local health departments to respond to the citizens of the state and do something about the problem. The State Health Department and the State Department of Environmental Protection provided training programs for local health departments. We are poised to do confirmatory testing, acting as an arm of the Department of Environmental Protection. We are really happy that things have progressed the way they have. Before it was mentioned about the stress of students in school; well there certainly should be a study done on the stress of local officials. I can tell you from my point of view that it has been a stressful experience. But I think we are on our way, and I'm proud to be in New Jersey and proud to be working with these people up here.

COORDINATING THE NEW JERSEY RADON RESPONSE

JUDITH KLOTZ*** (Discussant): I think the summation of what's going on in New Jersey is the teamwork, the coordination, the partnership that everybody before me has talked about. The Department of Health is definitely an important component of what's happening, but we are also definitely in a supporting role to the DEP as far as the state program is concerned. We are working very closely with local health departments, with other local officials, and with citizens who ask us for any help we can give.

Let me very briefly illustrate what I mean. The Department of Health is conducting an epidemiological study in which the Cancer Epidemiology Program is the principal investigator. DEP is going to be doing the actual monitoring. Public and professional education on radon is a collaboration between DOH and the local health departments. We require help in creating a registry of individuals previously exposed to radon. Our efforts to let these people know what future medical and other preventive actions they might take must be coordinated with other organizations that can tell us who can be the beneficiaries of that program. One of our main activities is helping local health departments help DEP in the confirmatory monitoring and outreach to the public. We are doing a lot of promotion of further evaluation and research on the health risks in conjunction with every single other research organization, including federal and state agencies, as well as the academic community which gives us information and works with us. We issue recommendations in coordination with EPA, CDC and with DEP. We are suggesting guidelines for safety standards for state and local officials in their mitigation

***Judith Klotz is Coordinator of Radon Programs for the New Jersey Department of Health.

work and testing, and we are working with DEP and with the Department of Community Affairs on recommending new building codes. None of this is on our own, and I'm very happy to say that at least so far, this collaboration has been extremely gratifying.

It is terrible to find a problem as serious as radon, but I would agree with Mayor Nulman that the good side of what is happening is that we seem to be, at least in New Jersey, addressing it in an extremely positive fashion. If this is any indication of what will be happening over the next few years as we continue to work on this, it will be a very positive experience for a lot of people. We may very well be a model program for how a state should and can deal with this and other types of environmental health problems as well.

INTEGRATING A RADON RESPONSE WITH A TOTAL VIEW OF INDOOR AIR POLLUTION

LINDA STANSFIELD:**** I know that all of you know that the problem with radon is lung cancer, and I represent the American Lung Association. So this problem is one that's right in our province. However, we see the problem perhaps differently than you have heard for the last three days. Last year, our topic for Clean Air Week, which is always the first week in May, was air pollution in your home. This involves more than just radon. There are a number of air pollutants in homes, all of which endanger health to various degrees. This year's Clean Air Week, which is just ending, addressed toxic chemicals in the air, indoors and out. I think we need to put this whole thing in perspective for the public, for budgetary measures, and for a lot of health reasons.

Of course smoking has for a long time been the Lung Association's bete noire, and we still think that smoking endangers your health to a much greater extent than radon does. But I don't think that for the general public there has been enough publicity on the interaction between smoking and radon on the lifespan. Earlier, someone spoke about stress among children in school. Someone should go into the school and say, "If you don't smoke, then the impact of radon on you will be much less." Children need to be told what they can do to help their health. In a certain sense, sometimes these meetings underplay the health effects of radon. People need to know more about the relative risks of all things they are breathing. For instance, in some homes, chlordane is a great health risk. This chlordane was put in the basement or around the basement of the house because the banks required it for the termite inspection. There are other things that people import into their house, such as formaldehyde which comes in with wood paneling,r new kitchen

****Linda Stansfield represented the New Jersey Chapter of the American Lung Association.

cabinets and new carpeting, and they aren't very aware of this. But I think that if they are going to be spending government funds and their private funds to remediate their home, then we've got to give them enough information so that they can use the method that will solve all their problems, not just one. There are many ways in which you can remediate for radon that do not involve ventilation and, therefore, do not improve your house for other pollutants. If your house has not been tested for other things, you probably are not going to think about it. I think government has a responsibility to tell people that there are more things that you should test for than just radon. I'm hoping that little kits will become available soon where you can test for a broad spectrum of the most likely pollutants.

Lung cancer, as you may or may not know, is a growing epidemic. We don't have a handle on the controls in lung cancer. We will have 126,000 deaths in the United States this year. We will have 144,000 new cases. At the moment, from what we know, we attribute 5,000 to 20,000 of those lung cancer deaths to radon. But those numbers -- it's a broad span -- and we're not sure just how much we should blame on radon. One of the roles that the Lung Association has is that we pressed the DEP and Department of Health to move more quickly in this area which we consider a significant health threat. We pressed them very hard on confidentiality. We knew that Pennsylvania was offering confidentiality, and we felt that they could make their lawyers overrule their original decision which said that under the Freedom of Information Act, we can't keep this confidential. I think it's nice that they have moved in that direction. We try to see that the information that is given to the public is accurate and up-to-date. I've chided the DEP about some materials that were distributed from EPA that talk about very simple measures that we don't believe are effective for more than six months. If someone takes those measures and thinks they have protected themselves, their lungs will know the difference, but they won't.

We are getting to the point where many government workers have gone into enough homes and done enough testing, that they have been exposed or have the potential to have been exposed to a great deal of radiation. Of course, the department would like to avoid putting protective clothing on these workers because of the fact that they may frighten the homeowner. We think that the worker has the right to protective clothing, and that his life should not be endangered. We have all seen the pesticide company that comes around and sprays all the workers as well as all the trees and says, "see it doesn't hurt me." Of course, it doesn't hurt him right then, but nobody is around when it hurts a lot. We think that there should be new construction and building regulations to take this into account and to take into account that we do have radon in New Jersey. We will be working with the Department of Community Affairs, which is not represented here but is dealing with another aspect of this whole complicated problem. The problem is complicated because indoor air is not a government responsibility. You should remember that and be very careful about trying to make it a government responsibility.

ADEQUACY OF THE 4 PCI/L GUIDELINE

L. Stansfield: I would like to know from the Mayor and Mr. Jordan if they feel that the levels that have been set are adequate to protect their constituents, or if they would like further examination of the levels that have been set for remedial action?

Mayor Nulman: I guess the important thing to remember is that I'm a lay person like many others here, and I rely very strongly on the recommendations of the experts. I know that there is some mild controversy over the recommended levels which are considered acceptable, and I, like everyone else, am really at the mercy of the scientific community as far as the determination. It's been worked out that 4 pCi/l is the acceptable level. I want that number to be as specific and as meaningful as can be, and that's all I know.

D. Jordan: I, too, am at the mercy of the professionals. There is no way that as a health professional I can set standards. We work in standards with water pollution, air pollution, and we are now working with standards with radon. I'm at the mercy of the experts to set those levels. It's as simple as that. In situations where there are questions about the specific level, we tell the public exactly that. That there is argument about those levels, and as far as we know, right now, these standards are in effect, and we will enforce those standards -- guidelines.

Question: Our concern, of course is that one in a hundred is not acceptable on any level. That means that if you take a hundred children in a classroom and you pick out one of those kids, and that kid happens to be your kid, do you want them dead in so many years? I don't think so. Do you want them to have a half life of 25 to 30 years? I don't know. So it has to be brought down to your children or my children.

M. Edelstein: Let me come back to Linda's question, which I think is what you are asking. Is there some consideration within the state in looking beyond the 4 pCi/l guideline?

J. Klotz: I don't think I've heard that one in a hundred is an acceptable risk for anything. But let's put it in some perspective. If one wants to get down to what some people might consider an acceptable risk from a particular radon exposure, one would have to get down to several orders of magnitude below what the current average exposure is in the

United States. We are never, in the near future, going to be able to get our exposures below what we would generally tolerate from something like a Superfund site or water contamination or air contamination. What I think is important is that we, in our state and federal programs, extend as much public health preventive action to the people who need it most as quickly as possible, and at the same time, carry out a program of prospective protection. The first part, the discovery of which homes are highest, requires that some kind of prioritizing be done. We've set our guideline at the lowest level that is recommended in any country or state that I know of, which is the 4 pCi/l level. Nobody is suggesting that people stop remediating at that level if they can remediate any further. It has been shown that it's very, very difficult, if you start at such a level, to get much improvement with any confidence. People could easily spend thousands of dollars finding out if they've even succeeded. They then might wish to have conducted some other kind of health protective activity, if this is their concern. In our work with public information, mitigation research and, from a prospective point of view in our work in building codes, we are hoping that ways of reducing radon to much lower levels will eventually be found. These can be put into all new housing regulations or guidelines for construction. So we are hoping to do two things at once. In the long run, lower the average U.S. exposure as low as possible, and in the short run, find those areas which need immediate remediation as quickly as we can find them, and to do that by prioritizing.

L. Stansfield: The point I was trying to make is that people are going to believe that the .02 WL (4 pCi/l) is a safe level, and that people at .021 WL (4.2 pCi/l) are not. For those who are concerned about health effects, I think if they knew that there is still significant risk at that level, that they would take action. I agree that remedial action probably would not be cost effective. But I think if they stop smoking, then they would benefit. I don't think they are getting enough information, although the Department of Health has certainly tried. Somehow the press is not picking up that the connection between cigarette smoking and death from radon is great. It is not a multiplying effect, but is more than an additive effect.

G. Nicholls: I would just like to comment on a couple of things. The DEP originally pressed EPA to provide a number that was "acceptable". After long periods of discussion with them, we came around to their way of thinking that the best thing to do was to provide people with realistic risk estimates, and then have the individual judge for him or herself what the risk was that they were willing to endure. We also agreed with EPA that 4 pCi/l was a reasonable number to shoot for in most situations. So that is the position that seems to be evolving. Not that there is any safe level for exposure to radon, but that there is a range of risk associated with that exposure, and it's up to the individual to decide. The benchmark is the 4 pCi/l number. There is a link between radon and smoking in most of the literature. The risks are at least additive. One of the things that you have to look at realistically, and we say this whenever we make a public presentation on radon, is to stop smoking, because smoking is at least 4 times as bad as radon exposure. That's in the statistics. Secondly, if you have radon in the house, probably smoking makes it worse, although we are not absolutely positive of that.

I was told by a newspaper reporter that he did not want me to couch risk in terms of packs of cigarette smoke, because the cigarette industry was a heavy advertiser in his newspaper. I've been told that twice by two different reporters. There are also magazines that have a heavy advertising component from the cigarette industry. They don't want to see risk couched in that term. Until you break that lobby, in both the advertising area and in the federal government, where we are subsidizing the tobacco industry, we are going to have a serious problem with regards to smoking. Until you break those two things, you don't stand a real good chance of addressing risk. There is a range of risk. We are trying to inform people about the range. Smoking makes the risk worse, we believe. Smoking is a more horrendous risk than even radon is.

There are indeed other pollutants in your home, and under certain circumstances it is possible to take measures against radon, if you have a radon problem, that will also reduce the risk associated with the other pollutants. Those cases where that is practical, we advise that. But the risk estimates associated with the exposure to the Urea Formaldehyde resins in homes, to the perchlorethylene that is on dry cleaned clothes when they come into your house, to the benzalfapyrenes that are in your kitchen after you grill a steak, all those things are orders of magnitude less than radon.

The last point that I want to make is that if you can get below the 4 pCi/l by remediating your house, you may not be able to measure it successfully. The measuring techniques are not good below one picoCurie/liter. If you have a 1 pCi/l house the natural variation in the radon concentration is probably going to be an order of magnitude greater than the measuring capability is. So if you are reading 1, you are probably somewhere between .1 pCi/l and 10 pCi/l. That's the reality of the measuring devices. As that improves though, maybe the standards will come down. But it's foolish to set a standard or acceptable level which you can't measure.

TESTING IN CLINTON

N. Weinstein: Mayor Nulman, could you give us a sense of what portion of the people in Clinton have now monitored their homes? Of people who may not have monitored, why? Are they unconcerned, or are they scared about the problem?

Mayor Nulman: We don't have an exact handle on the number of homes that have tested. I can tell you that of the canisters that we made available through the bulk purchases, we have sold hundreds. We are talking about a town that has somewhere in the neighborhood of 700-800 residential units and we have sold somewhere in the neighborhood of say, 350-400 detectors. Some homes have more than one detector. Others have sought to get their houses checked through independent sources or by mailing away in some cases to the University of Pittsburgh or have come to some of the other consultants. In those latter cases, people didn't want there to be any chance that others would find out what their level is. The latest Terradex detector that we offered was due to be back in the municipal office yesterday so that we can get it sent out and analyzed and get the results back. When the results come in, our town staff notifies the individuals; they come in and get the reading based on the serial number of the detector that they had. There are still some people who pooh-pooh the idea. There are skeptics all over the place who say, "Oh, this is just another scare. As a child, in elementary school, I was in the school basement covering up during civil defense alerts, and nothing ever came of it." A lot of people feel that this is just another scare. At one of the Department of Health meetings that was offered in the town, a lot of people were somewhat critical, saying, "We know what you are doing. You are just trying to get us scared again. You don't even have enough research data to make an intelligent determination, and you are just trying to frighten us. Come back to us when you have some hard data. You can't just come with some uranium miner results and extrapolate that and try to scare us." That kind of thing. It really runs the gamut. Believe it or not there are some people that think that the radon problem in Clinton stops at the municipal boundary and does not extend over into the next township. They say, "Boy, good thing I don't live in Clinton." Well, that's not the case. Let's get with it folks. It's cheap enough. The detectors we were offering were of two types, the charcoal canister at $12 and the Terradex detector at $25. I don't know if there is anyone in the Town of Clinton who would really be that hard pressed to come up with the $12 to get a charcoal canister and test. I thought it was kind of irresponsible for anybody to say, "Forget about it."

TRAINING A RADON WORKFORCE

P. Theiss: We have two states that have now had a rather serious problem, and there is a tremendous manpower need in both cases. We have a very, very limited supply of young people that are trained. Gerry knows this, because we have gone through a rather serious period in the last few years, and we have another industry that can offer a lot more money that needs about six people for every one of those that graduate right now. Have you thought at all about how, as we proceed up and down this east coast, we will find the people to do the monitoring, and who can be trained to do the mitigation?

G. Nicholls: I wish I had a coherent plan. We do know that we will not have the number of qualified individuals in any one of the three components of an overall radon program right away. The three components that I think of are the testing, the diagnosis, and the mitigation. One thing that has happened on the testing end is that the private sector has grown to fill that particular need. There is a question as to how well those firms perform. That is something that will be addressed, at least in New Jersey, and probably in New York, by a certification program. In New Jersey, we pin that certification to the diagnosis and mitigation end of things also. It will be far easier for a state to do training and testing of people who perform that work than for the state to do it itself. This time last year, I don't think there were any firms, to my knowledge, offering radon testing commercially in New Jersey other than Bernie Cohen doing his research project. By June there were three. By September there were 7, then 10, 17. We are up to 44 on the current list of firms that are doing testing. A total of 76 firms have requested to be on the list. So the private sector is filling in the numbers. What we have to do is work with the private sector to fill in the quality. On the diagnosis end of it, there are very, very few people involved. EPA is addressing that problem by retaining contractors to work on the development and the delivery of a training course. We are going to support EPA's training of people in diagnosis. We are also supporting an internship program with EPA where people would go and work with EPA and its contractors for a while to learn how to do this. Finally, we are going to work on the training and certification of the people who actually perform the work. How do they know if they install a vacuum system on a sump pump, that it actually does work, that it does reduce the radon level? How do they install the pipe? How do they install the covers over the French Drain system? You are absolutely right, Paul, we are not going to have the people right away, there is no question about that.

Question: What about using existing parts of the educational system?

G. Nicholls: We're looking at a lot of different options. We're looking into using Industrial Arts teachers while they are in training. They understand construction, and are sufficiently educated that they could take a program in radiation. They could learn diagnosis and mitigation. On the other hand, we are looking into training those who are schooled in radiation in basic construction techniques. There are a number of ideas we are kicking around. I would say it is going to be probably two months before the first training courses are in finished form and functioning regularly, although probably we will have the first session within a month. I would say it's probably going to be three years before there is a full training certification evaluation process in place for those three components. I don't think radon is going to go away.

PREVENTING RADON PROBLEMS

H. Horowitz: As the geological map fills in more and we get more sophisticated in terms of being able to predict where we have high levels of radon, has there been any thought given to pre-emptive purchasing of undeveloped parcels of land for green acres purposes in very high radon locations that are not currently developed?

G. Nicholls: It's a very good question. We have purchased some land on a couple of occasions to reduce radiation hazard. One example is the only uranium mine that has existed in New Jersey over near Lake Hopatcong. It was bought with Green Acres funds and turned into a park. We filled it in because we were concerned about kids falling into the two shafts that were there. In the long term, as the geology becomes better known, it may be necessary to develop local use ordinances. I think we have some fair amount of experience in Clinton right now. I know that their council is considering that kind of thing. But the entire "hot area" in Clinton is not a heck of a lot bigger than the college parking lot. If you find houses on high radon sites to start with, then you are going to have to work on the houses. If you don't find houses, then maybe you can develop a local use ordinance that would at least cause you to be thoughtful about the houses you place on them. We have been giving the names of certain consultants to people who are putting up developments in areas with known radon problems. We are building houses right now faster than we can remediate them down the line.

APPENDIX 1: GLOSSARY OF TECHNICAL TERMS

Activated Charcoal Canister: A type of time-integrated sampling detector which absorbs radon in carbon. Average radon concentration is subsequently found by detecting gamma rays from the radon daughters or radon decay products in the charcoal. Typical sampling times are on the order of 3-7 days

Activity: The number of disintegrations or decays per second. Both the number of atoms and the activity follow an exponential decay law for a given radionuclide.

Alpha particle: The nucleus of a helium atom which is emitted from some radioactive nuclei during radioactive decay.

Alpha track detector: A type of time-integrated sampling detector that records alpha particles from radon decay products or progeny by looking at radiation damage on plastic film. After exposure times of typically from 30 days to 1 year, the film is etched with sodium hydroxide solution revealing the tracks which are then counted manually or automatically under a microscope.

Basal cell: The target cell at risk from radiation in lung cancer production. They are the dividing stem cells that produce replacement cells for those normally lost from the bronchial epithelium.

Beta particle: A particle, either an electron or positron, emitted from some radioactive nuclei during radioactive decay. The electron and positron have the same mass but opposite charges.

Bronchial epithelium: The surface layer of cells lining the conducting airways.

Continuous sampling: A type of sampling method that determines real-time variations in radon concentrations by either continuous monitoring or by taking a series of closely-spaced grab samples. It reveals short-term fluctuations in radon concentrations rather than long-term average behavior as found from time-integrated measurements.

Convection: The movement of a fluid due to pressure or temperature differences.

Curie(Ci): A measure of activity or the rate of decay of radioactive nuclei. One curie equals 3.7×10^{10} decays per second. A picocurie is 10^{-12} curies or 0.037 decays per second.

Daughter: Another name for decay product or progeny.

Decay product: A nuclide formed from the radioactive decay of a radioactive nucleus. It refers to a nucleus, either radioactive or stable, that is formed directly from a decay, or one that results from successive transformations in a radioactive series.

Decay series: All members of a radioactive family of elements. A complete series begins with a long-lived parent and ends up with a stable element. Radon is one of the decay products found in the U-238 decay series which ultimately decays to stable Pb-206.

Diagnostic measurements: Measurements typically done after screening and follow-up tests to help determine the dynamics of radon gas entry and to help decide on a mitigation or remedial approach.

Dosimetry: The calculation or measurement of the energy absorbed by matter.

Drain tile: Perforated pipes typically placed in gravel around the foundation of a house in order to drain water away.

Electron: A negatively charged particle that orbits around the nucleus of an atom. The electron can also be created and emitted during the decay of a radioactive nucleus. See beta radiation.

Emanation rate: The rate of flow or flux of radon across a surface.

Epidemiology: A science dealing with the study of health and illness in human populations.

Equilibrium: Equilibrium is achieved when the activity of all the short-lived radon products or daughters is equal to the parent radon activity. Since this is typically not achieved, radon product activity is usually less than the associated radon activity.

Exposure: The quantity of radiation in a given environment representative of the potential health damage to an individual in that environment. Environmental exposure generally means exposure to radiation in non-occupational settings.

Follow-up measurement: Measurements made, often after screening measurements have indicated a potential radon problem, in order to find the average radon or radon decay product concentrations in a house. This is also sometimes referred to as a house characterization measurement.

Flux measurement: Measurements made to determine the rate at which radon is emanating from a surface or source into some part of the house. These diagnostic measurements allow the radon source strengths to be estimated and help in planning the most effective remediation procedures.

Gamma radiation: Energy released from some radioactive nuclei in the form of electromagnetic radiation. One of the three forms of radiation (alpha, beta and gamma) given off during the decay of a radioactive nucleus.

Grab sample: A type of sampling method which collects a small sample of air in a scintillation cell over a relatively short time span. The sample is subsequently analyzed for alpha decay in order to determine either concentrations of radon gas or radon decay products, depending on the method used. This approach is often used in screening and in diagnostic measurements.

Granular activated carbon (GAC): A material derived from typically wood, coal or other carbon base that has been made into charcoal and activated. Such material has a large internal pore structure and absorption capacity. It is used quite successfully in removing radon from water supplies.

Half-life: The time required for half of the atoms of a radioactive nuclei to decay. Each type of nucleus has its own unique half-life. For radon (Rn-222), it is 3.82 days.

Heat recovery ventilator (Air-to-air heat exchanger): A device which transfers the heat in the warm air being exhausted to the incoming cooler air.

Ionizing radiation: Radiation of sufficient energy to break apart or ionize atoms and molecules. Typically, alpha, beta and gamma radiation and neutrons are capable of ionizing tissue.

Lognormal distribution: The distribution of a random variable in which the logarithm follows the normal or Gaussian law of probability. The distribution of radon levels in houses seems to follow such a law.

MeV(Million Electron Volts): A unit of energy defined as the energy gained by an electron accelerated through a potential difference of one million electron volts.

Neutron: An electrically neutral particle in the nucleus of an atom with about the same mass as a proton.

Nuclide: A type of atom specified or characterized by the particular makeup of neutrons and protons in its nucleus.

Occupational exposure: Exposure of a worker during a period of work. For radon, it usually refers to miners, but may also refer to workers involved in radon mitigation.

Parent nucleus: A radionuclide which decays to yield a specified nuclide or nuclides. These nuclei may be produced directly through a single decay of the parent or may be the result of multiple decays through a radioactive decay chain. The resulting nuclei are often referred to as daughters, decay products or progeny.

Permeability: The property of a substance which measures its ability to allow passage or penetration of fluids through its pore spaces.

Picocuries per liter (pCi/l): A measurement of activity per volume of air. For radon, it implies a concentration of radon equivalent to 0.037 decays per second in a liter of air. See Curie.

Plate-out: The process in which radon decay products or progeny attach to materials and objects such as dust, walls, furniture, lungs, etc. due to electrostatic attraction.

Porosity: The property of a substance which is a measure of the volume of pore or open spaces compared to the total volume.

Proton: An electrically positive particle in the nucleus of an atom with about the same mass as a neutron. The number of protons in an atom is called the atomic number of the element and determines its chemical properties.

Radioactive decay: The spontaneous decay of a radionuclide by alpha, beta or gamma emission.

Radioactivity: The spontaneous release of energy by a nucleus which always results in a change in mass and often results in a different element being formed.

Radon: The radioactive noble gas Rn-222 whose immediate parent is radium or Ra-226. It is one element in the radioactive decay series starting with U-238.

Radon daughters: The short-lived radionuclides formed as a result of the decay of radon gas. They typically include Po-218, Pb-214, Bi-214 and Po-214.

Risk coefficient: The mortality rate due to lung cancer per unit of exposure after a suitable latent interval.

Screening measurement: A measurement that attempts to identify quickly those houses with high radon concentrations that require further investigation.

Time integrated: A type of sampling instrument that measures a single average concentration value over an extended time interval.

Uranium: A naturally-occurring radioactive nucleus with a very long half-life. About 99.3% of naturally-occurring uranium is U-238 and the remaining 0.7% is U-235. U-238 is the ultimate parent of radon gas.

Working level (WL): A measure of radon daughter or decay product concentration which is defined as any combination of the short-lived daughters in one liter of air which has a potential alpha energy release of 130,000 MeV. See MeV.

Working level month (WLM): An exposure or radiation dose which is found as the product of the average working level and the number of 170-hour working months. One WLM is a cumulative exposure equivalent to 1 WL for one working month or 170 hours.

SOME USEFUL CONVERSION FACTORS AND NUMBERS

1 WL is equivalent to 200 pCi/l at an equilibrium factor of 0.5.

1 pCi/l = 37 Bq/m^3

Based on very limited data, average radon exposure for the U.S. population is estimated to be about 0.8-1 pCi/l. Average outdoor concentrations of radon are around 0.1-0.2 pCi/l.

The U.S. occupational limit for exposure to radon progeny is 4 WLM/year.

APPENDIX 2: BIBLIOGRAPHY OF RADON REFERENCES*

EPA DOCUMENTS

"A Citizen's Guide to Radon. What It Is And What To Do About It." United States Environmental Protection Agency, OPA-86-004, August, 1986.

"Methods and Results of EPA's Study of Radon in Drinking Water." United States Environmental Protection Agency, Office of Radiation Programs, EPA 520/5-83-027, December, 1983.

"Radon Reduction Techniques for Detached Houses. Technical Guidance." United States Environmental Protection Agency, EPA/625/5-86/019, June, 1986.

Ronca-Battista M., P. Magno, S. Windham, E. Sensintaffar, "Interim Indoor Radon and Radon Decay Product Measurement Protocols." United States Environmental Protection Agency, Office of Radiation Programs, EPA 520/1-86-04, February, 1986.

Ronca-Battista M., P. Magno, P. Nyberg, "Interim Protocols for Screening and Follow-Up Radon and Radon Decay Product Measurements." United States Environmental Protection Agency, Office of Radiation Programs, Draft Copy, June, 1986.

"Radon Reduction Methods: A Homeowner's Guide." United States Environmental Protection Agency, Office of Research and Development, July, 1986.

AIR-TO-AIR HEAT EXCHANGERS

<u>Consumer Reports, 1986 Buying Guide Issue</u>. Mount Vernon, New York, December, 1985.

MITIGATION

Also see selected EPA documents, above.

---

*The reader is also referred to the numerous footnotes and references provided at the end of many of the individual papers throughout the proceedings.

Bliss, Steve, "The Importance of Ventilation, Part II," Solar Age. March, 1986.

Brennan, Terry, and Bill Turner, "Defeating Radon," Solar Age, March, 1986.

Turner, William, and Terry Brennan, "Radon's Threat Can Be Subdued", Solar Age, May, 1985.

HEALTH EFFECTS**

Cliff, K. D., A. D. Wrison, B. M. R. Green, and J. C. H. Miles, "Radon Daughter Exposures in the UK," Health Physics 45, No. 2, 1983, pp. 323-330.

Edling, C., and O. Axelson, "Quantitative Aspects of Radon Daughter Exposure and Lung Cancer in Underground Miners," British Journal of Industrial Medicine 40, 1983, pp. 182-187.

Gottlieb, Leon S., and Luverne A. Husen, "Lung Cancer Among Navajo Uranium Miners," CHEST 81:4, April, 1982.

Harley, Naomi H., and Bernard S. Pasternack, "A Model for Predicting Lung Cancer Risks Induced by Environmental Levels of Radon Daughters," Health Physics 40, March, 1981, pp. 307-316.

Radford, Edward P. and K. G. St. Clair Renard, "Lung Cancer in Swedish Iron Miners Exposed to Low Doses of Radon Daughters," The New England Journal of Medicine 310, No. 23, June 7, 1984.

"Radon and Lung Cancer in Mines and Homes," The New England Journal of Medicine 310, No. 23, June 7, 1984.

Samet, Jonathan M., et. al., "Uranium Mining and Lung Cancer in Navajo Men," The New England Journal of Medicine 310, No. 23, June 7, 1984.

Sevc, J., E. Kunz, and V. Placek, "Lung Cancer in Uranium Miners and Long-Term Exposures to Radon Daughter Products," Health Physics 30, June, 1976.

TECHNICAL REVIEW ARTICLES AND BOOKS

"Evaluation of Occupational and Environmental Exposures to Radon and Radon Daughters in the U.S." NCRP Report, No. 78, 1984.

**The entire issue of Health Physics 45, August, 1983 is devoted to radon.

"Exposures from the Uranium Series with Emphasis on Radon and Its Daughters." NCRP Report, No. 77, March, 1984.

Gesell, Thomas F., "Background Atmospheric 222-Rn Concentrations Outdoors and Indoors: A Review." Health Physics 45, No. 2, August, 1983, pp. 289-302.

Hess, C. T., J. Michel, T.R. Horton, H. M. Prichard, W. A. Coniglio, "The Occurrence of Radioactivity in Public Water Supplies in the United States." Health Physics 48, No. 5, May, 1985, pp. 553-586.

Nero, A. V., "Airborne Radionuclides and Radiation in Buildings: A Review." Health Physics 45, No. 2, August, 1983, pp. 303-322.

Nero, A. V., et. al., "Radon Concentrations and Infiltration Rate Measured in Conventional and Energy Efficient Houses." Health Physics 45, 1983, pp. 401-405.

Budnitz, R. J., et. al., "Human Disease from Radon Exposures: The Impact of Energy Conservation in Buildings." Energy and Environmental Division, Lawrence Radiation Laboratory, LBL-7809, August, 1978.

Hernandez, Thomas L., James W. Ring, Harvey M. Sachs, "The Variatiom of Basement Radon Concentration with Barometric Pressure." Health Physics 46, No. 2, February, 1981, pp. 440-445.

LESS TECHNICAL ARTICLES

"Indoor Air Pollution: Radon." Consumer Reports 50, October, 1985, pp. 601-602.

Lafavore, Michael, "The Radon Report." New Shelter January, 1986.

May, Harold, "Ionizing Radiation Levels in Energy-Conserving Structures." Underground Space 5, 1981, pp. 384-391.

"Radon Calling." The Mother Earth News March, 1986.

"Radon in Water and Air: Health Risks and Control Measures." The Land and Water Resources Center, University of Maine at Orono and the Division of Health Engineering, Maine Department of Human Services, February, 1983.

Smay, Elaine V., "Radon Exclusive." Popular Science 227, November, 1985, pp. 76-79.

APPENDIX 3: STATE RADIOLOGICAL HEALTH PROGRAM CONTACTS—AVAILABLE TECHNICAL ASSISTANCE*

For readers who wish to contact their state officials concerned with radon, we are providing the following list of agency contacts.

<u>Alabama</u>

Godwin, Aubrey V., Director
Division of Radiological Health
State Department of Public Health
State Office Building
Montgomery, Alabama 36130
Business: 205/261-5315

<u>Alaska</u>

Heidersdorf, Sidney D., Chief
Radiological Health Program
Department of Health and Social Service
Pouch H-06F
Juneau, Alaska 99811-9976
Business: 907/465-3019

<u>Arizona</u>

Tedford, Charles F., Director
Arizona Radiation Regulatory Agency
925 South 52nd Street, Suite 2
Tempe, Arizona 85281
Business: 602/255-4845

<u>Arkansas</u>

Wilson, E. Frank, Director
Division of Radiation Control & Emergency Management
Department of Health
4815 West Markham Street
Little Rock, Arkansas 72201
Business: 501-661-2301

*This listing was taken from the Directory of Personnel Responsible, Radiological Health Programs, January 1986, published by the Conference of Radiation Control Program Directors, Inc., 71 Fountain Place, Frankfort, Kentucky 40601.

California

Ward, Joseph O., Chief
Radiological Health Branch
State Department of health Services
714 P Street, Office Bldg. 8
Sacramento, California 95814
Business: 916/322-2073

Colorado

Hazle, A. J., Director
Radiation Control Division
Department of Health
4210 East 11th Avenue
Denver, Colorado 80220
Business: 303/320-8333, Ext. 6246

Connecticut

McCarthy, Kevin T.A., Director
Radiation Control Unit
Department of Environment Protection
State Office Building
165 Capital Avenue
Hartford, Connecticut 06106
Business: 203/566-5668

Delaware

Tapert, Allan C., Program Administrator
Office of Radiation Control
Division of Public Health
Department of Health & Social Services
Cooper Building, Cooper Square
Post Office Box 637
Dover, Delaware 19901
Business: 302/736-4731

District of Columbia

Bowie, Frances A., Administrator
Department of Consumer & Regulator Affairs
Service Facility Regulation Administration
614 H Street, N.W., Room 1014
Washington, D.C. 20004
Business: 202/727-7190

Florida

Jerrett, Lyle E., Director
Office of Radiation Control
Department of Health & Rehabilitative Services
1317 Winewood Boulevard
Tallahassee, Florida 32301
Business: 904/487-1004

Georgia

Rutledge, Bobby G., Director
Radiological Health Section
Department of Human Resources
878 Peachtree Street, Room 600
Atlanta, Georgia 30309
Business: 404/894-5795

Hawaii

Anamizu, Thomas, Chief
Noise and Radiation Branch
Environmental Protection & Health Services Div.
Department of Health
591 Ala Moana Boulevard
Honolulu, Hawaii 96813
Business: 808/548-4383

Idaho

Funderburg, Robert, Program Manager
Radiation Control Section
Department of Health and Welfare
Statehouse Mall
Boise, Idaho 83720
Business: 208/334-4107

Illinois

Lash, Terry, Director
Office of Environmental Safety
Department of Nuclear Safety
1035 Outer Park Drive
Springfield, Illinois 62704
Business: 217/546-8100

Indiana

Stocks, Hal S., Chief
Radiological Health Section
State Board of Health
1330 West Michigan Street
Post Office Box 1964
Indianapolis, Indiana 46206
Business: 317/633-0152

Iowa

Eure, John A., Director
Environmental Health Section
Iowa Department of Health
Lucas State Office Building
Des Moines, Iowa 50319
Business: 515/281-4928

Kansas

Romano, David J., Manager
Bureau of Air Quality and Radiation Control
Department of Health and Environment
Forbes Field, Building 321
Topeka, Kansas 66620
Business: 913/862-9360

Kentucky

Hughes, Donald R., Manager
Radiation Control Branch
Cabinet for Human Resources
275 East Main Street
Frankfort, Kentucky

Louisiana

Spell, William H., Administrator
Nuclear Energy Division
Office of Air Quality & Nuclear Energy
Department of Environmental Quality
Post Office Box 14690
Baton Rouge, Louisiana 70898-4690
Business: 504/925-4518

Maine

Hinckley, Wallace, Assistant Director
Division of Health Engineering
157 Capitol Street
Augusta, Maine 04333
Mailing Address: State House, Station 10
Augusta, Maine 04333
Business: 207/289-3826

Maryland

Resh, David L., Administrator
Community Health Management Program
Department of Health and Mental Hygiene
O'Conor Office Building
201 West Preston Street
Baltimore, Maryland 21201
Business: 301/225-6031

Massachusetts

Hallisey, Robert M., Director
Radiation Control Program
Department of Public Health
150 Tremont Street, 7th Floor
Boston, Massachusetts 02111
Business: 617/727-6214

<u>Michigan</u>

Bruchmann, George W. Chief
Division of Radiological Health
Bureau of Environmental and Occupational Health
Department of Public Health
3500 North Logan Street
Post Office Box 30035
Lansing, Michigan 48909
Business: 517/373-1578

<u>Minnesota</u>

Hennigan, Alice T. Dolezal, Chief
Section of Radiation Control
Environmental Health Division
Minnesota Department of Health
717 Delaware Street, S.E.
Post Office Box 9441
Minneapolis, Minnesota 55440
Business: 612/623-5323

<u>Mississippi</u>

Fuente, Eddie S., Director
Division of Radiological Health
State Department of Health
3150 Lawson Street
Post Office Box 1700
Jackson, Mississippi 39215-1700
Business: 601/354-6657

<u>Missouri</u>

Miller, Kenneth V., Chief
Bureau of Radiological Health
1730 East Elm Plaza
Post Office Box 570
Jefferson City, Missouri 65102
Business: 314/751-8208

<u>Montana</u>

Lloyd, Larry L., Chief
Occupational Health Bureau
Deparment of Health & Environmental Sciences
Cogswell Building
Helena, Montana 59620
Business: 406/444-3671

<u>Nebraska</u>

Borchert, Harold R., Director
Division of Radiological Health
301 Centennial Mall, S.
Post Office Box 95007
Lincoln, Nebraska 68509
Business: 402/471-2168

Nevada

Vaden, John D., Supervisor
Radiological Health Section, Health Division
Department of Human Resources
505 East King Street
Carson City, Nevada 89710
Business: 702/885-5394

New Hampshire

Tefft, Diane E., Program Manager
Radiological Health Program
Post Office Box 148
Concord, New Hampshire 03301
Business: 603/271-4588

New Jersey

Nicholls, Gerald P., Acting Chief
Bureau of Radiation Protection
Division of Environmental Quality
Department of Environmental Protection
380 Scotch Road
Trenton, New Jersey 08628
Business: 609/292-8392

New Mexico

Browen, Michael F., Acting Chief
Radiation Protection Bureau
Environmental Improvement Division
Department of Health and Environment
Post Office Box 968
Santa Fe, New Mexico 87504-0968
Business: 505/827-0020

New York

Rimawi, Karim, Director
Bureau of Environmental Radiation Protection
State Health Department
Empire State Plaza, Corning Tower
Albany, New York 12237
Business: 518/473-3613

North Carolina

Brown, Dayne H., Chief
Radiation Protection Section
Division of Facility Services
Department of Human Resources
Post Office Box 12200
Raleigh, North Carolina 27605-2200
Business: 919/733-4283

North Dakota

Mount, Dana K., Director
Division of Environmental Engineering
Department of Health
Missouri Office Building
1200 Missouri Avenue
Bismarck, North Dakota 58501
Business: 701/224-2348

Ohio

Quillan, Robert M., C.H.P., Director
Radiological Health Program
Department of Health
246 North High Street
Post Office Box 118
Columbus, Ohio 43216
Business: 614/466-1380

Oklahoma

McHard, J. Dale, Chief
Radiation & Special Hazards Service
State Department of Health
Post Office Box 53551
Oklahoma City, Oklahoma 73152
Business: 405/271-5221

Oregon

Paris, Ray D., Manager
Radiation Control Section
State Health Division
Department of Human Resources
1400 Southwest Fifth Avenue
Portland, Oregon 97201
Mailing Address:
State Health Division
Post Office Box 231
Portland, Oregon 97207
Business: 503/229-5797

Pennsylvania

Gerusky, Thomas M., Director
Bureau of Radiation Protection
Department of Environmental Resources
Fulton Building, 16th Floor
Third and Locust Street
Harrisburg, Pennsylvania 17120
Mailing Address:
Post Office box 2063
Harrisburg, Pennsylvania 17120
Business: 717/787-2480
800/237-2366 (Toll Free in State)

Puerto Rico

Saldana, David, Director
Radiological Health Division
G.P.O. Call Box 70184
Rio Piedras, Puerto Rico 00936
Business: 809/767-3563

Rhode Island

Hickey, James E., Chief
Division of Occupational Health & Radiation Control
Cannon Building, Davis Street
Providence, Rhode Island 02908
Business: 401/277-2438

South Carolina

Shealy, Heyward G., Chief
Bureau of Radiological health
South Carolina Dept. of Health & Environmental Control
2600 Bull STreet6
Columbia, South Carolina 29201
Business: 803/758-8354

South Dakota

Platt, C. James, Director
Licensure and Certification Program
State Department of Health
Joe Foss Office Building
523 East Capital
Pierre, South Dakota 57501
Business: 605/773-3364

Tennessee

Mobley, Michael H., Director
Division of Radiological Health
TERRA Building
150 9th Avenue, N.
Nashville, Tennessee 37203
Business: 615/741-7812

Texas

Lacker, David K., Chief
Bureau of Radiation Control
Department of Health
1100 West 49th Street
Austin, Texas 78756-3189
Business: 512/835-7000

Utah

Anderson, Larry, Director
Bureau of Radiation Control
State Department of Health
State Office Building, Box 45500
Salt Lake City, Utah 84145
Business: 801/533-6734

Vermont

McCandless, Raymond N., Director
Division of Occupational and Radiological Health
Department of Health
Administration Building
10 Baldwin Street
Montpelier, Vermont 05602
Business: 802/828-2886

Virginia

Price, Charles R., Director
Bureau of Radiological Health
Division of Health Hazard Control
Department of Health
109 Governor Street
Richmond, Virginia 23219
Business: 804/786-5932

Washington

Strong, T.R., Chief
Office of Radiation Protection
Department of Social & Health Services
Mail Stop LE-13
Olympia, Washington 98504
Business: 206/753-3468

West Virginia

Aaroe, William H., Director
Industrial Hygiene Division
151 11th Avenue
South Charleston, West Virginia 25303
Business: 304/348-3526

Wisconsin

McDonnell, Lawrence J., Chief
Radiation Protection Section
Divison of Health
Department of Health & Social Services
Post Office Box 309
Madison, Wisconsin 53701
Business: 608/273-5181

Wyoming

Haes, Julius E., Jr., Chief
Radiological Health Services
Division of Health & Medical Services
Hathaway Building
Cheyenne, Wyoming 82002-0717
Business: 307/777-6015